Torsion of
Reinforced Concrete

Torsion of
Reinforced Concrete

Thomas T. C. Hsu

Professor and Chairman
Department of Civil Engineering
University of Houston

VAN NOSTRAND REINHOLD COMPANY
NEW YORK   CINCINNATI   TORONTO   LONDON   MELBOURNE

Manufactured in the United States of America

Published by Van Nostrand Reinhold Company Inc.
135 West 50th Street
New York, New York 10020

Van Nostrand Reinhold Company Limited
Molly Millars Lane
Wokingham, Berkshire RG11 2PY, England

Van Nostrand Reinhold
480 Latrobe Street
Melbourne, Victoria 3000, Australia

Macmillan of Canada
Division of Gage Publishing Limited
164 Commander Boulevard
Agincourt, Ontario M1S 3C7, Canada

15 14 13 12 11 10 9 8 7 6 5 4 3 2 1

Library of Congress Cataloging in Publication Data

Hsu, Thomas T. C. (Thomas Tseng Chuang), 1933–
    Torsion of reinforced concrete.

    Includes bibliographical references and index.
    1. Reinforced concrete. 2. Torsion. I. Title.
TA440.H7   1983   620.1'37   82–24766
ISBN 0–442–26401–1

To

my wife and my family,
whose support was indispensable
in the preparation of this book.

Preface

Torsion is a major factor to consider in the design of many kinds of reinforced concrete structures, including space frames, beams that support cantilever slabs or balconies, horizontally curved beams, spandrel beams, beams next to floor openings, wall foundation beams, spiral staircases, skew bridges, and so on.

Although torsional stresses occur frequently in reinforced concrete buildings and structures, torsion was generally ignored by design engineers before the 1960s. It was assumed that torsional effects were minor and could be taken care of by the large safety factor used in flexural design. This assumption has been responsible for many cases of torsional distress and failure.

Torsion design began to arouse serious interest by the late 1950s because of three major stimuli. First, the ultimate strength design (USD) method was accepted as a replacement for the working stress design method then in use. In the new USD method, flexural analysis of reinforced concrete members was refined, and safety factors were more accurately defined so that negligence of torsional effects was no longer acceptable. Second, the rapid advances in electronic computer applications in structural analysis allowed engineers to consider many more design factors. It is now possible to take torsion into account with minimal effort. Third, in the post–World War II development of modern architectural concepts, buildings were considered as three-dimensional structures rather than flat-plane objects. Out-of-plane loadings, curved beams, skew structures, and many kinds of irregular shapes and forms were introduced. These new designs often required structural members to resist large torsional moments.

In response to the worldwide interest in torsion the American Concrete Institute (ACI) Committee 438 on Torsion was created in 1958 to study the torsion problem and to promote research in this area. This committee was also charged with the responsibility of preparing recommendations for suitable provisions in torsion design. As a result of the efforts of the committee, a new set of torsion design criteria was incorporated into the 1971 ACI Building Code. Later, provisions for the torsional limit design of spandrel beams were added to the 1977 and 1983 codes. At present, ACI Committee 445, Shear and Torsion (Committee 426, Shear and Committee 438, Torsion were merged in 1977) is working to establish torsional design criteria that would unify shear and torsion, as well as nonprestressed and prestressed concrete.

Interest in torsion and the world-wide codifications of torsional design criteria call for the publication of a comprehensive textbook on torsion of reinforced concrete. This textbook integrates all phases of the development of knowledge in torsion in a historical overview, giving the reader an intellectual appreciation of this aspect of engineering. A comprehensive presentation of modern research in torsion reveals the most current state-of-the-art. Moreover, this book has practical design value, providing substantial information in three important areas: the methods of analysis for various types of structures involving torsion, the background and derivation of the four building codes, and some actual examples of applications of the torsion design provisions.

The first chapter of this book deals with the classical theories of torsion for homogeneous members. These theories are applicable to plain concrete members as well as reinforced concrete members before cracking. Using knowledge obtained in Chapter 1, we investigate the torsional strength and behavior of plain concrete members in Chapter 2.

Chapter 3 studies the strength and behavior of reinforced concrete members subjected to torsion. Emphasis is placed on ultimate strength and post-cracking torsional stiffness. This study leads the way to the ACI torsion criteria presented in Chapter 4, which gives a detailed background of the 1971, 1977, and 1983 ACI Building Code.

Chapter 5 deals with prestressed concrete, including a new generalized torsion design criterion. The current ACI torsion criterion was systematically modified so that it is applicable to prestressed concrete as well as nonprestressed concrete. The latter is simply a special case of the former.

In Chapters 6 and 7 are shown studies of reinforced concrete members subjected to combined loadings (torsion, shear, bending, and axial load). Chapter 6 summarizes the development of skew-bending theories. It includes a description of the torsion provisions in the 1962 Russian Code and those in the 1973 Australian Code. Chapter 7 describes the development of the truss model method for shear, torsion, and combined loadings, and the background of the 1978 CEB-FIP Model Code. The most up-to-date information on the deformation behavior of torsional members is also covered.

Chapters 8 and 9 treat the analysis of structures involving torsion. The general analysis of space frames, including an in-depth study of symmetry, is given in Chapter 8. The very common design of spandrel beams is treated in Chapter 9. This chapter gives the background and application of the torsional limit design method incorporated in the 1977 and 1983 ACI codes.

The substance of this textbook was initially compiled as my lecture material for a graduate course at the University of Miami, Coral Gables, Florida. This course, offered six times during the past ten years, was well attended by graduate students as well as practicing engineers. This experience convinced me that a textbook on this subject would be very useful to the structural engineering profession. The book can be used either as a textbook for a graduate course or as a reference book for practicing engineers engaged in design.

In my experience, a three-credit-hour graduate course could cover Chapters 1 through 5, 8, and 9. These chapters should give the student a solid foundation in understanding the torsion phenomenon and the practical design of structures in torsion. For those who wish to have a broad view of the current state-of-the-art in torsion, Chapters 6 and 7 would be very useful. This material is particularly suitable for self study, or it can furnish sufficient substance for an additional two credit hours in the course.

T. T. C. Hsu

List of Plates

Plate 1: Coulomb's device for torsional oscillation tests of thin wires, 1784.

Plate 2: Barré de Saint-Venant, who produced rigorous solutions for elastic homogeneous noncircular members subjected to torsion. His name has been used to define the type of torsion, in which the shear stresses are flowing in a circulatory manner.

Plate 3: Torsion test rig in the Structural Laboratory of the Portland Cement Association, Skokie, Illinois, 1963.

Plate 4: American Hospital Association Buildings, Chicago. The continuous transfer girder with 5 ft by 9 ft cross section at the second floor level was the first to be designed by the ACI torsion criteria (Ref. 2 in Chapter 4), 1967.

Plate 5: Prestressed concrete curved beam used for the aerial guideway at Disney World, Orlando, Florida. Large torsional moment is induced by the curvature of the beam, centrifugal forces, and the wind load.

Plate 6: Collapse of a six-story reinforced concrete building due to the shear and torsion failure of the reinforced concrete ribbed raft foundation, 1979. One corner of the building has sunk 3.5 m.

Plate 7: Prestressed concrete double-tee girders in the Dade County Mass Rapid Transit aerial guideway, Florida, under construction in 1982. Large torsional moment in the girder is caused by the wind load and by the nosing/ lurching action of the vehicles.

Plate 8: Reinforced concrete wall beams supporting cantilever flower bins. Severe torsional cracking was observed in the beams and slabs, 1976.

Plate 9: A large torsional crack near the support of a spandrel beam in a reinforced concrete parking garage in South Florida, 1981.

Guide for Use of Book by Practicing Design Engineers

Although this book gives a historical overview of various theories on torsion, it also contains a great deal of material that is directly useful to practicing engineers in their design of structures involving torsion. Explanations of various torsion code provisions and many design example problems are included. This guide intends to help a practicing engineer find the materials that are required in a design.

There are two general steps in the design of a reinforced concrete structure. First, the structure is analyzed by the principles of mechanics to determine the axial loads, bending moments, shear forces, and torsional moments. Second, based on the actions obtained in the first step, each member in the structure is designed to determine its size, shape, and reinforcement.

ANALYSIS OF STRUCTURES

In the first step, the reader can start with Chapter 8, which presents three ways to analyze a space frame, namely, the flexibility method (Section 8.3), the stiffness method (Section 8.4), and the moment distribution method (Section 8.5). All these methods have been generalized to include the torsional action of the members in a structure. In addition to presentation of the principles of mechanics, this chapter includes ten example problems to illustrate the application of these three methods.

The analysis of space structures is often quite tedious, as it frequently involves the solving of many unknowns. However, space structures are often symmetrical, and the application of the theory of symmetry in such structures can drastically reduce the number of unknowns. This powerful theory of symmetry has been carefully presented in Section 8.1. To the author's knowledge, this is the first appearance of this theory in book form.

Since plane frames are very common structures, an understanding of the characteristics of plane frames can also be very useful in the analysis of such structures. Detailed discussions and proofs of these characteristics can be found in Section 8.2.

In the elastic analysis of space frames, the distribution of bending moments and torsional moments depends strongly on EI/GC, the ratio of the flexural rigidity to torsional rigidity of a member. ACI codes have always implied that this ratio can be calculated according to the uncracked sections. If this path is followed, EI can, of course, be found from any textbook on mechanics of solids, while GC can be determined from Chapter 1. Sections 1.1, 1.2, and 1.3 give the GC values for circular, noncircular, and thin-tube sections, respectively. The values for noncircular sections include rectangular sections (Section 1.2.4), thin-walled flanged sections (Section 1.2.6), and approximate solutions for arbitrary sections (1.2.5 and 1.4).

The EI/GC ratio can also be determined according to the cracked sections. It has been shown that, compared to the preceding method, such analysis will provide a more realistic picture of the stress distribution in a structure at ultimate load stage. If this path is followed, then the post-cracking torsional rigidity of a member, $G_{cr}C_{cr}$, can be determined from Section 3.3.4 of Chapter 3; and the post-cracking flexural rigidity can be obtained from regular textbooks in reinforced concrete design.

The elastic analysis of spandrel beams is included in Section 9.2 of Chapter 9. Torsional moments in such a spandrel beam are caused by loadings from the floor slab or closely spaced floor beams on one side only. The differential equations, which represent the equilibrium and compatibility conditions, have been solved, and design coefficients have been prepared for maximum torques of spandrel beams. These design coefficients are given in Appendix A.

Cracking of a spandrel beam will cause a redistribution of moments from the spandrel beam to the floor beams or floor slab. This could significantly reduce the torsional moment in the spandrel beam, resulting in great economy. A torsional limit design method has been developed to utilize this advantage. This method, which was first adopted by the 1977 ACI Code, is presented in Section 9.3.

DESIGN OF MEMBERS

The design of reinforced concrete members to resist shear and torsion has always been based on semi-empirical procedures. As a result, the design provisions are quite different in various countries. The four most influential design codes are included in this book. They are summarized in the Table below. These code provisions for shear and torsion are important because they have been derived from some sort of recognized theory and have been supported by some type of independent test results. Each one of them also has influenced design codes of other nations. Because of their significance, design examples have been included to illustrate the applications of these four code provisions. These example problems are also listed in the Table.

CODES	CHAPTERS OR SECTIONS	EXAMPLES
(1) ACI 318, "Building Code Requirements for Reinforced Concrete" (ACI 318–77), 1977; (ACI 318–83), 1983.	Chapter 4	Examples 4.4, 9.1
(2) CEB-FIP, "Model Code for Concrete Structures," 3rd ed., 1978.	Sections 7.4 and 7.7	Example 7.1
(3) USSR Council of Ministers, "Structural Standards and Regulations," SNiP II-B, 1–62, 1962; SNiP II-21–75, 1976.	Section 6.2.3	Example 6.1
(4) Standard Association of Australia, "SAA Concrete Structures Code" AS CA2–1973; AS 1480–1974 (metric).	Section 6.3.6	Example 6.2

In the United States, the design for shear and torsion is, of course, based on the ACI Building Code. Therefore, all of Chapter 4 has been devoted to discussing the background and the application of this code. Four example problems (Examples 4.1–4.4) are included in this chapter, while another detailed example problem (Example 9.1) is given in Chapter 9 in connection with the torsional limit design method.

The ACI Code design procedure for shear and torsion has recently been generalized to include prestressed concrete. This generalization has been carefully presented in Chapter 5, and a recommendation entitled "Tentative Recommendations for the Design of Prestressed and Nonprestressed Members to Resist Torsion" is given in Appendix B. Example 4.4, which has been designed as a reinforced concrete member, has been redesigned as a prestressed concrete member. In this way, the advantage of prestressing is clearly illustrated.

As intended, the ACI Code provisions for torsion are quite satisfactory when applied to regular members normally used in buildings. However, this code could be quite awkward and overly conservative when applied to the large-size hollow members, with arbitrary cross section, that are normally employed in long-span bridges. To design such large structures, the CEB-FIP Code is more versatile. For this reason, the theory of the variable-angle truss model, which is the basis of the CEB-FIP Code, has been carefully presented in Chapter 7.

From a practical design point of view, the CEB-FIP Code has three important advantages over the ACI Code when applied to large bridge structures. First, the CEB-FIP Code is applicable to arbitrary bulky cross sections and is not restricted to members composed of rectangular cross sections as in the ACI Code. Second, the CEB-FIP Code allows a designer to provide different web reinforcement in each wall of a box section according to the distribution of shear and torsional stresses. Third, the CEB-FIP Code provides a more reasonable way to design longitudinal bars in the flexural compression zone.

Caution must be exercised, however, in the application of the CEB-FIP Code, particularly when it is applied to smaller members normally used in buildings. First, the "staggering concept" of shear design in the CEB-FIP Code is not conservative and should be abandoned (Sections 7.4.4.4 and 7.3.3). Shear reinforcement should be designed according to the conventional shear diagram. Next, in the so-called compatibility torsion condition, the torsional moment in a member should not be blindly neglected as permitted by the CEB-FIP Code (Section 7.7.1.1). The torsional limit design method as permitted by the ACI Code would be more appropriate (Sections 9.1 and 9.3). Finally, the equation in the CEB-FIP Code to prevent the compression failure of concrete struts is quite unreasonable for smaller members normally used in buildings. This point is discussed in Sections 7.7.2.3 and illustrated in Example 7.1. A more logical equation has been proposed.

Contents

Torsion of
Reinforced Concrete

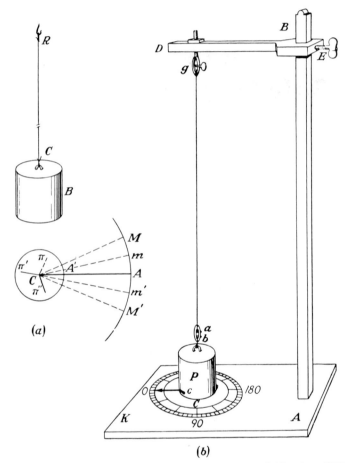

Plate 1: Coulomb's device for torsional oscillation tests of thin wires, 1784.

1
Torsion of Homogeneous Members

1.1 NAVIER'S EQUATION FOR ELASTIC CIRCULAR SECTIONS[2]

1.1.1 Derivation

The problem of torsion in an elastic circular member was first studied by Coulomb[1] in 1784. This study was a by-product of his investigation of the property of electric charges. To measure the relationship between an electric charge and a small magnetic force, Coulomb invented a very sensitive device. This device was simply a horizontal needle rod suspended from a long, thin metal wire. A small magnetic force at the end of the rod would cause a large twist of the wire, and consequently a large angular rotation of the rod.

To investigate the behavior of the wire, Coulomb first studied the torsional oscillation of a weight suspended on a wire. He found that torsional moment, T, is proportional to the twisting angle, ϕ. To determine the proportional constants between T and ϕ, he varied the diameter, length, and material of the wire and established the following equation:

$$T = \mu \frac{d^4}{l} \phi \qquad (1\text{--}1)$$

where

$d =$ diameter of the metal wire
$l =$ length of the wire
$\mu =$ a material constant (μ of steel was found to be $3\frac{1}{3}$ times that of brass.)

A theoretical equation for torsion of an elastic circular shaft, however, was derived about 40 years later by Navier in his famous lecture notes on

3

strength of materials. These lecture notes were collected and published in 1826.[2] The theoretical solution was made possible by his clear understanding of equilibrium, compatibility, and stress–strain relationships.

A circular shaft fixed at one end and subjected to a torque, T, at the other, free end is shown in Fig. 1.1 (a). The longitudinal axis is designated z, the diameter d, and the length l. A transverse slice of the shaft with the length dz was also isolated, as shown in Fig. 1.1 (b). The deformation of this slice under twisting is described by two assumptions of compatibility as follows:

1. The shape of a cross section must remain unchanged after twisting.
2. A plane section must remain plane after twisting; i.e., no warping.

Using these two assumptions, a radial line of the right face of the slice should rotate an angle $d\phi$ with respect to the left face which lies at a distance dz away. Then the shear strain γ at a distance r is:

$$\gamma = \frac{rd\phi}{dz} \qquad (1\text{--}2)$$

Let's define $\theta = \dfrac{d\phi}{dz}$ = angle of twist per unit length of the shaft. Then:

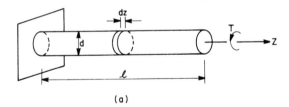

(a)

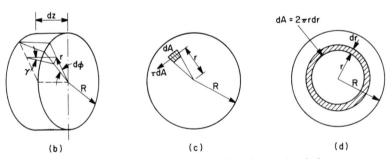

(b)   (c)   (d)

Fig. 1.1. Equilibrium and compatibility of a circular shaft.

$$\gamma = r\,\theta \tag{1-3}$$

Notice that the shear strain γ varies linearly with the distance r from the longitudinal axis. At the surface where $r = R$ the maximum shear strain $\gamma_{max} = R\theta$.

The shear stress at any distance from the longitudinal axis, τ, can be obtained from the stress–strain relationship:

$$\tau = G\gamma \tag{1-4}$$

where G is the modulus of rigidity. Substituting γ from Eq. (1–3) into Eq. (1–4) results in:

$$\tau = rG\theta \tag{1-5}$$

Equation (1–5) shows that shear stress is proportional to r, and the maximum shear stress at the surface becomes $\tau_{max} = RG\theta$.

The torque, T, is obtained from the equilibrium condition by equating external moment to internal moment. Referring to Fig. 1.1 (c):

$$T = \int \tau\, r\, dA \tag{1-6}$$

Substituting τ from Eq. (1–5) gives:

$$T = G\theta \int r^2 dA. \tag{1-7}$$

Let's define $I_p = \int r^2 dA$ = polar moment of inertia. Then:

$$T = GI_p\,\theta \tag{1-8}$$

Equation (1–8) shows that T is proportional to θ, and the proportional constant GI_p is called the torsional rigidity for a circular section. Substituting $G\theta$ from Eq. (1–8) into Eq. (1–5):

$$\tau = \frac{Tr}{I_p} \tag{1-9}$$

and the maximum shear stress $\tau_{max} = TR/I_p$ at the surface. Equations (1–5), (1–8), and (1–9) relate the quantities τ, θ, and T. They are the basic equations for elastic torsion of circular shafts.

The polar moment of inertia, I_p, can easily be calculated for a circular section. Referring to Fig. 1.1 (d):

$$I_p = \int r^2 dA = 2\pi \int_{r=0}^{r=R} r^3 dr = 2\pi \frac{R^4}{4} = \frac{\pi d^4}{32} \tag{1–10}$$

Substituting Eq. (1–10) into Eq. (1–8):

$$T = \frac{\pi}{32} Gd^4\theta \tag{1–11}$$

For a uniform torsion along the shaft, $\theta = d\phi/dz = \phi/\ell$, where ϕ is the twisting angle or torsion angle at the end of the shaft. Therefore,

$$T = \left(\frac{\pi}{32} G\right)\frac{d^4}{\ell}\phi \tag{1–12}$$

Equation (1–12) is identical to Coulomb's Eq. (1–1) if the material constant μ is taken as $(\pi/32)G$. Since Coulomb's equation is obtained from tests, it can be construed as an experimental confirmation of Navier's theoretical equation. It also implies that the assumptions of compatibility used in Navier's derivation are correct.

1.1.2 Example Problems

EXAMPLE 1.1

An automobile engine has a rated power of 230 hp at 4,400 rpm. Determine the diameter of the transmission shaft having a wall thickness of $\frac{3}{16}$ in. Use an allowable stress τ of 3 ksi, and note that 1 hp = 550 ft-lb/sec.

Solution

Let
   $T =$ torque supplied by the automobile engine
   $d_o =$ outer diameter of the transmission shaft
   $d_i =$ inner diameter of the transmission shaft

The work done by the torque T per revolution is $T(2\pi)$, and the work done by the torque T per second would be $T(2\pi)$ (4,400/60). This should be equal to the rated power of 230 (550) lb-ft/sec:

$$T(2\pi)\left(\frac{4,400}{60}\right) = 230(550)$$

$$T = 274.5 \text{ lb-ft} = 3.295 \text{ in.-k}$$

According to Eq. (1–9):

$$I_p = \frac{T \cdot r}{\tau} = \frac{3.295}{3}\left(\frac{d_o}{2}\right) = 0.5492 \, d_o \tag{a}$$

From Eq. (1–10) the polar moment of inertia of a circular tube can be expressed as:

$$I_p = \frac{\pi}{32}(d_o^4 - d_i^4) = \frac{\pi}{32}\left[d_o^4 - \left(d_o - \frac{3}{8}\right)^4\right]$$

$$= \frac{\pi}{32}[1.5 \, d_o^3 - 0.844 \, d_o^2 + 0.211 \, d_o - 0.0198] \tag{b}$$

Equating Eqs. (a) and (b), we have:

$$d_o^2 - 0.563 \, d_o - 3.588 - 0.0132\,\frac{1}{d_o} = 0 \tag{c}$$

Neglecting the small term of $(0.0132/d_o)$, then:

$$d_o = \frac{1}{2}(0.563 + \sqrt{0.563^2 + 4(3.588)}) = 2.20 \text{ in.}$$

Use a 2¼-in.-diameter transmission shaft.

EXAMPLE 1.2

As shown in Fig. 1.2, a 15-ft stepped shaft AC is made up of a 9-ft steel shaft AB of 4 in. diameter and a 6-ft aluminum shaft BC of 3 in. diameter. The moduli of elasticity, the shear rigidities, and the allowable stresses are also given in the figure for both the steel and the aluminum. The stepped shaft is fixed torsionally at both ends A and C, and a torque T is applied

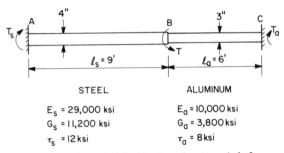

Fig. 1.2. Example 1.2—Torsion of a stepped shaft.

at the joint B. Find the allowable torque, T. Neglect the weight of the shaft.

Solution—Method 1

From Eq. (1–5), $\tau = rG\theta = rG\phi/\ell$ so that:

$$\phi = \frac{\tau\ell}{Gr}$$

For the steel shaft:

$$\phi_s = \frac{12(9)(12)}{11,200(2)} = 0.0579 \text{ rad.} \qquad \text{governs}$$

For the aluminum shaft:

$$\phi_a = \frac{8(6)(12)}{3,800(1.5)} = 0.1011 \text{ rad.}$$

For compatibility:

$$\phi_s = \phi_a = 0.0579 \text{ rad.}$$

From equilibrium, and Eq. (1–8):

$$T = T_s + T_a$$
$$= G_s I_{ps} \frac{\phi_s}{\ell_s} + G_a I_{pa} \frac{\phi_a}{\ell_a}$$

$$= 11,200 \left(\frac{\pi 4^4}{32}\right) \frac{0.0579}{9(12)} + 3,800 \left(\frac{\pi 3^4}{32}\right) \frac{0.0579}{6(12)}$$

$$= 150.8 + 24.3 = 175.1 \text{ in.-k } (19.79 \text{ kN-m})$$

The allowable torque is 175.1 in.-k (19.79 kN-m).

Solution—Method 2

Let's first calculate the polar moments of inertia:

$$I_{ps} = \frac{\pi(4)^4}{32} = 25.13 \text{ in.}^4 \qquad I_{pa} = \frac{\pi(3)^4}{32} = 7.95 \text{ in.}^4$$

From compatibility $\phi_s = \phi_a$ and using Eq. (1–8):

$$\frac{T_s l_s}{G_s I_{ps}} = \frac{T_a l_a}{G_a I_{pa}}$$

$$\frac{T_s(9)(12)}{11,200(25.13)} = \frac{T_a(6)(12)}{3,800(7.95)}$$

$$T_s = 6.21 \, T_a \qquad\qquad\qquad \text{(d)}$$

Assume steel governs and from Eq. (1–9):

$$T_s = \frac{I_{ps}\tau_s}{r_s} = \frac{25.13(12)}{2} = 150.8 \text{ in.-k } (17.04 \text{ kN-m})$$

From Eq. (d):

$$T_a = \frac{150.8}{6.21} = 24.3 \text{ in.-k } (2.75 \text{ kN-m})$$

$$T = T_s + T_a = 150.8 + 24.3 = 175.1 \text{ in.-k } (19.79 \text{ kN-m}) \qquad \text{governs}$$

Assume aluminum governs:

$$T_a = \frac{I_{pa}\tau_a}{r_a} = \frac{7.95(8)}{1.5} = 42.4 \text{ in.-k } (4.79 \text{ kN-m})$$

From Eq. (d):

$$T_s = 42.4 \, (6.21) = 263.3 \text{ in.-k } (29.75 \text{ kN-m})$$
$$T = T_a + T_s = 42.4 + 263.3 = 305.7 \text{ in.-k } (34.54 \text{ kN-m})$$

Since the assumption of steel governing gives a smaller allowable torque, the allowable $T = 175.1$ in.-k (19.79 kN-m).

1.2   ST. VENANT'S SOLUTION FOR ELASTIC NONCIRCULAR SECTIONS

1.2.1   Duleau's Experiment[3]

In Navier's book, equations for rectangular sections were also derived similar to circular sections. It was assumed that torsional stresses are proportional to the distance from the axis of twist. Therefore, Eqs. (1–8) and (1–9) should be applicable to rectangular sections if I_p is taken as the polar moment of inertia for rectangular sections:

$$I_p = \frac{ab(a^2 + b^2)}{12}$$

where a and b are the shorter and longer sides of the rectangular section. Unfortunately these simple extrapolations could not be supported by Duleau's tests.

In 1820, six years before the publication of Navier's book, Duleau made torsion tests on iron members with circular and square cross sections. He applied a torque, T, at the end of a cantilever member and then measured the resulting torsional angle, ϕ. Substituting T and $\theta = \phi/\ell$ into Eq. (1–8), the average moduli of rigidity, G, were calculated to be:

$$G = 6613 \times 10^6 \text{ kg/m}^2 \, (9.39 \times 10^6 \text{ psi}) \text{ for circular sections}$$
$$G = 5511 \times 10^6 \text{ kg/m}^2 \, (7.83 \times 10^6 \text{ psi}) \text{ for square sections}$$

To explain this 20% difference between the moduli of rigidity for a circular section and a square section, Navier wrote: "The difference of the two results is without doubt almost entirely due to the diversity of the qualities of iron. Perhaps one can also attribute a part (of this difference in results) to the fact that the previous formulae for square members do not depict as accurately the behavior as those (equations) for circular members." In other words, Navier attempted to explain the inconsistency of test results by the quality diversity of the material, but at the same time, he had doubts regarding the correctness of Eq. (1–8) for a rectangular section.

The puzzle regarding torsion of rectangular members was solved three decades later by St. Venant, in 1855. It was discovered that the polar moment of inertia, I_p, in Eq. (1-8) should be substituted by St. Venant's torsional constant C. This torsional constant is 18% less than the polar moment of inertia in the case of a square section. This correction, therefore, adequately explains the 20% difference between the moduli of rigidity for circular and square sections in Navier's analysis of Duleau's tests.

A period of three decades had passed before St. Venant was able to solve the torsion problem for rectangular sections because the solution had to await the development of the necessary mathematical tools. These tools included the Fourier series and the theory of elasticity. The equations for theory of elasticity are summarized in the following section.

1.2.2 Cauchy's Equations for Theory of Elasticity[4]

The technique of stress analysis in an elastic solid body was first developed by Cauchy[4] in 1828. He derived a total of 15 equations: 3 differential equations from the equilibrium condition, 6 differential relationships from the compatibility condition, and 6 algebraic equations from the stress–strain relationship. These 15 equations are used to solve the 15 unknowns: 6 stress components, 6 strain components, and 3 displacements of an element in a solid body.

A detailed discussion of Cauchy's equations can be found in any text on theory of elasticity such as Ref. 6. Hence, only a brief derivation will be presented here.

Stress components (referring to Fig. 1.3a):

σ_x = normal stress in the x-direction
σ_y = normal stress in the y-direction
σ_z = normal stress in the z-direction
τ_{xy} = shear stress on the x-face and in the y-direction
τ_{yz} = shear stress on the y-face and in the z-direction
τ_{zx} = shear stress on the z-face and in the x-direction

Notice that the first subscript of shear stress τ indicates the face on which τ is acting. A face is defined by the direction of its outward normal. The second subscript of τ shows the direction of the shear stress.

Strain components:

ϵ_x = elongation strain in the x-direction

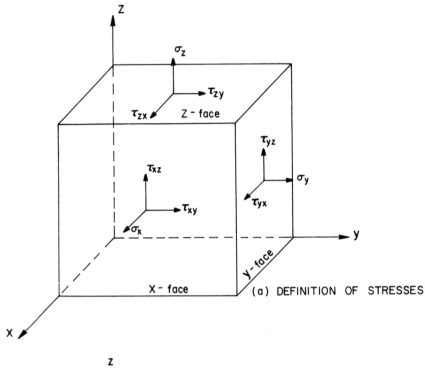

(a) DEFINITION OF STRESSES

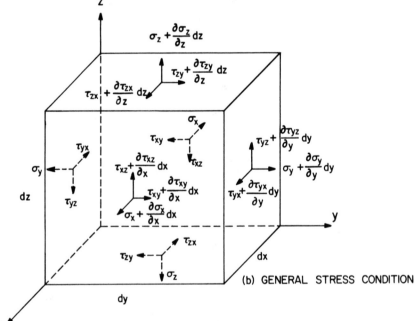

(b) GENERAL STRESS CONDITION

Fig. 1.3. Stresses on a differential interior element.

12

ϵ_y = elongation strain in the y-direction
ϵ_z = elongation strain in the z-direction
γ_{xy} = distortion strain in the x-y plane
γ_{yz} = distortion strain in the y-z plane
γ_{zx} = distortion strain in the z-x plane

Displacement components:

u = displacement in the x-direction
v = displacement in the y-direction
w = displacement in the z-direction

Equilibrium Equations. Figure 1.3 (b) shows all the stresses acting on a differential element in the interior of an elastic body. Taking the equilibrium of forces in the x-direction:

$$\left(\sigma_x + \frac{\partial \sigma_x}{\partial x} dx - \sigma_x\right) dydz + \left(\tau_{yx} + \frac{\partial \tau_{yx}}{\partial y} dy - \tau_{yx}\right) dzdx$$

$$+ \left(\tau_{zx} + \frac{\partial \tau_{zx}}{\partial z} dz - \tau_{zx}\right) dxdy = -Xdxdydz$$

Therefore:

$$\Sigma F_x = 0 \qquad \frac{\partial \sigma_x}{\partial x} + \frac{\partial \tau_{yx}}{\partial y} + \frac{\partial \tau_{zx}}{\partial z} = -X \qquad (1\text{--}13)$$

Similarly:

$$\Sigma F_y = 0 \qquad \frac{\partial \sigma_y}{\partial y} + \frac{\partial \tau_{zy}}{\partial z} + \frac{\partial \tau_{xy}}{\partial x} = -Y \qquad (1\text{--}14)$$

$$\Sigma F_z = 0 \qquad \frac{\partial \sigma_z}{\partial z} + \frac{\partial \tau_{xz}}{\partial x} + \frac{\partial \tau_{yz}}{\partial y} = -Z \qquad (1\text{--}15)$$

In Eqs. (1–13) through (1–15), X, Y, and Z designate the body forces per unit volume in the x, y, and z-directions, respectively. Taking summation of moment about x, y, and z-axes, it can be easily shown that:

$$\tau_{yz} = \tau_{zy}, \tau_{xz} = \tau_{zx}, \text{ and } \tau_{xy} = \tau_{yx}$$

This means that the two subscripts for τ can be interchanged.

Compatibility Equations. If the three displacement components u, v, and w of an elastic body can be described as continuous mathematical functions, then the three strain components for elongation can be defined as follows (refer to Fig. 1.4):

$$\epsilon_x = \frac{\left(u + \dfrac{\partial u}{\partial x} dx - u\right)}{dx} = \frac{\partial u}{\partial x} \tag{1-16}$$

$$\epsilon_y = \frac{\left(v + \dfrac{\partial v}{\partial y} dy - v\right)}{dy} = \frac{\partial v}{\partial y} \tag{1-17}$$

$$\epsilon_z = \frac{\left(w + \dfrac{\partial w}{\partial z} dz - w\right)}{dz} = \frac{\partial w}{\partial z} \tag{1-18}$$

Referring also to Fig. 1.4, the shear strain, γ_{xy}, can be expressed as a function of u and v:

$$\gamma_{xy} = \gamma_1 + \gamma_2 = \frac{\left(u + \dfrac{\partial u}{\partial y} dy - u\right)}{dy} + \frac{\left(v + \dfrac{\partial v}{\partial x} dx - v\right)}{dx}$$

Therefore:

$$\gamma_{xy} = \frac{\partial u}{\partial y} + \frac{\partial v}{\partial x} \tag{1-19}$$

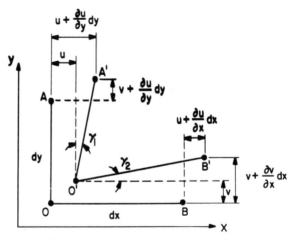

Fig. 1.4. Elongation and shear deformations.

Similarly:

$$\gamma_{yz} = \frac{\partial v}{\partial z} + \frac{\partial w}{\partial y} \tag{1-20}$$

$$\gamma_{zx} = \frac{\partial w}{\partial x} + \frac{\partial u}{\partial z} \tag{1-21}$$

Stress–Strain Relationship. Using the generalized form of Hooke's law in three dimensions for isotropic materials:

$$\epsilon_x = \frac{1}{E} [\sigma_x - \nu (\sigma_y + \sigma_z)] \tag{1-22}$$

$$\epsilon_y = \frac{1}{E} [\sigma_y - \nu (\sigma_z + \sigma_x)] \tag{1-23}$$

$$\epsilon_z = \frac{1}{E} [\sigma_z - \nu (\sigma_x + \sigma_y)] \tag{1-24}$$

$$\gamma_{xy} = \frac{1}{G} \tau_{xy} \tag{1-25}$$

$$\gamma_{yz} = \frac{1}{G} \tau_{yz} \tag{1-26}$$

$$\gamma_{zx} = \frac{1}{G} \tau_{zx} \tag{1-27}$$

In Eqs. (1–22) through (1–24), E is the modulus of elasticity, and ν is Poisson's ratio.

Equations (1–13) through (1–27) constitute the 15 equations necessary to solve the 15 unknowns of stress, strain, and displacement components.

Boundary Conditions. The six stress components should also satisfy the equilibrium condition at the surface of an elastic body, which is called the boundary condition. A tetrahedron was isolated at the surface as shown in Fig. 1.5. The direction of the surface is represented by its normal, N. The cosine of an angle between the normal and a coordinate axis is called the directional cosine. The three directional cosines are:

$$\cos (N,x) = l \quad \text{for angle between } N \text{ and } x\text{-axis}$$
$$\cos (N,y) = m \quad \text{for angle between } N \text{ and } y\text{-axis}$$
$$\cos (N,z) = n \quad \text{for angle between } N \text{ and } z\text{-axis}$$

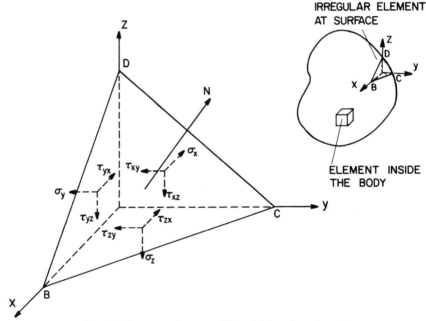

Fig. 1.5. Stresses acting on a differential boundary element.

If the area of the triangular surface is designated A, the areas of the three other faces of the tetrahedron become: $A l$, $A m$, and $A n$ on the x, y, and z-faces, respectively. Summation of forces in the three directions gives:

$$\Sigma F_x = 0 \qquad \sigma_x l + \tau_{yx} m + \tau_{zx} n = \overline{X} \qquad (1\text{--}28)$$

$$\Sigma F_y = 0 \qquad \tau_{xy} l + \sigma_y m + \tau_{zy} n = \overline{Y} \qquad (1\text{--}29)$$

$$\Sigma F_z = 0 \qquad \tau_{xz} l + \tau_{yz} m + \sigma_z n = \overline{Z} \qquad (1\text{--}30)$$

In Eqs. (1–28) through (1–30) $\overline{X}$, $\overline{Y}$, and $\overline{Z}$ designate the components of external stresses on area A in the x, y, and z-directions, respectively.

1.2.3 St. Venant's Semi-inverse Method[5]

There are two general approaches to the solution of stresses in the theory of elasticity.

(a) *Direct Approach.* Integrate the equilibrium equations to obtain stresses. The constant of integration should be chosen to satisfy the boundary conditions and compatibility conditions. The lack of a general method of integration precludes the use of this approach except in very special cases.

(b) *Inverse Approach.* Assume the displacement components u, v, and w, and substitute them into the compatibility equations to obtain the six strain components ϵ_x, ϵ_y, ϵ_z, γ_{xy}, γ_{yz}, and γ_{zx}. Then substitute these strain components into the stress–strain equations to get the six stress components σ_x, σ_y, σ_z, τ_{xy}, τ_{yz}, and τ_{zx}. With the stress components known, the body forces can be obtained from equilibrium equations and the surface forces from the boundary equations. This inverse approach is quite straightforward, but there is little chance of producing the surface stresses and body forces of practical interest.

To solve the torsion problem for noncircular cross sections, St. Venant invented a "semi-inverse" method.[5] In this semi-inverse method some features of the displacements u, v, and w are first assumed. The remaining features are then determined to satisfy all the equations of the theory of elasticity. Guided by the solutions of elementary strength of materials and the observation of deformation, it is possible to produce rigorous solutions of practical importance.

In Navier's derivation of torsion formulae for circular sections, two deformation assumptions were made and were found to be correct. These assumptions must be reevaluated according to the laboratory observation of torsional deformation for noncircular members. Such observation shows that warping of cross sections occurs after twisting. Therefore, the second assumption, that a plane section remains plane, used so successfully by Navier for circular sections, appears not to be applicable to noncircular sections. On the other hand, Navier's first assumption, that the cross-sectional shape remains unchanged, seems to be valid for noncircular sections as well as circular sections. Consequently, the following two assumptions are made to describe the displacement components for noncircular sections:

1. The shape of the cross sections remains unchanged after twisting.
2. Warping of the cross section is identical throughout the length of the noncircular member.

Based on the first assumption and referring to Fig. 1.6:

$$u = \theta z r \,(\sin \alpha) = -\theta z r \left(\frac{y}{r}\right) = -\theta z y \qquad (1\text{–}31a)$$

$$v = \theta z r \,(\cos \alpha) = +\theta z r \left(\frac{x}{r}\right) = +\theta z x \qquad (1\text{–}31b)$$

Using the second assumption:

$$w = \theta \psi(x,y) \qquad (1\text{–}31c)$$

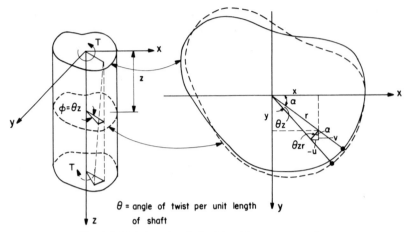

θ = angle of twist per unit length of shaft

Fig. 1.6. Twisting of a shaft with arbitrary cross section.

where $\psi(x,y)$ is the warping function and is not a function of z. $\psi(x,y)$ is unknown and will be determined later to satisfy all the equations of the theory of elasticity.

Substituting the derivatives of u, v, and w into the compatibility Eqs. (1–16) through (1–21) gives $\epsilon_x = \epsilon_y = \epsilon_z = \gamma_{xy} = 0$ and:

$$\gamma_{yz} = \theta \left(\frac{\partial \psi}{\partial y} + x \right) \tag{1–32a}$$

$$\gamma_{zx} = \theta \left(\frac{\partial \psi}{\partial x} - y \right) \tag{1–32b}$$

Substituting these six strain components into the equations of the stress–strain relationship, Eqs. (1–22) through (1–27), $\sigma_x = \sigma_y = \sigma_z = \tau_{xy} = 0$ and:

$$\tau_{yz} = G\theta \left(\frac{\partial \psi}{\partial y} + x \right) \tag{1–33}$$

$$\tau_{zx} = G\theta \left(\frac{\partial \psi}{\partial x} - y \right) \tag{1–34}$$

Substitute these six stress components into the equilibrium Eqs. (1–13) through (1–15), and assume the body forces X, Y, and Z to be zero (weightless body). It can be seen that Eqs. (1–13) and (1–14) are satisfied because τ_{zx} and τ_{zy} are not a function of z. Equation (1–15), however, becomes:

$$\frac{\partial \tau_{xz}}{\partial x} + \frac{\partial \tau_{yz}}{\partial y} = 0 \qquad (1\text{--}35)$$

Equations (1–33) through (1–35) give three equations for three unknowns, τ_{xz}, τ_{yz}, and ψ. To simplify these equations we observe that Eq. (1–35) can be satisfied by a stress function $\Phi(x,y)$ such that:

$$\tau_{xz} = \frac{\partial \Phi}{\partial y} \qquad \tau_{yz} = -\frac{\partial \Phi}{\partial x} \qquad (1\text{--}36)$$

Substitute τ_{xz} and τ_{yz} into Eqs. (1–33) and (1–34):

$$\frac{\partial \Phi}{\partial x} = -G\theta \left(\frac{\partial \psi}{\partial y} + x\right) \qquad (1\text{--}37)$$

$$\frac{\partial \Phi}{\partial y} = G\theta \left(\frac{\partial \psi}{\partial x} - y\right) \qquad (1\text{--}38)$$

Equations (1–37) and (1–38) give two equations for two unknowns, Φ and ψ. Differentiating Eqs. (1–37) and (1–38) with respect to x and y respectively, and adding the two equations, we can eliminate ψ to get the following equation with one unknown, Φ:

$$\frac{\partial^2 \Phi}{\partial x^2} + \frac{\partial^2 \Phi}{\partial y^2} = -2G\theta \qquad (1\text{--}39)$$

Equation (1–39) shows that the stress function Φ must satisfy a harmonic differential equation commonly known as Laplace's equation. It may be noted that St. Venant's original equation was expressed in terms of ψ. The use of the stress function Φ in Eq. (1–39) was introduced by Prandtl because it gave simpler equations for the boundary condition and for the torsional moment.

The boundary condition should now be established. At the lateral surface (Fig. 1.6) the surface stresses $\overline{X} = \overline{Y} = \overline{Z} = 0$, since only torsional moments are applied at the ends of the shaft. Also, the directional cosine with respect to the z-axis, $n = 0$. By substituting these values and $\sigma_x = \sigma_y = \sigma_z = \tau_{xy} = 0$ into Eqs. (1–28) through (1–30), it can be seen that the first two equations are satisfied, while the third gives:

$$\tau_{xz} l + \tau_{yz} m = 0$$

In this equation, $l = \cos(N,x) = dy/ds$ and $m = \cos(N,y) = -dx/ds$ (see Fig. 1.7). Also notice that τ_{xz} and τ_{yz} can be expressed in terms of Φ in Eq. (1–36); we get:

$$\frac{\partial \Phi}{\partial y}\frac{dy}{ds} + \frac{\partial \Phi}{\partial x}\frac{dx}{ds} = 0$$

From the chain rule:

$$\frac{d\Phi}{ds} = 0$$

This means that Φ must be a constant along the boundary of the cross section. This constant can be chosen arbitrarily for singly connected boundaries (i.e., solid bar) because the stresses are derivative of Φ and will not be affected. Hence, we choose:

$$\Phi = 0 \text{ along the boundary} \tag{1–40}$$

At the end surfaces of the shaft (Fig. 1.6), $l = m = 0$, $n = \pm1$ and $\sigma_x = \sigma_y = \sigma_z = \tau_{xy} = 0$. Substituting these values into the three boundary

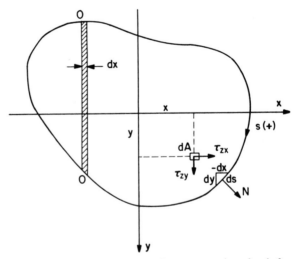

Fig. 1.7. Stresses and geometry in a cross section of a shaft.

conditions, we observe that Eq. (1–30) is satisfied, while Eqs. (1–28) and (1–29) become:

$$\tau_{zx} = \pm \overline{X} \qquad \tau_{zy} = \pm \overline{Y}$$

The positive and negative signs are for bottom and top end faces, respectively. Physically these two equations state that the stress distribution at the end faces ($\overline{X}$ and $\overline{Y}$) should be identical to that at any other section along the length of the shaft (τ_{zx} and τ_{zy}).

The force resultant on the end surfaces in the x-direction is:

$$\iint \overline{X} \, dxdy = \iint \tau_{zx} \, dxdy = \iint \frac{\partial \Phi}{\partial y} \, dxdy = \int dx \int \frac{\partial \Phi}{\partial y} \, dy$$

$\int (\partial \Phi / \partial y) dy$ is integrated in the y-direction from boundary to boundary (Fig. 1.7). It should be equal to zero according to Eq. (1–40). Therefore, the force resultant in the x-direction vanishes. Similarly, the force resultant in the y-direction will vanish, and the resultant of the shear stress at each end face must produce only a couple, T.

The couple at each end face can be expressed by:

$$T = \iint (\tau_{zy}x - \tau_{zx}y)dxdy = -\iint \frac{\partial \Phi}{\partial x} xdxdy - \iint \frac{\partial \Phi}{\partial y} ydxdy$$

Each of the two terms can be integrated by parts:

$$-\iint \frac{\partial \Phi}{\partial x} xdxdy = -\int dy \int x \frac{\partial \Phi}{\partial x} dx = -\int dy \int xd\Phi$$

$$= -\int dy \left[x\Phi \Big|_0^0 - \int \Phi dx \right] = \iint \Phi \, dxdy$$

Similarly:

$$-\iint \frac{\partial \Phi}{\partial y} ydxdy = \iint \Phi dxdy$$

Hence:

$$T = 2 \iint \Phi dxdy \qquad (1\text{–}41)$$

To summarize our progress up to this point, we note that the solution for a noncircular shaft is reduced to finding a stress function Φ that satisfies the harmonic differential Eq. (1–39) and the boundary condition expressed by Eq. (1–40). Then the torque, T, can be calculated by Eq. (1–41) and the shear stresses, τ_{xz} and τ_{yz}, by Eq. (1–36). It is interesting to note that an expression for warping function ψ is unnecessary in the analysis of stresses. ψ can be obtained, of course, from either Eq. (1–37) or Eq. (1–38), once the stress function Φ is found.

1.2.4 Solution for Rectangular Sections

Using St. Venant's semi-inverse method, the solution of stresses in a rectangular section is reduced to finding a stress function Φ that satisfies Eq. (1–39) and the boundary conditions, Eq. (1–40). For the rectangular section shown in Fig. 1.8, the boundary conditions are:

$$\Phi = 0 \text{ at } y = \pm b \tag{1–42}$$

$$\Phi = 0 \text{ at } x = \pm a \tag{1–43}$$

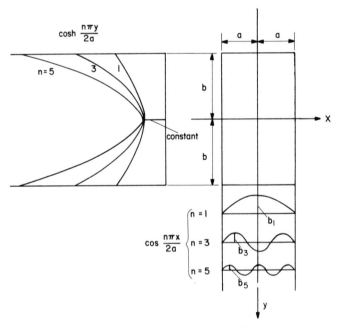

Fig. 1.8. Fourier series applied to rectangular sections.

Both Eqs. (1–39) and (1–40) can be satisfied by a cosine Fourier series:

$$\Phi = \sum_{n=1,3,5\ldots}^{\infty} b_n \cos \frac{n\pi x}{2a} \, Y_n \qquad -a < x < a \qquad (1\text{–}44)$$

where b_1, b_3, b_5, etc. are constant coefficients and Y_1, Y_3, Y_5, etc. are functions of y only. Φ in the form of the cosine series will satisfy Eq. (1–43), while Y_n will be determined to satisfy Eqs. (1–39) and (1–42).

Before we substitute Φ into Eq. (1–39), the constant $2G\theta$ on the right-hand side of the equation should also be expressed in terms of a Fourier series:

$$2G\theta = \sum_{n=1,3,5\ldots}^{\infty} 2G\theta \, \frac{4}{n\pi} \, (-1)^{(n-1)/2} \cos \frac{n\pi x}{2a} \qquad (1\text{–}45)$$

Now substituting Φ from Eq. (1–44) and $2G\theta$ from Eq. (1–45) into Eq. (1–39) and equating each term on both sides of the equation, we obtain for the typical nth term

$$\frac{\partial^2 Y_n}{\partial y^2} - \frac{n^2\pi^2}{4a^2} Y_n = -2G\theta \, \frac{4}{n\pi b_n} \, (-1)^{(n-1)/2} \qquad (1\text{–}46)$$

This is a second-order linear differential equation with constant coefficient and constant right-hand term. The general solution is:

$$Y_n = A \sinh \frac{n\pi y}{2a} + B \cosh \frac{n\pi y}{2a} + \frac{32G\theta a^2}{n^3\pi^3 b_n} (-1)^{(n-1)/2} \qquad (1\text{–}47)$$

Owing to symmetry about the x-axis, the constant $A = 0$. From the boundary Eq. (1–42), and keeping in mind the expression of Φ in Eq. (1–44), we deduce that $Y_n = 0$ when $y = \pm b$. Therefore:

$$B = -\frac{32G\theta a^2}{n^3\pi^3 b_n} (-1)^{(n-1)/2} \frac{1}{\cosh \dfrac{n\pi b}{2a}}$$

Substituting A and B into Eq. (1–47) gives:

$$Y_n = \frac{32G\theta a^2}{n^3\pi^3 b_n} (-1)^{(n-1)/2} \left[1 - \frac{\cosh \dfrac{n\pi y}{2a}}{\cosh \dfrac{n\pi b}{2a}} \right] \qquad (1\text{–}48)$$

and substituting Y_n into Eq. (1–44) results in:

$$\Phi = \frac{32G\theta a^2}{\pi^3} \sum_{n=1,3,5\ldots}^{\infty} \frac{1}{n^3} (-1)^{(n-1)/2} \left[1 - \frac{\cosh \dfrac{n\pi y}{2a}}{\cosh \dfrac{n\pi b}{2a}} \right] \cos \frac{n\pi x}{2a} \quad (1\text{–}49)$$

Substituting Φ from Eq. (1–49) into Eq. (1–37) or Eq. (1–38) and noticing the mathematical expression:

$$\frac{8a}{\pi^2} \sum_{n=1,3,5\ldots}^{\infty} \frac{1}{n^2} (-1)^{(n-1)/2} \sin \frac{n\pi x}{2a} = x \quad -a < x < a$$

we can get an expression for warping function ψ:

$$\psi = xy - \frac{32a^2}{\pi^3} \sum_{n=1,3,5\ldots}^{\infty} \frac{(-1)^{(n-1)/2} \sinh \dfrac{n\pi y}{2a}}{n^3 \cosh \dfrac{n\pi b}{2a}} \sin \frac{n\pi x}{2a} \quad (1\text{–}50)$$

This warping function ψ is not required in the following derivation of equations relating τ, θ, and T, but it will be used in Chapter 3, Section 3.2.2, in the derivation of Cowan's efficiency coefficient.

Now that Φ is found for rectangular sections, the shear stress τ_{yz} can be calculated from Eq. (1–36):

$$\tau_{yz} = \frac{16G\theta a}{\pi^2} \sum_{n=1,3,5\ldots}^{\infty} \frac{1}{n^2} (-1)^{(n-1)/2} \left[1 - \frac{\cosh \dfrac{n\pi y}{2a}}{\cosh \dfrac{n\pi b}{2a}} \right] \sin \frac{n\pi x}{2a} \quad (1\text{–}51)$$

The maximum shear stress, τ_{max}, will occur at the midpoint of the wider face. Let $b > a$; then τ_{max} will occur at $x = \pm a$ and $y = 0$. Keeping in mind that:

$$\sin \frac{n\pi}{2} = (-1)^{(n-1)/2}$$

τ_{max} is expressed from Eq. (1–51) as:

$$\tau_{max} = \frac{16G\theta a}{\pi^2} \sum_{n=1,3,5\ldots}^{\infty} \frac{1}{n^2} \left[1 - \frac{1}{\cosh \dfrac{n\pi b}{2a}} \right]$$

Observing that the first series can be summed:

$$\sum_{n=1,3,5\ldots}^{\infty} \frac{1}{n^2} = 1 + \frac{1}{3^2} + \frac{1}{5^2} + \cdots = \frac{\pi^2}{8}$$

Then:

$$\tau_{max} = G\theta \,(2a) \left[1 - \frac{8}{\pi^2} \sum_{n=1,3,5\ldots}^{\infty} \frac{1}{n^2 \cosh \dfrac{n\pi b}{2a}} \right] \tag{1-52}$$

Let $x = 2a =$ the shorter dimension of a rectangular cross section, $y = 2b =$ the large dimension of a rectangular cross section, and:

$$k = \left[1 - \frac{8}{\pi^2} \sum_{n=1,3,5\ldots}^{\infty} \frac{1}{n^2 \cosh \dfrac{n\pi y}{2x}} \right] \tag{1-53}$$

Then Eq. (1–52) can be written as

$$\tau_{max} = kG\theta x \tag{1-54}$$

The coefficient k is a function of y/x and is tabulated in Table 1.1.

The torsional moment T can also be calculated by substituting Φ from Eq. (1–49) into Eq. (1–41) and integrating:

$$T = \frac{32G\theta(2a)^3(2b)}{\pi^4} \left[\sum_{n=1,3,5\ldots}^{\infty} \frac{1}{n^4} - \frac{2a}{\pi b} \sum_{n=1,3,5\ldots}^{\infty} \frac{1}{n^5} \tanh \frac{n\pi b}{2a} \right] \tag{1-55}$$

Observing that:

$$\sum_{n=1,3,5\ldots}^{\infty} \frac{1}{n^4} = \frac{1}{1^4} + \frac{1}{3^4} + \frac{1}{5^4} + \cdots = \frac{\pi^4}{96}$$

Eq. (1–55) becomes:

$$T = G\theta(2a)^3(2b)\frac{1}{3}\left[1 - \frac{192}{\pi^5}\frac{a}{b}\sum_{n=1,3,5\ldots}^{\infty}\frac{1}{n^5}\tanh\frac{n\pi b}{2a}\right] \qquad (1\text{–}56)$$

Again, taking $x = 2a$, $y = 2b$, and:

$$\beta = \frac{1}{3}\left[1 - \frac{192}{\pi^5}\frac{x}{y}\sum_{n=1,3,5\ldots}^{\infty}\frac{1}{n^5}\tanh\frac{n\pi y}{2x}\right] \qquad (1\text{–}57)$$

T can be expressed by a simple equation:

$$T = \beta x^3 y\, G\theta \qquad (1\text{–}58)$$

The coefficient β is a function of y/x and is also tabulated in Table 1.1.

**Table 1.1 St. Venant's coefficients
for rectangular sections**

y/x	k	β	α	α_2
1.0	0.675	0.141	0.208	0.208
1.2	0.759	0.166	0.219	0.196
1.4	0.822	0.187	0.227	0.185
1.6	0.869	0.204	0.234	0.174
1.8	0.904	0.217	0.240	0.164
2.0	0.930	0.229	0.246	0.155
2.5	0.968	0.249	0.258	0.135
3.0	0.985	0.264	0.267	0.118
4.0	0.997	0.281	0.282	0.0945
5.0	0.999	0.291	0.291	0.0782
10.0	1.00	0.312	0.312	0.0397
100	1.00	0.331	0.331	0.00217
∞	1.00	0.333	0.333	0

Note: α_2 was calculated by electronic computer using eight terms and eight significant figures.

Let us define:

$$C = \beta x^3 y \qquad (1\text{–}59)$$

where C is termed the torsional constant and represents the geometric parameter in Eq. (1–58). Then Eq. (1–58) can be written as:

$$T = CG\theta \qquad (1\text{–}60)$$

The product CG is the proportional constant between T and θ, and is called the torsional rigidity.

Equation (1–54) relates τ_{max} and θ, while Eq. (1–58) relates T and θ. Eliminating θ from Eq. (1–54) and Eq. (1–58), we obtain the relationship between T and τ_{max}:

$$T = \frac{\beta}{k} x^2 y \, \tau_{max} \tag{1-61}$$

Let $\alpha = \beta/k$, which is a function of y/x and is also tabulated in Table 1.1, and we have

$$\tau_{max} = \frac{T}{\alpha x^2 y} \tag{1-62}$$

Equations (1–54), (1–58), and (1–62) are the three basic equations relating T, θ, and τ_{max}.

Using a similar derivation, the maximum shear stress at the center of the shorter face ($y = \pm b$, $x = 0$) can be expressed as:

$$\tau_{x,\,max} = \frac{T}{\alpha_2 x y^2} \tag{1-63}$$

where α_2 is also given in Table 1.1.

The distributions of shear stresses along the edges of the longer and shorter faces are shown in Fig. 1.9 for y/x ratios of 1, 2, 3, 4, 5, and 10. The distributions of shear stresses along three radial lines are given in Fig. 1.10 for these same y/x ratios. The shear stress at each point was calculated by $\sqrt{\tau_{zx}^2 + \tau_{zy}^2}$. It will be seen later, in Eq. (1–89), that this is actually the maximum stress at a point. The numbers in Figs. 1.9 and 1.10 have been computed by electronic computer, each using 500 terms of a series.

EXAMPLE 1.3

A 12 in. by 20 in. rectangular concrete section is subjected to a torque of 15 ft-k. Find the torsional constant and the maximum shear stresses on both the wider and the shorter faces.

Solution

$$\frac{y}{x} = \frac{20}{12} = 1.667$$

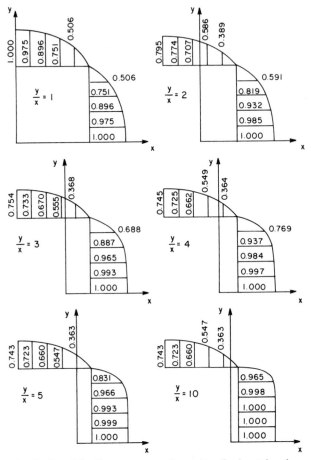

Fig. 1.9. Distribution of St. Venant stresses along edges (horizontal scale expanded).

From Table 1.1 and by linear interpolation:

$$\beta = 0.208 \qquad \alpha = 0.236 \qquad \alpha_2 = 0.171$$

The torsional constant is calculated from Eq. (1–59):

$$C = \beta x^3 y = 0.208(12)^3(20) = 7,188 \text{ in.}^4 \ (299,200 \text{ cm}^4)$$

The maximum shear stress on a wider face is:

$$\tau_{max} = \frac{T}{\alpha x^2 y} = \frac{15(12,000)}{0.236(12)^2(20)} = 265 \text{ psi } (1,827 \text{ kN/m}^2)$$

and the maximum shear stress on a shorter face is:

$$\tau_{x,\,max} = \frac{T}{\alpha_2 x y^2} = \frac{15(12,000)}{0.171(12)(20)^2} = 219 \text{ psi } (1,513 \text{ kN/m}^2)$$

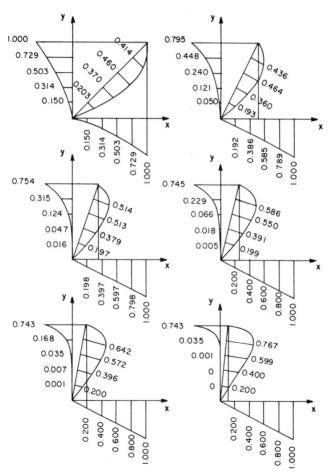

Fig. 1.10. Distribution of St. Venant stresses along three radial lines (x-scale expanded).

1.2.5 Solution for Other Cross Sections

In addition to the rectangular shape, St. Venant also obtained rigorous solutions for elliptical, triangular, and other cross sections. The torsional constant for an elliptical cross section was found to be:

$$C = \frac{1}{4\pi^2} \frac{A^4}{I_p} \qquad (1\text{-}64)$$

where $A = \pi ab$, $I_p = \pi ab^3/4 + \pi ba^3/4$, and a and b are the longer and shorter principal radii, respectively, of an ellipse. For any bulky cross sections without re-entrant corners, St. Venant found that the torsional constant can be approximately expressed by Eq. (1-64) if A and I_p are taken as the area and the polar moment of inertia, respectively, of the given cross section.

1.2.6 Bach's Equation for Thin-Walled Flanged Sections

For flanged sections such as T, L, and I sections, shown in Fig. 1.11, approximate torsion equations were suggested by Bach[7] in 1911. These equations are based on the following two assumptions:

1. The width of each rectangular component is thin as compared to the overall dimension.
2. The shape of the cross section remains unchanged after twisting. In other words, the angle of twist is the same for all rectangular components.

According to the first assumption, the torque for each thin component can be approximated by:

$$T = \frac{1}{3} x^3 y \, G\theta \qquad (1\text{-}65)$$

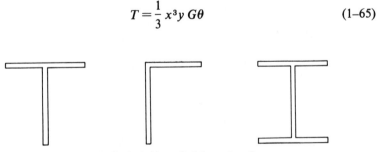

Fig. 1.11. Typical thin-walled flanged sections.

where the constant $\frac{1}{3}$ is an approximation of St. Venant's coefficient β when the y/x ratio approaches infinity (see Table 1.1). Using the second assumption, the torque for a thin-walled flanged section becomes

$$T = G\theta \underbrace{\sum \frac{1}{3} x^3 y}_{C} \qquad (1\text{–}66)$$

The geometric parameter $\sum \frac{1}{3} x^3 y$ is the torsional constant C for thin-walled flanged sections.

The maximum torsional stress along the face of the thickest rectangular component can be obtained from Eq. (1–54), in which $k = 1$ when $y/x = \infty$.

$$\tau_{max} = G\theta\, x_{max} \qquad (1\text{–}67)$$

In Eq. (1–67) x_{max} is the largest thickness among the rectangular components. Eliminating $G\theta$ from Eqs. (1–66) and (1–67), we obtain the relationship between T and τ_{max}:

$$\tau_{max} = \frac{T x_{max}}{\sum \frac{1}{3} x^3 y} \qquad (1\text{–}68)$$

Equations (1–66), (1–67), and (1–68) are Bach's three basic equations for torsion of thin-walled flanged sections. They were originally intended for application to structural steel sections where the thicknesses of rectangular components are smaller than the overall dimensions by an order of magnitude. Recent experiments, however, indicate that these equations can also be reasonably applied to concrete flanged members with rather bulky sections. This will be discussed later in Chapter 2, Section 2.4.

It should be mentioned that Bach's Eqs. (1–66), (1–67), and (1–68) apply only to *free* torsion of thin-walled open sections, where warping of the section is free to occur. If warping deformation is restrained by supports, diaphragms, or the requirement of symmetry, another type of torsion phenomenon, called warping torsion, will occur. Warping torsion will drastically modify the torsional behavior and stresses of a thin-walled open section. A detailed discussion of warping torsion is beyond the scope of this book; interested readers are referred to Refs. 8 and 9.

EXAMPLE 1.4

A double-tee girder used as the standard aerial guideways in the Miami Rapid Transit System is shown in Fig. 1.12 (a). Find the torsional constant C for this cross section. Neglect fillets in the calculation.

Solution

The double-tee cross section is divided into five parts with three shapes ①, ②, ③ as shown in Fig. 1.12 (b), (c), and (d), respectively. According to Eq. (1–66), the torsional constant of the whole double-tee is the sum of the torsional constants of the five parts. In order to maximize the torsional constant of the whole double-tee, the 8 in. by 13.25 in. pieces at the intersections of the webs and flange are included in the webs.

Shape 1—rectangle:

$$C_1 = \frac{1}{3} x^3 y = \frac{1}{3} (8)^3 (46.75) = 7{,}979 \text{ in.}^4$$

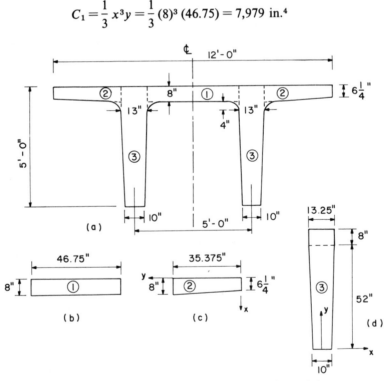

Fig. 1.12. Example 1.4—Cross section of a double-tee girder.

Shape 2—trapezoid: The thickness x can be expressed in terms of the length y as shown by the x–y coordinates in Fig. 1.12 (c).

$$x = 6.25 + \frac{8 - 6.25}{35.375} \, y = 6.25 + 0.04947 \, y$$

$$C_2 = \frac{1}{3} \int_0^{35.375} (6.25 + 0.04947 \, y)^3 \, dy$$

$$= \frac{1}{3} \left[\frac{(6.25 + 0.04947 \, y)^4}{4(0.04947)} \right]_0^{35.375}$$

$$= 4,329 \text{ in.}^4$$

Shape 3—a trapezoid plus a rectangle: The thickness x of the trapezoid can be expressed in terms of length y as shown in Fig. 1.12 (d).

$$x = 10 + \frac{13.25 - 10}{52} \, y = 10 + 0.0625 \, y$$

$$C_3 = \frac{1}{3} \int_0^{52} (10 + 0.0625 \, y)^3 \, dy + \frac{1}{3} (13.25)^3 (8)$$

$$= \frac{1}{3} \left[\frac{(10 + 0.0625 \, y)^4}{4(0.0625)} \right]_0^{52} + 6,203$$

$$= 27,763 + 6,203 = 33,966 \text{ in.}^4$$

Total:

$$C = C_1 + 2C_2 + 2C_3$$
$$= 7,979 + 2(4,329) + 2(33,966) = 84,569 \text{ in.}^4$$

1.3 BREDT'S THIN-TUBE THEORY

According to St. Venant, a prismatic member subjected to torsion will produce only circulating shear stresses. The shear stress will be zero at the axis of twist and will increase linearly to a maximum at the surface (except at sharp corners). From a practical point of view, the most efficient cross section to resist torsion would then be a thin tube, where all the materials will be fully stressed. Very simple equations were derived for thin tubes by Bredt[10] in 1896. These equations are also very useful for torsion of reinforced concrete members, which will be discussed in Chapters 3, 4 and 7.

As in St. Venant's assumption for solid members, the cross-sectional shape of a prismatic thin tube will remain unchanged after twisting. The warping deformation perpendicular to the cross section will also be uniform throughout the length of the tube. Such deformations imply that all components of stresses should vanish except the shear components τ_{xz} and τ_{yz}. For a thin tube, the direction of the resultant shear stress τ must follow the boundary as shown in Fig. 1.13.

A small length of a thin tube with variable wall thickness is shown in Fig. 1.13. The tube has a longitudinal axis z and is subjected to a torque T about the z-axis. An element $ABCD$ is isolated with the shear stresses shown. The shear stress on face AD is denoted τ_1 and that on face BC is τ_2. The thicknesses at faces AD and BC are designated h_1 and h_2, respectively. Taking the equilibrium of forces of the element in the longitudinal z-direction and canceling out dz:

$$\tau_1 h_1 = \tau_2 h_2 \tag{1–69}$$

Since shear stresses on mutually perpendicular planes must be equal, the shear stresses on face AB must be τ_1 at point A and τ_2 at point B. Equation (1–69), therefore, means that τh on face AB must be equal at points A and B. For convenience we define $q = \tau h$ as the shear flow, which is the shear force per unit length. Then q must be equal at points A and B. Notice also that the two longitudinal planes at AD and BC are selected at an arbitrary distance apart. It follows that the shear flow q must be constant around the cross sections of a thin tube.

The relationship between T and τ can be derived directly from the equilibrium of moment about the z-axis. As shown in Fig. 1.13, the shear force along a length of wall ds is qds. The contribution of this element to torsional resistance is $qds(r)$, where r is the distance from the center of twist (z-

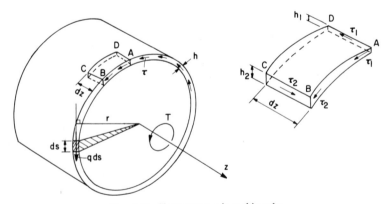

Fig. 1.13. Shear stresses in a thin tube.

axis) to the shear force qds. Since q is a constant, integration along the wall of the whole cross section gives the total torsional resistance:

$$T = q \oint rds \qquad (1\text{-}70)$$

From Fig. 1.13 it can be seen that rds is equal to twice the area of the shaded triangle formed by r and ds. Summing these areas around the whole cross section results in:

$$\oint rds = 2A \qquad (1\text{-}71)$$

where A is the whole area bounded by the center line of the wall. Substituting Eq. (1–71) into Eq. (1–70) gives:

$$q = \tau h = \frac{T}{2A} \qquad \text{or} \qquad \tau = \frac{T}{2Ah} \qquad (1\text{-}72)$$

Incidentally, the resultant of the shear flow q integrated over the whole cross section in any direction perpendicular to the z-axis must be zero.

The relationship between T and θ can be derived from the compatibility condition of warping deformation. In Fig. 1.14, a longitudinal cut is made in an infinitesimal length dz of a thin tube. Progressing from one side of the cut along the perimeter to the other side of the cut, the differential warping displacement (in the z-direction) integrating from the beginning to the end must be zero.

The warping displacement of a differential element A as shown in Fig. 1.14 (a) and (b) is induced by the rotation $d\phi$ and by the shear deformation γ. Figure 1.14 (a) shows the differential warping displacement of element A due to rotation. Under torsion, element A rotates an angle $rd\phi/dz$ in the tangential direction of the perimeter, and the differential warping displacement is:

$$dw = -r\frac{d\phi}{dz}\,ds = -r\theta\,ds \qquad (1\text{-}73)$$

The differential warping displacement of element A due to shear deformation is shown in Fig. 1.14 (b):

$$dw = \gamma\,ds \qquad (1\text{-}74)$$

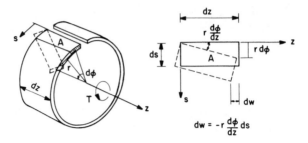

(a) WARPING DISPLACEMENT (IN Z-DIRECTION)
DUE TO ANGLE OF TWIST

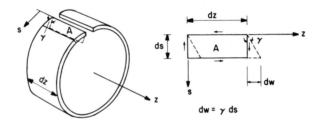

(b) WARPING DISPLACEMENT (IN Z-DIRECTION)
DUE TO SHEAR DEFORMATION

Fig. 1.14. Warping displacement in a thin tube.

Adding Eqs. (1–73) and (1–74) gives the differential warping displacement due to both the rotation and the shear deformation:

$$dw = -r\theta ds + \gamma ds \qquad (1\text{–}75)$$

For a closed section the total differential warping displacement integrating around the whole perimeter must be equal to zero:

$$\oint dw = -\theta \oint r ds + \oint \gamma ds = 0 \qquad (1\text{–}76)$$

Recalling $\oint r ds = 2A$ from Eq. (1–71), Eq. (1–76) becomes:

$$\oint \gamma ds = 2\theta A \qquad (1\text{–}77)$$

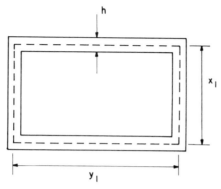

Fig. 1.15. Rectangular box section.

Substituting $\gamma = \tau/G$ from the stress–strain relationship into Eq. (1–77):

$$\oint \tau \, ds = 2G\theta A \qquad (1\text{--}78)$$

Replacing τ by $T/2Ah$ according to Eq. (1–72):

$$T = \frac{4A^2}{\underbrace{\oint \dfrac{ds}{h}}_{C}} G\theta \qquad (1\text{--}79)$$

where $4A^2/\left(\oint ds/h\right)$ is the torsional constant C for a thin tube with variable wall thickness. Equations (1–72), (1–78), and (1–79) are the three basic equations relating T, θ, and τ for thin tubes with variable wall thickness.

For thin tubes with uniform wall thickness, h is constant, and the torsional constant is:

$$C = \frac{4A^2 h}{u} \qquad (1\text{--}80)$$

where $u = \oint ds$ = the circumference of the center line of the wall. For a rectangular box section, shown in Fig. 1.15, $A = x_1 y_1$, $u = 2(x_1 + y_1)$, and Eq. (1–80) becomes:

$$C = \frac{2x_1^2 \, y_1^2 h}{(x_1 + y_1)} \qquad (1\text{--}81)$$

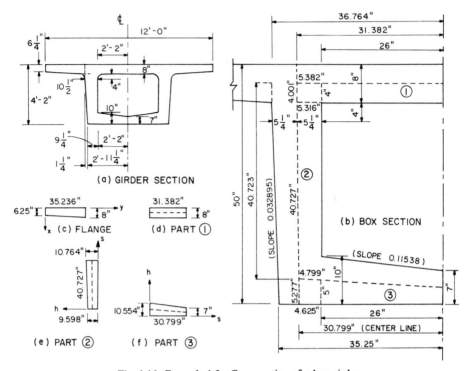

Fig. 1.16. Example 1.5—Cross section of a box girder.

EXAMPLE 1.5

A box girder with overhanging flanges, designed as an alternative to the double-tee girder (see Example 1.4) used in the Miami Rapid Transit System, is shown in Fig. 1.16 (a). Find the torsional constant for this cross section. Neglect the fillets in the calculation.

Solution

The cross section of the girder is divided into a box, shown in Fig. 1.16 (b), and two flanges, shown in Fig. 1.16 (c). The torsional constant C of the girder is assumed to be the sum of the torsional constant of the box and that of the flanges. The torsional constant of the box is calculated from Eq. (1–79):

$$C = \frac{4A^2}{\oint \frac{ds}{h}}$$

Note that A and s are measured from the center line of the box wall thickness. Detailed geometry of the center line and the wall thickness are given in Fig. 1.16 (b).

$$A = 30.799(40.723)(2) + 30.799(5.277 - 3.5) + (5.382 - 4.799)(40.723)$$
$$= 2,508.5 + 54.7 + 23.7 = 2,587 \text{ in.}^2$$

The calculation of $\oint (ds/h)$ can be divided into three parts ①, ②, and ③ as shown in Fig. 1.16 (d), (e) and (f), respectively.

Part ①:

$$\int_1 \frac{ds}{h} = \frac{31.382}{8} = 3.923$$

Part ②:

$$h = 9.598 + 0.03289s$$

$$\int_2 \frac{ds}{h} = \int_0^{40.727} \frac{ds}{(9.598 + 0.03289s)}$$
$$= \left[\frac{\ln(9.598 + 0.03289s)}{0.03289} \right]_0^{40.727} = 3.973$$

Part ③:

$$h = 10.554 - 0.11538s$$

$$\int_3 \frac{ds}{h} = \int_0^{30.799} \frac{ds}{(10.554 - 0.11538s)}$$
$$= \left[\frac{\ln(10.554 - 0.11538s)}{0.11538} \right]_0^{30.799} = 3.559$$

$$\oint \frac{ds}{h} = 2\left(\int_1 \frac{ds}{h} + \int_2 \frac{ds}{h} + \int_3 \frac{ds}{h} \right)$$
$$= 2(3.923 + 3.973 + 3.559) = 22.91$$

$$C_{box} = \frac{4(2,587)^2}{22.91} = 1,168,500 \text{ in.}^4 \ (48.64 \times 10^6 \text{ cm}^4)$$

The torsional constant for an overhanging flange is shown in Fig. 1.16 (c). It can be calculated as illustrated in Example 1.4.

$$x = 6.25 + 0.04967\ y$$

$$C_{fl} = \frac{1}{3}\ x^3 y = \frac{1}{3} \int_0^{35.236} (6.25 + 0.04967\ y)^3\ dy$$

$$= \frac{1}{3} \left[\frac{(6.25 + 0.04967\ y)^4}{4(0.04967)} \right]_0^{35.236}$$

$$= 4{,}313\ \text{in.}^4\ (0.1795 \times 10^6\ \text{cm}^4)$$

The total torsional constant of the girder is:

$$C = C_{box} + 2C_{fl} = 1{,}168{,}500 + 2(4{,}313)$$
$$= 1{,}177{,}100\ \text{in.}^4\ (49.00 \times 10^6\ \text{cm}^2)$$

Comparing this torsional constant with the torsional constant of the double-tee section in Example 1.4 (84,569 in.^{4}), it can be seen that a box girder is torsionally stiffer than a double-tee girder by an order of magnitude.

1.4 PRANDTL'S MEMBRANE ANALOGY

In 1903 Prandtl[11] discovered an interesting analogy between the stress function in the torsion problem and the deflection of a membrane under uniform loading. The analogy stems from the observation that both the stress function and the deflection are governed by Laplace's harmonic differential equation, and must also satisfy the same boundary conditions. This analogy provides a convenient tool to visualize the direction and the magnitude of shear stresses on any cross section subjected to torsion. Based on this analogy, devices have also been invented to determine the torsional rigidity and shear stresses for various complicated cross sections, which were difficult to solve by mathematical methods.

Figure 1.17 shows a homogeneous membrane, such as a soap film, stretched horizontally within a boundary having a shape similar to the cross section of a given torsion beam. The uniform tensile force per unit length of the membrane is denoted S. The membrane is also subjected to a uniform load per unit area, q, that produces a deflection, w. Taking the coordinate system as shown, q and w are positive when pointing downward. Isolate an infinitesimal element A and take the equilibrium of forces in the vertical direction. It can be seen that the membrane stresses in the x and y directions will produce upward forces $S(\partial^2 w/\partial x^2)dxdy$ and $S(\partial^2 w/\partial y^2)dxdy$, respectively, due to the curvature of the membrane. Hence:

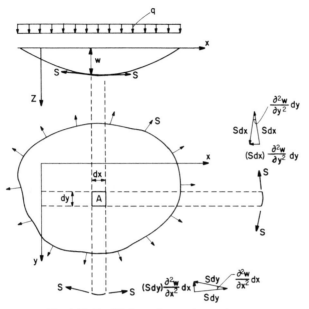

Fig. 1.17. Equilibrium of a membrane element.

$$q\,dxdy + S\frac{\partial^2 w}{\partial x^2}\,dxdy + S\frac{\partial^2 w}{\partial y^2}\,dxdy = 0$$

$$\frac{\partial^2 w}{\partial x^2} + \frac{\partial^2 w}{\partial y^2} = -\frac{q}{S} \qquad\qquad (1\text{--}82)$$

Equation (1–82) states that the deflection w must satisfy Laplace's harmonic differential equation. It must also be zero along the boundary.

The equations for the deflection w in the membrane analogy and those for the stress function Φ in the torsion solution are compared as follows:

TORSION	MEMBRANE
$\dfrac{\partial^2 \Phi}{\partial x^2} + \dfrac{\partial^2 \Phi}{\partial y^2} = -2G\theta$	$\dfrac{\partial^2 w}{\partial x^2} + \dfrac{\partial^2 w}{\partial y^2} = -\dfrac{q}{S}$
$\Phi = 0$ along the boundary	$w = 0$ along the boundary
$\tau_{zy} = -\dfrac{\partial \Phi}{\partial x}$	$(\text{slope})_x = \dfrac{\partial w}{\partial x}$
$\tau_{zx} = \dfrac{\partial \Phi}{\partial y}$	$(\text{slope})_y = \dfrac{\partial w}{\partial y}$
$T = 2\iint\Phi\,dxdy$	$\text{Volume} = \iint w\,dxdy$

Comparison of the first two pairs of equations shows that both w and Φ must satisfy Laplace's harmonic equation and the same boundary condition. Therefore, if w and q/s are replaced by Φ and $2G\theta$, respectively, a solution for the membrane can be converted into a solution for torsion. If the stress function Φ in the torsion problem is imagined as the deflection of a membrane, the third and fourth pair of equations imply that the shear stresses τ_{zy} and τ_{zx} can be visualized as the slope of the deflected membrane in the x and y directions. The last pair of equations also shows that the torque T can be taken as two times the volume bounded by the deflected membrane and the x–y plane.

The direction and magnitude of shear stresses can be visualized by the contour lines of the deflected surface. A contour line is a curve connecting all the points that have the same deflection. Using this definition and denoting s as the length along the contour line:

$$\frac{dw}{ds} = 0 \text{ along the contour line} \tag{1-83}$$

Similarly we can set $\partial\Phi/\partial s = 0$ in terms of the chain rule:

$$\frac{d\Phi}{ds} = \frac{\partial\Phi}{\partial y}\frac{dy}{ds} + \frac{\partial\Phi}{\partial x}\frac{dx}{ds} = 0 \tag{1-84}$$

Using the definition of the stress function, Eq. (1–36), and referring to Fig. 1.18, Eq. (1–84) becomes:

$$\tau_{zx} \cos\alpha - \tau_{zy} \sin\alpha = 0 \tag{1-85}$$

where α is the angle between the normal and the x-axis. Equation (1–85) states that the sum of the projection of τ_{zx} and $-\tau_{zy}$ on the normal to the contour line, n, must be equal to zero. In other words, the resultant shear stress τ must be tangential to the contour line.

The magnitude of the resultant shear stress τ can be obtained by projecting τ_{zy} and $-\tau_{zx}$ at a point onto the tangent of the contour line:

$$\tau = \tau_{zy} \cos\alpha - \tau_{zx} \sin\alpha \tag{1-86}$$

The maximum shear stress τ can be obtained by taking the derivative of τ with respect to α in Eq. (1–86) and equating it to zero:

$$-\tau_{zy} \sin\alpha - \tau_{zx} \cos\alpha = 0$$

$$\tan\alpha = \frac{-\tau_{zx}}{\tau_{zy}} \tag{1-87}$$

from which:

$$\sin \alpha = \frac{-\tau_{zx}}{\sqrt{\tau_{zx}^2 + \tau_{zy}^2}} \qquad (1\text{--}88a)$$

and:

$$\cos \alpha = \frac{\tau_{zy}}{\sqrt{\tau_{zx}^2 + \tau_{zy}^2}} \qquad (1\text{--}88b)$$

Substituting $\sin \alpha$ and $\cos \alpha$ from Eqs. (1–88a) and (1–88b) into Eq. (1–86) results in:

$$\tau_{\max} = \sqrt{\tau_{zx}^2 + \tau_{zy}^2} \qquad (1\text{--}89)$$

Thus the maximum shear stress at a point is the resultant of the shear stresses in the x and y directions, τ_{zx} and τ_{zy}.

τ can also be expressed in terms of Φ by substituting Eq. (1–36) into Eq. (1–86) and recalling the definition of α in Fig. 1.18:

$$\tau = -\frac{\partial \Phi}{\partial x} \frac{dx}{dn} - \frac{\partial \Phi}{\partial y} \frac{dy}{dn} = -\frac{d\Phi}{dn} \qquad (1\text{--}90)$$

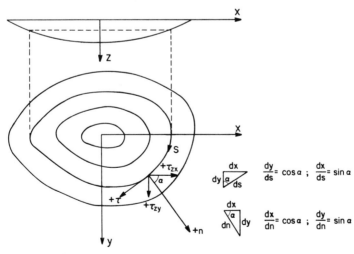

Fig. 1.18. Contour lines and direction of shear stresses.

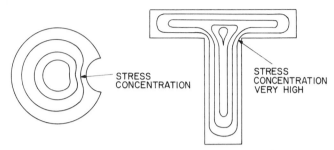

Fig. 1.19. Contour lines showing stress concentration.

Comparing with the slope of a deflected membrane along the normal, dw/dn, it can be seen that the magnitude of the resultant shear stress τ at a point can be visualized as the slope of a deflected membrane in the normal direction at that point. Hence the closer the contour lines near a point, the larger the shear stress.

The contour lines are often used to illustrate the phenomenon of stress concentration. Figure 1.19 shows the contour lines for a circular cross section with a notch and those for a tee cross section. Large stress concentration occurs near the notch of the circular section and at the re-entrant corners of the tee section because the contour lines are most congested at these points.

1.5 NADAI'S SAND-HEAP ANALOGY (PLASTIC TORSION)

Prandtl's membrane analogy was extended to the case of plastic material by Nadai[12] in 1923. In the plasticity problem, the solution must satisfy the plasticity condition in addition to the equilibrium and boundary conditions. The compatibility conditions are automatically satisfied because the material will yield and will adjust itself to maintain continuity.

The stress–strain curve of a material in shear can be idealized into two straight lines as shown in Fig. 1.20. In the plastic region:

$$\tau = \tau_{\text{max}} = k \tag{1-91}$$

where k is the yield constant. According to the Huber-Mises-Hencky distortion energy criterion for yielding, $k = \sigma_y/\sqrt{3}$, where σ_y is the uniaxial yield stress. For Tresca's maximum shear stress criterion of yielding, $k = \sigma_y/2$.

The solution for plastic torsion follows the same procedures as in St. Venant's elastic torsion. To summarize, we first assume that the shape of the cross section remains unchanged after twisting of the member, and that the

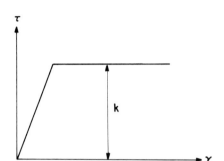

Fig. 1.20. Stress–strain curve in shear showing plastic region.

warping of the cross section is identical throughout the length of the member. From these assumptions we arrive at the conclusion that all stress components except τ_{zx} and τ_{zy} must vanish. Second, we substitute τ_{zx} and τ_{zy} into Cauchy's three equilibrium equations and show that τ_{zx} and τ_{zy} must satisfy Eq. (1–35), i.e.:

$$\frac{\partial \tau_{zx}}{\partial x} + \frac{\partial \tau_{zy}}{\partial y} = 0$$

This equation can be satisfied by a stress function F defined by:

$$\tau_{zx} = \frac{\partial F}{\partial y} \quad \text{and} \quad \tau_{zy} = -\frac{\partial F}{\partial x} \tag{1–92}$$

The plasticity condition of Eq. (1–91) can be written as follows using Eq. (1–89):

$$\sqrt{\tau_{zx}^2 + \tau_{zy}^2} = k \tag{1–93}$$

Substituting Eq. (1–92) into Eq. (1–93), we obtain the governing equation

$$\sqrt{\left(\frac{\partial F}{\partial x}\right)^2 + \left(\frac{\partial F}{\partial y}\right)^2} = k \tag{1–94}$$

A stress function F that satisfies Eq. (1–94) is called a plastic stress function. In Eq. (1–89) the term $\sqrt{\tau_{zx}^2 + \tau_{zy}^2}$ was shown to be the maximum shear stress at a point. Similarly, $\sqrt{(\partial F/\partial y)^2 + (\partial F/\partial y)^2}$ can be shown to be the maximum slope of F. Thus Eq. (1–94) can also be written as:

$$\text{max slope of } F = k \tag{1–95}$$

In other words, the plastic stress function F is a surface of constant maximum slope. Substituting Eq. (1–91) into Eq. (1–95):

$$\tau = \text{max slope of } F \tag{1–96}$$

The boundary condition for the plastic stress function F can be derived in the same manner as the elastic stress function, which results in

$$F = 0 \text{ along the boundary of lateral surface} \tag{1–97}$$

$$T = 2\!\iint\! F \, dxdy \text{ at the end surface} \tag{1–98}$$

Summarizing the characteristics of the plastic stress function F from Eqs. (1–95) through (1–98), it can be concluded that F is a surface of constant maximum slope, which can be constructed over the cross-sectional boundary of a twisted member. The maximum slope of the surface represents the resultant shear stress, and twice the volume enclosed by the surface and the plane of the boundary gives the end torque.

The plastic stress function F can be visualized by a sand-heap analogy. If a piece of stiff paper is cut into the shape of the cross section of a twisted member and is covered with sand while lying horizontally, the natural slope of the sand heap gives a picture of the surface F. Figure 1.21 shows the

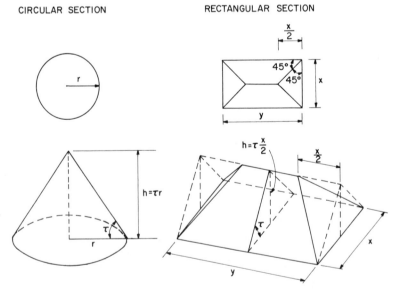

Fig. 1.21. Sand-heap analogy for circular and rectangular sections.

sand heaps on a circular and a rectangular shape. Both surfaces of the sand heaps satisfy Eqs. (1-95) and (1-97). The remaining two equations, (1-96) and (1-98), are used to derive the relationship between T and τ for these two shapes as follows:

Circular sections:

From Eq. (1-96), $h = \tau r$

From Eq. (1-98), $T = 2$ Volume $= \dfrac{2\pi r^2 h}{3}$ (1-99)

$$T = \frac{2\pi r^3}{3}\, \tau$$

Rectangular section:

From Eq. (1-96), $h = \tau \dfrac{x}{2}$

From Eq. (1-98), $T = 2$ Volume $= 2\left(\dfrac{xh}{2}\,y - 2\dfrac{1}{3}\dfrac{xh}{2}\dfrac{x}{2}\right)$ (1-100)

$$T = \left(\frac{1}{2} - \frac{1}{6}\frac{x}{y}\right)x^2 y\,\tau$$

or $T = \alpha_p\, x^2 y\,\tau$

where $\alpha_p = [0.5 - (\frac{1}{6})(x/y)]$ is called the plastic coefficient. α_p varies from $\frac{1}{3}$ when $y/x = 1$ to $\frac{1}{2}$ when $y/x = \infty$.

The relationship between T and τ is also quite useful for flanged sections in the fully plastic stage. Referring to Fig. 1.22, equations for T, L, and I sections are derived:

T and L sections:

When $b_w > h_f$,

$$T = \tau\left[b_w^2 h\left(\frac{1}{2} - \frac{1}{6}\frac{b_w}{h}\right) + \frac{1}{2}h_f^2\,(b - b_w)\right]$$ (1-101)

When $b_w < h_f$,

$$T = \tau\left[h_f^2 b\left(\frac{1}{2} - \frac{1}{6}\frac{h_f}{b}\right) + \frac{1}{2}b_w^2\,(h - h_f)\right]$$ (1-102)

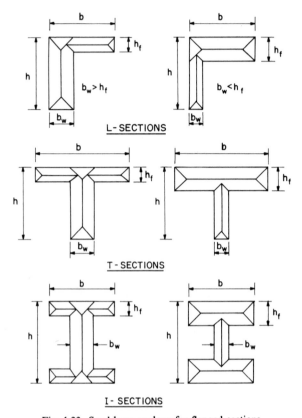

Fig. 1.22. Sand-heap analogy for flanged sections.

I section:

When $b_w > h_f$,

$$T = \tau \left[b_w^2 h \left(\frac{1}{2} - \frac{1}{6} \frac{b_w}{h} \right) + h_f^2 (b - b_w) \right] \qquad (1\text{--}103)$$

When $b_w < h_f$,

$$T = \tau \left[2 h_f^2 b \left(\frac{1}{2} - \frac{1}{6} \frac{h_f}{b} \right) + \frac{1}{2} b_w^2 \left(h - 2 h_f + \frac{b_w}{3} \right) \right] \qquad (1\text{--}104)$$

The uses of Eqs. (1–100) and (1–104) are given in Examples 2.1 and 2.2, respectively, in Chapter 2.

REFERENCES

1. Coulomb, C. A., "Recherches théoriques et experimentales sur la force de torsion et sur l'élasticité des fils de métal," *Histoire de l'Academie Royale des Sciences,* Paris, 1784; L'Imprimerie Royale, Paris, 1787, pp. 229–269.
2. Navier, C. L., *Résumé des leçons données à l'école des ponts et chaussées sur l'application de la mécanique à l'éstablissement des constructions et des machines.* Première partie, "continant les leçons sur la résistance des matériaux et sur l'établissement des constructions en terre, en maçonnerie et en charpente," Firmin Didet, Paris, 1826. (Article V, "de la résistance d'un corps prismatique à la torsion," pp. 71–76. Article VI, "de la resistance d'un corps prismatique à la rupture causée par la torsion," pp. 76–80.)
3. Duleau, A., "Essai théorique et experimental sur la résistance du fer forgé," Paris, 1820.
4. Cauchy, A., "Sur les équations qui expérimentent les conditions d'équilibre ou les lois de mouvement intérieur d'un corps solide," *Exercices de mathématique,* Paris, 1828.
5. Saint-Venant, B. de, Mémoire sur la torsion des prismes (lu à l'Académie le 13 juin 1853). *Mémoires des savants étrangers,* Mémoires présentés par divers savants à l'Académie des Sciences, de l'Institut Imperial de France et imprimé par son ordre, V. 14, Imprimerie Impériale, Paris, 1856, pp. 233–560.
6. Timoshenko, S. and J. N. Goodier, *Theory of Elasticity,* 2nd ed., McGraw-Hill Book Co., N.Y., 1951, p. 506.
7. Bach, B., *Elasticitat und Festigkeit,* 6th ed., 1911.
8. Vlasov, V. Z., *Thin-walled Elastic Beams,* 2nd ed., 1959 (in Russian). Translation available as publication TT61–11400, Office of Technical Services, U.S. Department of Commerce, Washington, D.C.
9. Kollbrunner, C. F. and K. Basler, *Torsion in Structures,* Springer-Verlag, New York, 1969.
10. Bredt, R., "Kritische Bemerkungen zur Drehungselastizitat," *Zeitschrift des Vereines Deutscher Ingenieure,* Band 40, No. 28, July 11, 1896, pp. 785–790; No. 29, July 18, 1896, pp. 813–817.
11. Prandtl, L., "Zur Torsion von prismatischen Staben," *Physik Zeitschrift* 4, 1903, p. 758.
12. Nadai, A., *Theory of Flow and Fracture of Solids,* Vol. 1, McGraw-Hill Book Co., New York, 1950, Chapter 35.

Plate 2: Barré de Saint-Venant (1797–1886), who has produced the rigorous solutions for homogeneous non-circular members subjected to torsion. His name has been used to define the type of torsion, in which the shear stresses are flowing in a circulatory manner.

2
Torsion of Plain Concrete Members

2.1 BEHAVIOR OF RECTANGULAR SECTIONS

The torsional behavior of plain concrete members can be demonstrated by a series of torque-twist curves, shown in Fig. 2.1 (a). The angle of twist was obtained by measuring the rotation between two cross sections and then dividing the rotation by the distance between the sections. It can be seen that the torque-twist curves behave elastically at low torque and become somewhat curved at higher load. Failure is brittle and is always accompanied by a loud noise.

The slope of a torque-twist curve at a point, $dT/d\theta$, representing the torsional rigidity, can be related to the stress–strain curve of concrete in uniaxial compression and in uniaxial tension. To show this relationship a typical stress–strain curve, including both compression and tension, is shown in Fig. 2.2. This curve has the following characteristics:

1. A compressive stress–strain curve is approximately straight up to about one-half of the ultimate compressive strength ($0.5 f'_c$). It becomes curved thereafter and reaches a maximum at a strain of about 0.002. Beyond this strain, the curve has a descending branch and terminates at a strain of 0.003 to 0.01, depending on the strength of concrete.
2. A tensile stress–strain curve is approximately straight up to failure. Failure occurs at a strain of approximately 0.0001. No descending branch can be discerned without special test set-up and control.
3. The tensile strength of concrete is roughly one-tenth of the compressive strength. The initial moduli of elasticity are very close for compressive and tensile tests.

Using St. Venant's Eq. (1–58) and expressing G_c as $E_c/2(1 + \nu)$, torsional rigidity for rectangular sections at a point on a torque-twist curve can be written as:

$$\frac{dT}{d\theta} = \beta x^3 y \frac{E_c}{2(1 + \nu)} \qquad (2-1)$$

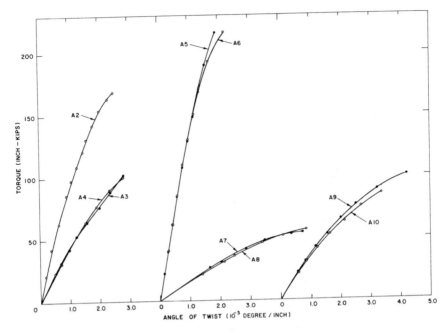

(a)   Torque-twist curves for series A (Ref. 1)

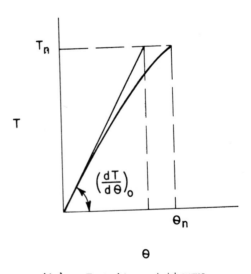

(b)   Typical torque-twist curve

Fig. 2.1. Torque-twist curve of plain concrete beams.

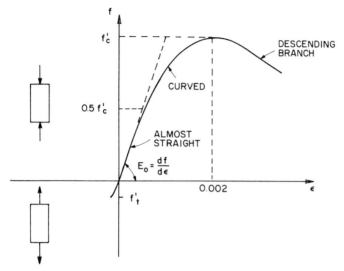

Fig. 2.2. Stress–strain curve of concrete including uniaxial compression and uniaxial tension.

At zero load T/θ can be written as $(dT/d\theta)_0$ (see Fig. 2.1b), and E_c becomes the initial modulus of elasticity, E_0 (see Fig. 2.2). Hence, E_0 can be expressed by:

$$E_0 = \frac{2(1 + \nu)}{\beta x^3 y} \left(\frac{dT}{d\theta}\right)_0 \qquad (2\text{--}2)$$

Using the measurement of $(dT/d\theta)_0$ from torsion tests, E_0 was found[1] from Eq. (2–2) to be very close to the E_0 obtained from the stress–strain curve of axial compression and axial tension tests. It can, therefore, be stated that St. Venant's theory can accurately describe the torsion behavior of plain concrete members at low torque.

At high load, the torque-twist curve deviates from the straight line representing the initial torsional rigidity. However, the curve can still be approximated by a straight line if some kind of secant modulus of elasticity is used. The secant modulus of elasticity at the stress level of about 0.5 f'_c can be estimated by the equation recommended by the ACI Code:

$$E_c = 33 \; w^{1.5} \sqrt{f'_c} \qquad (2\text{--}3)$$

where w is the unit weight of concrete. This equation, when used in Eq. (2–1), appears to give a good approximation of the whole range of torsional rigidity.

The applicability of St. Venant's theory can also be confirmed by the measurement of the maximum tensile strains ϵ_{max} at the center of the longer face of a torsional member.[1] ϵ_{max} was measured by electric strain gages located at an angle inclined at 45° to the longitudinal axis of the beam. Figure 2.3 shows that the T vs. ϵ_{max} curve is almost a straight line at lower range of torque. The slope of this straight line can be derived from St. Venant's Eq. (1–62), $\tau_{max} = T/\alpha x^2 y$. In this case of pure shear, the maximum shear stress is equal to the maximum tensile stress, $\tau_{max} = \sigma_{max}$. Noting also that $\sigma_{max} = E_c \epsilon_{max}$, then the slope T/ϵ_{max} is:

$$\frac{T}{\epsilon_{max}} = \alpha x^2 y\, E_c \qquad (2\text{–}4)$$

At zero load T/ϵ_{max} can be written as $(dT/d\epsilon_{max})_0$, and the initial modulus of elasticity E_0 is:

$$E_0 = \frac{1}{\alpha x^2 y} \left(\frac{dT}{d\epsilon_{max}} \right)_0 \qquad (2\text{–}5)$$

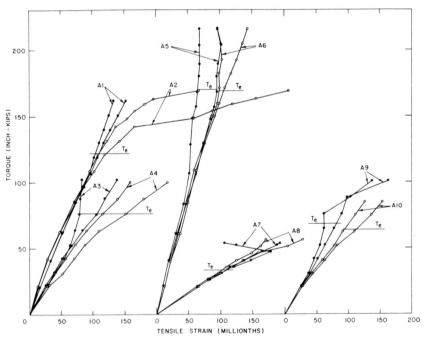

Fig. 2.3. Principal tensile strains measured at the center of the wider face (Ref. 1).

Measurements of $(dT/d\epsilon_{max})_0$ from torsion tests show that E_0 is very close to that obtained from the stress–strain curve of axial tensile and compressive tests.[1] Again, St. Venant's theory was proved to be very accurate at low load.

At high torque the T vs. ϵ_{max} curve behaves quite erratically. Some gages record a sudden increase of tensile strains, whereas others may show reversal of strain. This phenomenon can be explained by the development of microcracks. A gage mounted across a microcrack would show a sudden increase of strain, whereas one located between two microcracks might record a decrease of strain. With the interference of microcracks, the behavior of torsional members gradually deviates from St. Venant's theory at high loads.

2.2 TORSIONAL STRENGTH OF RECTANGULAR SECTIONS

Three theories have been developed to predict the torsional strength of plain concrete members: elastic theory, plastic theory, and skew-bending theory.

2.2.1 Elastic Theory

As discussed in the previous section, the behavior of a torsional member is reasonably well described by St. Venant's theory, even with the development of microcracks at high torque. In view of this success, St. Venant's theory has also been extended to the prediction of torsional strength. In applying this theory, the following failure criterion is assumed: Torsional failure of a plain concrete member occurs when the maximum principal tensile stress, σ_{max}, equals to the tensile strength of concrete, f'_t. Since $\sigma_{max} = \tau_{max}$ in pure shear, the elastic failure torque, T_e, can be derived from Eq. (1–62) as:

$$T_e = \alpha x^2 y f'_t \qquad (2–6)$$

where α is St. Venant's coefficient (Table 1.1) and f'_t is the tensile strength of concrete obtained from a uniaxial tension test. A resonable value of $f'_t = 5\sqrt{f'_c}$ has been suggested by the author.*

The elastic theory was first used by Bach and Graf[2] as early as 1912. It was widely adopted later by Young et al.[3] (1922), Andersen[4] (1937), Cowan[5] (1951), Humphreys[6] (1957), and Zia[7] (1961). However, tests have shown that this theory consistently underestimates the failure strength of a plain concrete beam. The actual test strength is roughly 50% greater than that predicted by the theory.

---

* Unpublished report at Portland Cement Association, Skokie, Illinois.

2.2.2 Plastic Theory

Since elastic theory consistently underestimates torsional strength, Nylander[8] surmised that the extra strength may be contributed by the plastic property of concrete. In other words, concrete may develop plasticity and thus increase the ultimate strength. Similarly to elastic theory, failure was assumed to occur when the maximum principal tensile stress reaches the tensile strength of concrete, $\tau = f'_t$. The plastic failure torque, T_p, can therefore be expressed by Eq. (1–100), assuming full plasticity:

$$T_p = \alpha_p \, x^2 y f'_t \qquad (2\text{–}7)$$

where $\alpha_p = (0.5 - x/6y)$. This plastic coefficient varies from $\frac{1}{3}$ to $\frac{1}{2}$, about 50% greater than α used in the elastic theory. Hence, plastic theory can roughly account for the extra strength that cannot be explained by the elastic theory.

Plastic theory, however, has three weaknesses. First, it is theoretically unsatisfactory because principal tension is the cause of torsional beam failure, but no significant plastic behavior has been observed in tension of concrete as shown in the stress–strain curve of Fig. 2.2. Second, torsional failure of plain concrete members is quite brittle. There is no sign of a plastic rotation as shown in Fig. 2.1. Third, the theory cannot account for a size effect. Tests indicated that for small torsional specimens the calculated plastic torques are usually smaller than the test values, whereas the opposite is true for large specimens.

2.2.3 Skew-Bending Theory[1]

In view of the difficulties in using the classical elastic and plastic theories to predict the ultimate strengths of plain concrete torsional members, the author surmised that the failure criterion used in these theories may be incorrect. Consequently, the mechanism of failure was reexamined with the aid of a high-speed movie camera at a speed of 1,200 frames per second. By projecting the film at 20 frames per second, the process of failure was slowed down so that the characteristics of torsion failure could be clearly observed.

The failure process is shown in Fig. 2.4, a series of drawings prepared from the motion picture. The face of the beam away from the camera was photographed in a mirror placed behind the beam. The movie clearly showed that the first crack, which is inclined at 45° to the axis of the beam, appeared on the front face. It gradually widened and progressed across the top of

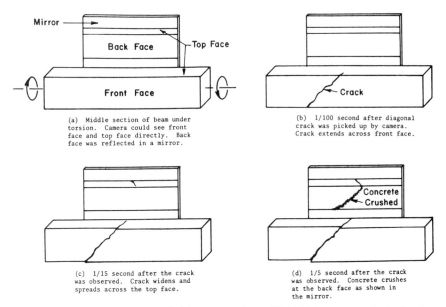

(a) Middle section of beam under torsion. Camera could see front face and top face directly. Back face was reflected in a mirror.

(b) 1/100 second after diagonal crack was picked up by camera. Crack extends across front face.

(c) 1/15 second after the crack was observed. Crack widens and spreads across the top face.

(d) 1/5 second after the crack was observed. Concrete crushes at the back face as shown in the mirror.

Fig. 2.4. Torsion failure process of plain concrete beam (10 × 15-in. cross section) taken by movie camera with speed of 1,200 frames/sec (Ref. 1).

the beam until, finally, the concrete crushed on the back face. This failure process was similar to that of a plain concrete flexural beam and, therefore, revealed a bending-type failure.

The failure surface of a torsional beam is shown in Fig. 2.5. The front longer edge of the failure surface is inclined at approximately 45° to the axis of the beam. The two shorter edges are curves that start out almost perpendicular to the front wider face and then turn gradually to 45° as they approach the opposite wider face. This longer edge on the other side is approximately a straight line connecting the ends of the two short curves. Examination of the fracture surface further reveals that the failure surface close to the 45° edge in the foreground is approximately a plane and appears to have been broken by tension. The concrete at the edge of the opposite wider face, however, has a jagged appearance, suggesting a compression failure. This observation again confirms a bending-type failure for torsional rectangular members.

Based on the bending mechanism of torsional failure, an equation for nominal torsional strength of plain concrete rectangular members can be derived. Referring to Fig. 2.6, the applied torque can be resolved into two components acting on the failure surface: the bending component, T_b, and the twisting

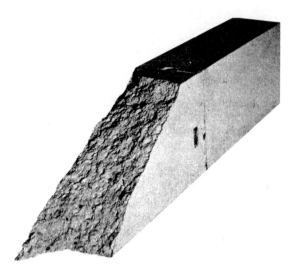

Fig. 2.5. Failure surface of 10 × 15-in. beam A1 (Ref. 1).

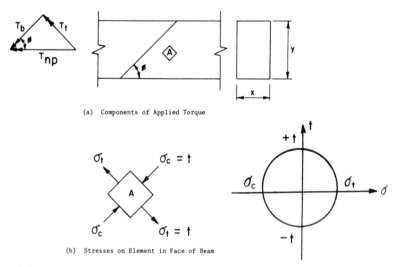

(a) Components of Applied Torque

(b) Stresses on Element in Face of Beam

Fig. 2.6. Bending component (T_b) and torsional component (T_t) on a beam (right-hand-screw convention).

component, T_t. The former is responsible for the observed bending-type failure.

According to the well-known elastic bending theory:

$$T_b = T_{np} \cos \phi = \frac{x^2 y}{6} \csc \phi f_r \qquad (2\text{–}8)$$

where

T_{np} = nominal torsional strength of a plain concrete beam
ϕ = angle between the tensile crack on the wider face and the axis of beam
f_r = modulus of rupture of concrete

Rearranging the terms in Eq. (2–8) leads to:

$$T_{np} = \left(\frac{x^2 y}{3} f_r\right) \csc 2\phi \qquad (2\text{–}9)$$

To find ϕ that gives the minimum torsional strength, we differentiate Eq. (2–9) with respect to ϕ and equate it to zero:

$$\frac{dT_{np}}{d\phi} = \frac{x^2 y}{3} f_r \, (2 \cot 2\phi \csc 2\phi) = 0$$

$$\phi = 45° \qquad (2\text{–}10)$$

Substituting $\phi = 45°$ into Eq. (2–9):

$$T_{np} = \frac{x^2 y}{3} f_r \qquad (2\text{–}11)$$

The effect of the twisting component, T_t, should also be considered. An element A taken from the wider face of the beam is subjected not only to a 45° tensile stress due to T_b but also to a perpendicular compressive stress of equal magnitude due to T_t. According to tests by McHenry and Karni,[9] this perpendicular compression will reduce the tensile strength of concrete by 15%. Since bending failure in plain concrete is due to tension, the modulus of rupture in Eq. (2–11) should also be reduced by a factor of 0.85. Thus Eq. (2–11) becomes:

$$T_{np} = \frac{x^2 y}{3} (0.85 \, f_r) \qquad (2\text{–}12)$$

Equation (2–12) provides a new failure criterion; namely, the failure of a torsional beam is reached when the tensile stress induced by a 45° bending component of torque on the wider face reaches a reduced modulus of rupture of concrete.

Since modulus of rupture is not often available for analysis and design, it is desirable to express f_r in terms of f'_c by the author's empirical equation for $x \geqslant 4$ in.:

$$f_r = 21 \left(1 + \frac{10}{x^2} \right) \sqrt[3]{f'_c} \qquad (2\text{–}13)$$

In this empirical equation the dimensions are not consistent. f_r and f'_c must be in psi, and x in inches. Substituting Eq. (2–13) into Eq. (2–12) gives:

$$T_{np} = 6 \, (x^2 + 10) y \sqrt[3]{f'_c} \qquad (2\text{–}14)$$

for $x \geqslant 4$ in. Equation (2–14) was substantiated by 55 beam tests of eight groups of researchers.[1]

Comparison. Comparing the elastic theory (Eq. 2–6), plastic theory (Eq. 2–7), and skew-bending theory (Eq. 2–12), it can be seen that they all have the same parameter, $x^2 y$. The difference lies only in the nondimensional coefficients and in the material constants.

In the elastic and plastic theories, the material constant is the direct tensile strength of concrete, f'_t. In the skew-bending theory, it is the reduced modulus of rupture, $0.85 f_r$. Although f_r represents a type of indirect tensile strength, it is strongly affected by the tensile strain gradient and, therefore, the size of the specimen. Since Eq. (2–13) expresses f_r as a function of x, the skew-bending theory adequately explains the size effect observed in torsion tests.

A comparison of the coefficients is shown in Fig. 2.7. It can be seen that St. Venant's coefficients in the elastic theory and Nadai's coefficients in the plastic theory are functions of y/x. The coefficient in the skew-bending theory, however, is a constant of $\frac{1}{3}$. This constant always lies between St. Venant's and Nadai's coefficients.

Because of the difficulties encountered in obtaining a reliable tensile strength for concrete, comparisons of theories for torsional tests of plain concrete members were often inconclusive. Analysis of all available results by the author, however, seems to indicate that the actual coefficient is slightly affected by the ratio y/x, but this effect is definitely not as strong as that predicted by St. Venant's or Nadai's coefficients.

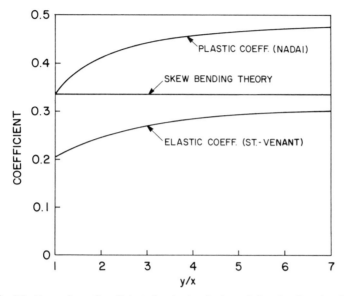

Fig. 2.7. Comparison of coefficients for elastic, plastic, and skew-bending theories.

The mild effect of the y/x ratio can be explained by the skew-bending theory, considering the nonuniform strain distribution on the wider face of a torsional beam. Such stress distribution is shown in Fig. 2.8 for a 10 in. by 20 in. cross section. Figure 2.8 (a) and (b) gives the stress distributions, respectively, at the elastic torque and at a load stage just prior to failure. At elastic torque, the measured strains along the wider face are found to be very close to St. Venant's stress distribution. The principal tensile strain at mid-depth reaches 0.0001, the estimated tensile failure strain of concrete.

Beyond the elastic torque, Fig. 2.8 (b) shows that the strains continue to increase according to St. Venant's stress distribution. The strains near the mid-depth, however, exceed the tensile strain of concrete, and microcracks can be detected in this region. These microcracks apparently will weaken the tensile strength of concrete near the mid-depth and, in turn, the torsional strength of the beam. According to St. Venant's stress distributions, shown in Fig. 1.9 for various ratios of y/x, nonuniformity of stress distribution along the wider face becomes more severe for a smaller y/x ratio. The weakening effect of microcracks near mid-depth should therefore increase, and the torsional coefficient should decrease, when the y/x ratio is decreased.

Since the effect of the ratio y/x on the torsional strength of plain concrete members is small, the inclusion of this effect in the skew-bending theory is unwarranted.

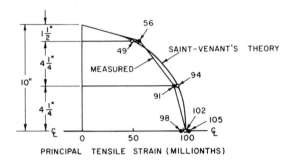

(a) AT ELASTIC TORQUE

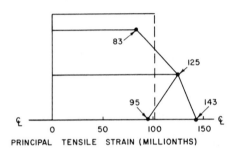

(b) AT ULTIMATE TORQUE

Fig. 2.8. Stress distribution on wider face of beam A6 (Ref. 1).

2.2.4 Example 2.1

A plain concrete beam with 6 in. by 12 in. cross section is subjected to pure torsion. $f'_c = 5,000$ psi. Find the torsional strength by elastic theory, plastic theory, and skew-bending theory.

Solution

Elastic theory:

$$\frac{y}{x} = \frac{12}{6} = 2, \qquad \alpha = 0.246 \text{ from Table 1.1}$$

$$f'_t = 5\sqrt{5,000} = 354 \text{ psi}$$

$$T_e = \alpha x^2 y f'_t = 0.246 \ (6)^2(12)(354)$$

$$= 37,600 \text{ in.-lb} = 3.14 \text{ k-ft}$$

Plastic theory:

$$\frac{y}{x} = \frac{12}{6} = 2, \qquad \alpha_p = \frac{1}{2} - \frac{1}{6}\frac{x}{y} = \frac{1}{2} - \frac{1}{12} = 0.417$$
$$f_t' = 5\sqrt{5,000} = 354 \text{ psi}$$
$$T_p = \alpha_p\, x^2 y f_t' = 0.417\,(6)^2(12)(354)$$
$$= 63,800 \text{ in.-lb} = 5.31 \text{ k-ft}$$

Skew-bending theory:

$$f_r = 21\left(1 + \frac{10}{x^2}\right)\sqrt[3]{f_c'} = 21\left(1 + \frac{10}{6^2}\right)\sqrt[3]{5,000} = 459 \text{ psi}$$
$$T_{np} = \frac{1}{3} x^2 y\,(0.85\, f_r) = \frac{1}{3}\, 6^2\,(12)(0.85)(459)$$
$$= 56,200 \text{ in.-lb} = 4.68 \text{ k-ft}$$

Comparison of the three theories shows that the value predicted by skew-bending theory lies in between the elastic and the plastic theories, but closer to the latter.

2.3 TORSIONAL STRENGTH OF CIRCULAR SECTIONS

The torsional strength of prismatic members with circular sections can also be predicted by the three theories used in the previous section, namely, elastic, plastic, and skew-bending theories.

2.3.1 Elastic Theory

From Navier's Eq. (1–9), the elastic torque, T_e, is expressed by the maximum shear stress at the surface, τ_{max}:

$$T_e = \frac{I_p}{R}\, \tau_{max} = \frac{\pi d^3}{16}\, \tau_{max}$$

where

I_p = polar moment of inertia
R = the radius of the cross section
d = the diameter of the cross section
τ_{max} = maximum shear stress at the surface

Assuming that failure occurs when $\tau_{max} = f'_t$, the tensile strength of concrete:

$$T_e = \frac{\pi d^3}{16} f'_t \qquad (2\text{--}15)$$

Again f'_t can be taken approximately as $5\sqrt{f'_c}$.

2.3.2 Plastic Theory

From Nadai's Eq. (1–99), the plastic torque is:

$$T_p = \frac{\pi d^3}{12} \tau$$

where τ is the plastic shear stress over the whole cross section. Similarly, Assuming $\tau = f'_t$:

$$T_p = \frac{\pi d^3}{12} f'_t \qquad (2\text{--}16)$$

The plastic torque in Eq. (2–16) is always 33% higher than the elastic torque in Eq. (2–15).

2.3.3 Skew-Bending Theory

Similarly to the derivation of torsional strength for rectangular sections, the ultimate torque acting on members with circular sections can be resolved into two components, T_b and T_t. The bending component, T_b, is responsible for the observed bending-type failure. Refer to Fig. 2.6 and assume the cross section of the beam to be circular. Using the elastic bending formula:

$$T_b = T_{np} \cos \phi = \frac{I_x}{(d/2)} \csc \phi \, f_r$$

where

T_{np} = nominal torsional strength of a plain concrete beam
I_x = the moment of inertia about a diametrical axis $= \pi d^4/64$
d = diameter of cross section
f_r = modulus of rupture of a concrete circular member

Hence:

$$T_{np} = \frac{\pi d^3}{16} f_r \csc 2\phi$$

Differentiate T_{np} with respect to ϕ, and equate it to zero. ϕ is found to be $45°$. Thus:

$$T_{np} = \frac{\pi d^3}{16} f_r$$

Considering the effect of the twisting component (see the discussion of Eq. 2–12 for rectangular sections), f_r is reduced to $0.85 f_r$:

$$T_{np} = \frac{\pi d^3}{16} (0.85 f_r) \tag{2-17}$$

The torsional strength T_{np} in Eq. (2–17) is identical to the elastic torque in Eq. (2–15) except that the material constant $0.85 f_r$ is used instead of f'_t.

2.4 TORSIONAL STRENGTH OF FLANGED SECTIONS

The flanged sections commonly used in practice include T-sections, L-sections, and I-sections. The torsional strength of these sections can also be calculated by some kinds of approximation of the elastic, plastic, or skew-bending theories.

2.4.1 Bach's Equation

In Chapter 1 Bach's equation was derived from St. Venant's elastic theory based on two approximate assumptions. First, each rectangular component of a flanged section is very thin so that St. Venant's coefficient can be approximated by $\frac{1}{3}$. Second, the rotations of all rectangular components in a cross section are identical. Using these two assumptions, the torque T can be obtained in Eq. (1–68). Assuming that failure occurs when $\tau_{max} = f'_t$, the torsional strength becomes:

$$T_{np} = \sum \frac{x^3 y}{3x_{max}} f'_t \tag{2-18}$$

If all the rectangular components in a cross section have the same thickness, $x_{max} = x$, then:

$$T_{np} = \sum \frac{x^2 y}{3} f'_t \qquad (2\text{-}19)$$

2.4.2 Plastic Theory

For flanged sections the plastic torque can be calculated by Nadai's Eqs. (1-101) through (1-104). The shear stress in these equations is taken as $\tau = f'_t$, and f'_t may be taken as $5\sqrt{f'_c}$.

2.4.3 Skew-Bending Approximation

Although an attempt has been made to derive the torsional strength of flanged sections based on the skew-bending mechanism of failure, the mathematical complexity has rendered it useless in practice. To simplify the solution, an approximate assumption was made; namely, the torsional strength of a flanged section is the sum of the torsional strength of the component rectangles. Thus:

$$T_{np} = \sum \frac{x^2 y}{3} (0.85 f_r) \qquad (2\text{-}20)$$

where $f_r = 21 (1 + 10/x^2) \sqrt[3]{f'_c}$ for $x \geqslant 4$ in. Comparing Eq. (2-20) with Bach's Eq. (2-19) for uniform thickness of rectangular components, it can be seen that they are identical except that $0.85 f_r$ is used in the skew-bending theory instead of f'_t in the elastic theory. The parameter $\Sigma x^2 y/3$ in Eq. (2-20) has been adopted since the 1971 ACI Code.

Equation (2-20) has been substantiated by a series of ten beam tests,[10] shown in Fig. 2.9. This series of tests includes two rectangular beams, four L-beams, and four T-beams. The stems of all these beams are 10 in. by 15 in., and the flange thickness is 5 in. The only variable is the overhanging flange width. Plotting the torsional strength as a function of total overhanging flange width, Fig. 2.9 shows that two cut-off points must be imposed on Eq. (2-20). For the L-sections, the overhanging flange width should be limited to 15 in.; and for the T-sections, the total overhanging flange width should be limited to 30 in. This means that an effective flange width can be taken as approximately three times the overhanging flange thickness.

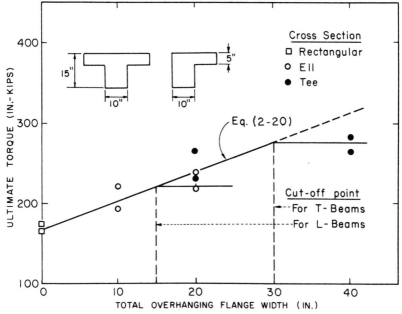

Fig. 2.9. Verification of Eq. (2–20) (Ref. 10).

2.4.4 Example 2.2

A plain concrete beam with a cross section shown in Fig. 2.10 is subjected to pure torsion. $f'_c = 4{,}000$ psi. Find the torsional strength using Bach's equation, plastic theory, and the skew-bending approximation.

Solution

Bach's equation (Eq. 2–18):

$$f'_t = 5\sqrt{f'_c} = 5\sqrt{4{,}000} = 317 \text{ psi}$$

$$T_{np} = \frac{\Sigma x^3 y}{3x_{max}} f'_t = \frac{317}{3(8)} [2(8)^3(18) + 6^3(20)]$$

$$= 300{,}000 \text{ in.-lb} = 25.0 \text{ k-ft}$$

Plastic theory (Eq. 1–104):

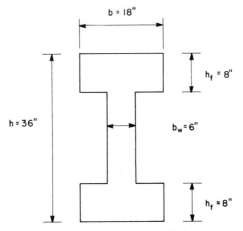

Fig. 2.10. Example 2.2—A concrete beam with flanged cross section.

$$\tau = f'_t = 5\sqrt{f'_c} = 317 \text{ psi}$$

$$T_p = \tau \left[h_f^2\, b \left(1 - \frac{1}{3}\frac{h_f}{b} \right) + \frac{1}{2}\, b_w^2 \left(h - 2h_f + \frac{1}{3}\, b_w \right) \right]$$

$$= 317 \left[8^2(18) \left(1 - \frac{1}{3}\frac{8}{18} \right) + \frac{1}{2}\,(6)^2 \left(36 - 16 + \frac{1}{3}\,6 \right) \right]$$

$$= 435{,}000 \text{ in.-lb} = 36.2 \text{ k-ft}$$

Skew-bending approximation (Eq. 2–20):

For flange: $f_r = 21 \left(1 + \dfrac{10}{x^2} \right) \sqrt[3]{f'_c} = 21 \left(1 + \dfrac{10}{8^2} \right) \sqrt[3]{4{,}000} = 385 \text{ psi}$

For web: $f_r = 21 \left(1 + \dfrac{10}{6^2} \right) \sqrt[3]{4{,}000} = 426 \text{ psi}$

$$T_{np} = \sum \frac{x^2 y}{3}\,(0.85\, f_r)$$

$$= \frac{1}{3}\,[2(8)^2(18)(0.85)(385) + 6^2(20)(0.85)(426)]$$

$$= 338{,}000 \text{ in.-lb} = 28.1 \text{ k-ft}$$

The torsional strength calculated by the skew-bending approximation falls in between those of Bach's equation and the plastic theories, but is much closer to the former. The prediction by the plastic theory appears to be too high.

REFERENCES

1. Hsu, T. T. C., "Torsion of Structural Concrete—Plain Concrete Rectangular Sections," *Torsion of Structural Concrete,* SP-18, American Concrete Institute, Detroit, 1968, pp. 203–238; Portland Cement Association Research and Development Laboratories, Development Dept. Bulletin D-134.
2. Bach, B. and O. Graf, "Versuche uber die Widerstands Fahigkeit von Beton und Eisenbeton gegen Verdrehung" (Investigation of Torsional Strength of Concrete and Reinforced Concrete), *Deutscher Ausschuss fur Eisenbeton,* Heft 16, Wilhelm Ernst, Berlin, 1912.
3. Young, C. R., W. L. Sagar, and C. A. Hughes, "Torsional Strength of Rectangular Sections of Concrete, Plain and Reinforced," University of Toronto, School of Engineering, Bulletin No. 9, 1922.
4. Andersen, P., "Rectangular Concrete Sections under Torsion," *Journal of the American Concrete Institute,* Proc., Vol. 34, No. 1. Sept.–Oct. 1937, pp. 1–11.
5. Cowan, H. J., "Tests of Torsional Strength and Deformation of Rectangular Reinforced Concrete Beams," *Concrete and Constructional Engineering,* London, Vol. 46, No. 2, Feb. 1951, pp. 51–59.
6. Humphreys, R., "Torsional Properties of Prestressed Concrete," *The Structural Engineer,* London, V. 35, No. 6, June 1957, pp. 213–224.
7. Zia, P., "Torsional Strength of Prestressed Concrete Members," *Journal of the American Concrete Institute,* Proc., Vol. 57, No. 10, April 1961, pp. 1337–1359.
8. Nylander, H., *Vridning och Vridningsinspanning vid Betong Konstruktiener* (Torsion and Torsional Restraint by Concrete Structures), Statens Kommitteé för Byggnadsforskning, Stockholm, Bulletin No. 3, 1945.
9. McHenry, D. and J. Karni, "Strength of Concrete under Combined Tensile and Compressive Stresses," *Journal of the American Concrete Institute,* Proc., Vol. 54, April 1958, pp. 829–839; Portland Cement Association Research and Development Laboratories, Development Dept. Bulletin D-19.
10. Hsu, T. T. C., "Torsion of Structural Concrete—A Summary of Pure Torsion," *Torsion of Structural Concrete,* SP-18, American Concrete Institute, Detroit, 1968, pp. 165–178; Portland Cement Association Research and Development Laboratories, Development Dept. Bulletin D-133.

Plate 3: Torsion test rig in the Structural Laboratory of the Portland Cement Association, Skokie, Illinois, 1963.

3
Torsion of Reinforced Concrete Members

3.1 BEHAVIOR OF TORSIONAL MEMBERS

3.1.1 Members with Longitudinal Steel Only

The behavior of torsional members with longitudinal steel only is depicted by the torque-twist curve in Fig. 3.1. Before cracking the torque-twist relationship is very close to that of a plain concrete member. In other words, the effect of the longitudinal steel can be neglected, and the torsional rigidity can be reasonably predicted by St. Venant's theory, presented in Chapter 1.

After cracking, the beam may collapse suddenly if the beam is reinforced with a light amount of steel. In the case of heavy reinforcement, the ultimate strength may exceed the cracking torque but seldom exceed it by more than 15%. This ineffectiveness of longitudinal steel in increasing the strength of the beams was thought by early investigators[3] to be the result of the location of longitudinal bars. Such longitudinal bars were always placed at the corners

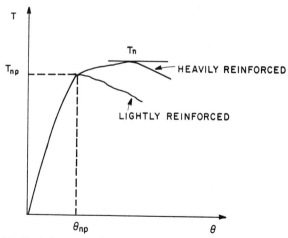

Fig. 3.1. Typical torque-twist curve for beams with longitudinal steel only.

of a beam where the shear stress is zero according to St. Venant's stress distribution. However, later tests with longitudinal bars at the center of the faces show that longitudinal steel alone is ineffective regardless of the location.

Neglecting the small effect of longitudinal steel, a concrete beam reinforced with longitudinal steel alone can be treated as a plain concrete beam in calculating the torsional rigidity as well as the torsional strength. The ultimate strength can be calculated by the equations in Chapter 2, particularly Eq. (2–12) based on skew-bending theory.

3.1.2 Members with Longitudinal Steel and Stirrups

The torque-twist curves of a series of specimens with a 10 in. by 15 in. cross section and reinforced with various amounts of torsional reinforcement (1.07 to 5.28%, including equal volumes of longitudinal steel and stirrups) are shown in Fig. 3.2. Each curve can be divided into two distinct regions—

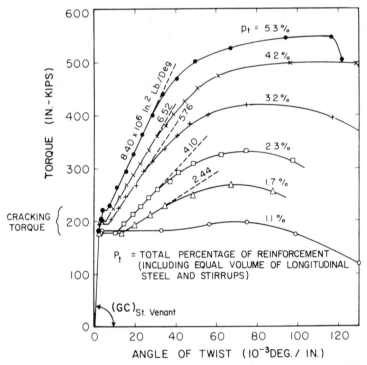

Fig. 3.2. Torque-twist curves of beams with various percentages of reinforcement (Series B in Ref. 1).

before and after cracking. A horizontal plateau exists at cracking, where a member continues to twist under a constant load.

The cracking torques are plotted as a function of total percentage of steel (including longitudinal steel and stirrups) in Fig. 3.3 for 55 beams.[1] It can be seen that cracking torque T_{cr} is a mild function of the total steel percentage ρ_t and can be expressed by the following equation:

$$T_{cr} = (1 + 4\,\rho_t)\,T_{np} \qquad (3\text{--}1)$$

In Eq. (3–1), ρ_t is expressed in terms of the total steel ratio. For example, for a member with 3% total steel, $T_{cr} = [1 + 4(0.03)]\,T_{np} = 1.12\,T_{np}$. In view of the small effect of ρ_t, it would be simpler and conservative in practical design to neglect the favorable effect of ρ_t and to take $T_{cr} = T_{np}$.

Before cracking, Fig. 3.2 shows that the percentage of steel has a negligible effect on the torsional rigidity of the member. In other words, all the members behave as plain concrete members. Therefore, St. Venant's torsional rigidity presented in Chapter 1, Eqs. (1–59) and (1–60), is applicable to members with longitudinal steel and stirrups before cracking.

After cracking, Fig. 3.2 shows that the behavior can no longer be predicted by St. Venant's theory. This is so because cracking terminates the basic premise

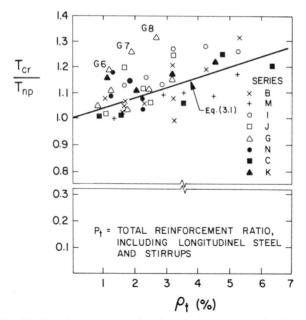

Fig. 3.3. Cracking torque as a function of reinforcement ratio (Ref. 1).

of the theory of elasticity that the material must be continuous. Hence a new equilibrium condition is established after cracking, in which the steel picks up the tensile stresses and the concrete carries the compression. The transition from the St. Venant's equilibrium condition to the new post-cracking equilibrium condition is manifested by the horizontal plateau of a torque-twist curve.

An interesting phenomenon occurs after cracking; i.e., the length of the beam increases with increasing torque. The unit lengthening of the beam is plotted against torque as the solid line in Fig. 3.4. It can be seen that the relationship is approximately linear. The strains in the longitudinal bars have also been measured, and the average strain for all the bars in a beam is also plotted as the dotted line in Fig. 3.4. The close resemblance of the solid and the dotted curves indicates that the lengthening of the beam is due to the stretching of the longitudinal bars. This stretching is required to produce tensile stresses in the longitudinal bars, so as to maintain the post-cracking equilibrium condition.

The ramification of the post-cracking lengthening of the torsional beam

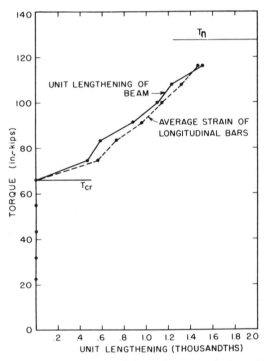

Fig. 3.4. Increase of beam length with increasing torque for beam N2 (Ref. 1).

is also very interesting. In a building a torsional beam is often not free to lengthen. If a beam is restrained longitudinally, a compression will be induced in the beam. This compression is equivalent to a longitudinal concentric prestress that will increase the torsional strength of a beam. This favorable, self-generated prestress may be a contributing factor in explaining the fact that complete torsional collapse is seldom observed in the interior spans of continuous beams.

Returning to Fig. 3.2, it can be seen that the ultimate strengths and the post-cracking torsional rigidity (slope of torque-twist curve after cracking) are very strongly a function of steel percentage. If the total steel percentage is less than about 1%, the beam will fail brittlely upon cracking because the reinforcement is insufficient to produce an ultimate torque greater than the cracking torque. At about 1% steel, the ultimate strength is about equal to the cracking torque, and the torque-twist curve develops a long horizontal plateau. Such ductile behavior is desirable in design. This minimum 1% of total torsional reinforcement to ensure ductility is about three times greater by volume than the minimum longitudinal steel required for flexure (ρ_{min} = $200/f_y$ according to the ACI Code), or about ten times greater than the minimum stirrups required for flexural shear ($50/f_y$).

If reinforcement is provided in an excessive amount, the steel will not yield at failure. This provides a maximum percentage of steel that limits the maximum ultimate torque to about 2.5 to 3.0 times the cracking torque.

When beams are reinforced with a steel percentage in between the minimum and the maximum, the ultimate strengths and the post-cracking torsional rigidity increase approximately linearly with steel percentage. A detailed study of ultimate strengths and post-cracking torsional rigidity is given below in Sections 3.2 and 3.3, respectively.

3.2 TORSIONAL STRENGTH

In recent years many theories have been developed for calculating the torsional strengths of members with both longitudinal steel and stirrups. These theories can be roughly divided into two types: the truss analogy type and the skew-bending type. In this chapter we shall introduce only three theories, by Rausch,[2] Cowan,[3] and the author.[4] These three theories deal with pure torsion and, therefore, provide a good insight into the failure mechanism of torsion. The first two theories belong to the truss analogy type and also have great historical value. The third theory belongs to the skew-bending type. It serves as the basis of torsion design since the 1971 ACI Codes. Other theories, by Lessig, Yudin, Collins et al., Lampert-Thurlimann, Elfgren et al., and Collins-

Mitchell, deal primarily with the interaction of torsion and bending or the interaction of torsion, bending and shear (pure torsion is only a special case in these theories). They will be presented later, in Chapters 6 and 7.

3.2.1 Rausch's Space Truss Analogy

The first theory for reinforced concrete subjected to torsion was proposed by Rausch in 1929 in the form of a Ph.D. thesis. A second edition, in the form of a technical book, was published in 1938 and a third edition in 1952. This third edition[2] is available in many libraries.

A short length of a reinforced concrete member subjected to torsion is shown in Fig. 3.5 (a). The cross section of the member has an arbitrary shape and is assumed to be hollow. After cracking, the concrete is separated by 45° cracks into a series of helical members. These helical concrete members are assumed to interact with the longitudinal steel bars and the hoop steel bars to form a space truss as shown in Fig. 3.5 (b). Each of the helical members is idealized into a series of 45° short straight struts connected at the joints. The compression force in the concrete struts will produce an

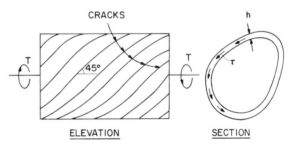

(a) DIAGONAL CRACKING IN A REINFORCED CONCRETE MEMBER

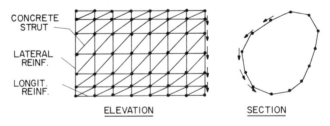

(b) SPACE TRUSS IDEALIZATION

Fig. 3.5. Space truss analogy.

outward radial force at each joint that will be resisted by lateral hoop reinforcement. These lateral hoop bars are also idealized as chains of short straight bars connected to the concrete strut at the joints. The chains of diagonal concrete struts and the chains of hoop bars thus form a mechanism that will lengthen under an infinitesimal external torque. This tendency to lengthen is resisted by longitudinal reinforcement. Each longitudinal bar is assumed to be a chain of short bars connecting at the joints to the diagonal struts and the hoop bars. In this way a space truss is formed that consists of 45° concrete struts in compression and longitudinal and hoop bars in tension (Fig. 3.5b). This space truss is able to resist large external torque.

A space struss thus formed implies the following four assumptions:

1. The space truss is made up of 45° diagonal concrete struts, longitudinal bars, and hoop bars connected at the joints by hinges.
2. A diagonal concrete member carries only axial compression; i.e., shear resistance is neglected.
3. Longitudinal and lateral bars carry only axial tension; i.e., dowel resistance is neglected.
4. For a solid section, the concrete core does not contribute to the ultimate torsional resistance.

Referring to Fig. 3.6, we shall now analyze the forces in all the members of the space truss. First, the internal forces in the longitudinal bars, the hoop bars, and the diagonal concrete struts are denoted X, Y, and D, respectively. The force at each joint representing the shear flow is designated F. Second, each force is labeled in sequence with subscripts from 1 to n along

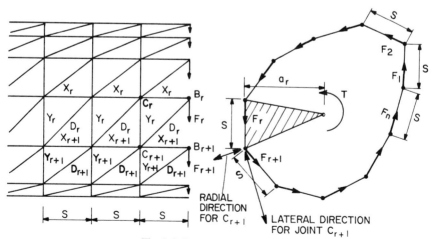

Fig. 3.6. Forces in a space truss.

the periphery of the cross section. A typical bar would have the subscript r between 1 and n. The bar next to the typical bar would have subscript $r + 1$, and so on. Third, we notice that the internal forces X, Y, and D must be identically distributed in each cell in the longitudinal direction. They are so labeled in Fig. 3.6. This requirement is similar to those derived in St. Venant's theory or Bredt's thin-tube theory, which state that the stress distribution must be identical at each cross section for a prismatic member subjected to end torsions.

Now let us first take the equilibrium of joint C_{r+1} in the *longitudinal* direction. Observing that the two X_{r+1} forces cancel each other, we can write:

$$D_r = D_{r+1}$$

Similarly, taking equilibrium successively for joints C_1, C_2, . . . C_n, we conclude that:

$$D_1 = D_2 = \cdots D_r = D_{r+1} = \cdots D_n = D \text{ (constant)} \qquad (3\text{--}2)$$

Second, we consider the equilibrium of joint C_{r+1} in the *lateral* (or tangential) direction:

$$Y_r = Y_{r+1}$$

Similarly, equilibrium of joints C_1, C_2, . . . C_n results in:

$$Y_1 = Y_2 = \cdots Y_r = Y_{r+1} = \cdots Y_n = Y \text{ (constant)} \qquad (3\text{--}3)$$

Third, we shall study the equilibrium of joint C_{r+1} in the *radial* direction, which is perpendicular to both the longitudinal and lateral directions. As shown in Fig. 3.7 (a) the forces D_r and D_{r+1} can each be resolved into the two components, one along the longitudinal direction and another along the direction of the shear flow. The longitudinal components of D_r and D_{r+1} balance each other exactly according to Eq. (3–2). However, the two components of forces D_r and D_{r+1} (i.e., $D/\sqrt{2}$) in the direction of the shear flow produce an outward radial force as shown in Fig. 3.7 (b). This outward radial force must be counterbalanced by the inward radial force provided by the forces Y_r and Y_{r+1} (Fig. 3.7c). Since $Y_r = Y_{r+1} = Y$ from Eq. (3–3), we conclude that:

$$Y = \frac{D}{\sqrt{2}} \qquad (3\text{--}4)$$

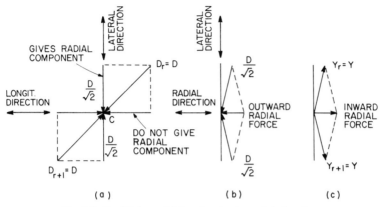

Fig. 3.7. Equilibrium of joint C_{r+1} in the radial direction.

Next, let us consider the equilibrium of joint B_r in the plane of the wall. From the force triangle shown in Fig. 3.8, we have:

$$X_r = \frac{D_r}{\sqrt{2}} = \frac{D}{\sqrt{2}}$$

and:

$$F_r = \frac{D_r}{\sqrt{2}} = \frac{D}{\sqrt{2}}$$

Similarly, taking the equilibrium of joints $B_1, B_2, \ldots B_n$, we have:

$$X_1 = X_2 = \cdots = X_r = X_{r+1} = \cdots X_n = X = \frac{D}{\sqrt{2}} \text{ (constant)} \quad (3\text{–}5)$$

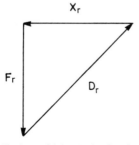

Fig. 3.8. Equilibrium of joint B_r in the plane of the wall.

$$F_1 = F_2 = \cdots = F_r = F_{r+1} = \cdots F_n = F = \frac{D}{\sqrt{2}} \text{ (constant)} \quad (3\text{-}6)$$

Combining Eqs. (3–2) through (3–6), we arrive at:

$$X = Y = \frac{D}{\sqrt{2}} = F = \text{const} \quad (3\text{-}7)$$

Finally we shall study the equilibrium of the whole cross section. The internal torques contributed by F_1 to F_n must be equal to the external torque T. Denoting in Fig. 3.6 the distance from the force F_r to the axis of twist as a_r, the internal torque due to F_r is $F_r a_r$. Summing all the internal torque for F_1 to F_n and noticing that $F_1 = F_2 = \cdots = F_n = F$ (constant) from Eq. (3–6), we have:

$$T = \sum_{r=1}^{n} F_r a_r = F \sum_{r=1}^{n} a_r \quad (3\text{-}8)$$

From geometry the area of the shaded triangle A_r can be expressed by $a_r s/2$, where s is the spacing of the hoop bars and of the longitudinal bars. Then:

$$a_r = \frac{2A_r}{s} \quad (3\text{-}9)$$

Substituting Eq. (3–9) into Eq. (3–8):

$$T = F \sum_{r=1}^{n} \frac{2A_r}{s} = \frac{2F}{s} \sum_{r=1}^{n} A_r \quad (3\text{-}10)$$

Notice that $\sum_{r=1}^{n} A_r = A =$ the total area within the truss. We have:

$$T = \frac{2AF}{s} \quad (3\text{-}11)$$

Substituting $F = Y = A_t f_s$ into Eq. (3–11) gives:

$$T = \frac{2AA_t f_s}{s} \quad (3\text{-}12)$$

where

A_t = cross-sectional area of one hoop bar
f_s = allowable stress of hoop steel

Equation (3–12), which was based on the working stress design method, was adopted by several codes in the 1950s. Using the current ultimate strength concept, the nominal torsional strength T_n can be expressed as:

$$T_n = \frac{2AA_t f_{sy}}{s} \tag{3-13}$$

where f_{sy} is the yield strength of hoop steel. Equation (3–13) can be used for the design of the hoop reinforcement.

If $F = X = A_l f_y$ is substituted into Eq. (3–11), we obtain:

$$T_n = \frac{2AA_l f_{ly}}{s} \tag{3-14}$$

where

A_l = area of one longitudinal bar
f_{ly} = yield strength of longitudinal steel

Let $\hat{A}_l = nA_l$ = the total area of longitudinal bars in the cross section of the member, and notice that $u = ns$ = the perimeter of the area bounded by the center line of a complete hoop bar. Equation (3–14) can be written as:

$$T_n = \frac{2A\hat{A}_l f_{ly}}{u} \tag{3-15}$$

Equation (3–15) can be used to design the total area of the longitudinal bars.

A more common method for the design of longitudinal bars is to relate the area of the longitudinal bars $\hat{A}_l$ to the area of a hoop bar A_t. Equating Eqs. (3–13) and (3–15), we have:

$$\frac{\hat{A}_l f_{ly}}{u} = \frac{A_t f_{sy}}{s} \tag{3-16}$$

If the yield strengths of longitudinal steel and hoop steel are equal, i.e., $f_{ly} = f_{sy} = f_y$, then:

$$\hat{A}_l s = A_t u \tag{3-17}$$

Equation (3-17) states that the volume of all longitudinal steel within the spacing s should be equal to the volume of one complete hoop bar. This is the so-called equal volume principle.

Taking, for example, a rectangular member reinforced with four longitudinal corner bars and closed stirrups with spacing s (see Fig. 3.9), the torsional strength can be obtained from Eq. (3-13):

$$T_n = \frac{2x_1 y_1 A_t f_{sy}}{s} \tag{3-18}$$

and assuming $f_{ly} = f_{sy} = f_y$, the total area of longitudinal bars from Eq. (3-17) is:

$$\hat{A}_l = A_t \frac{u}{s} = A_t \frac{2(x_1 + y_1)}{s} \tag{3-19}$$

If reinforcement is in the form of 45° helical bars, described in Fig. 3.10, it can be shown that the torsional strength is:

$$T_n = \frac{2\sqrt{2} A A_t f_{sy}}{s} \tag{3-20a}$$

where A_t = the cross-sectional area of the one helical bar. For a rectangular section when $A = x_1 y_1$:

$$T_n = \frac{2\sqrt{2} x_1 y_1 A_t f_{sy}}{s} \tag{3-20b}$$

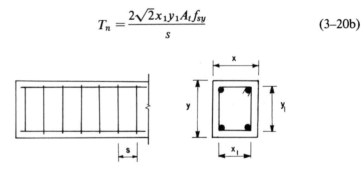

Fig. 3.9. Rectangular beam with longitudinal bars and closed stirrups.

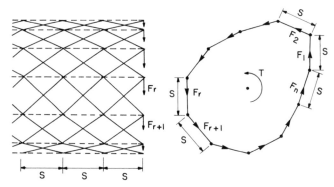

Fig. 3.10. Space truss analogy for members with 45° helical reinforcement.

In conclusion, Rausch's concept of the space truss analogy is an ingenious combination of Bredt's thin-tube theory with the plane truss analogy for flexural shear in reinforced concrete. It gives a very clear idea of the main function of reinforcement and concrete in resisting torsion. The equation so derived is also very simple and elegant. For these reasons the space truss analogy has provided very useful services right up to the present.

It will be pointed out later, in Fig. 3.13 and Section 3.2.3, however, that Rausch's equation for ultimate torsional strength is unconservative in many cases. The trend predicted by Rausch's equation also deviates to some degree from the test results. From a theoretical point of view, the space truss analogy could not take into account the effect of the shear resistance of the concrete struts and the dowel action of reinforcement. It could not explain the contribution of concrete observed in tests. For these reasons, various new theories have been developed, in the hope of improving the accuracy of Rausch's equations.

3.2.2 Cowan's Efficiency Coefficient

In 1935 Andersen[5] pointed out that Rausch's truss analogy assumed uniform stress along all the reinforcement in a member subjected to torsion. This assumption of uniform stress contradicts St. Venant's stress distribution for all types of cross sections except circular. In the case of a rectangular section, St. Venant's stress distribution requires that maximum stress occur at the middle of the wider face and decrease to zero at the corner (Fig. 1.9). In view of this nonuniform stress distribution in the steel, the torsional resistance of reinforcement should be less effective than that predicted by Rausch's Eq. (3–13). Consequently, Andersen suggested that Rausch's Eq. (3–13) should be modified by an efficiency coefficient that is less than unity. In addition, in accordance with American practices, Andersen suggested that the concrete in a reinforced concrete member should also contribute to the torsional resistance.* Hence, Andersen's equation is expressed as:

$$T_n = T_e + \lambda \frac{2AA_t f_{sy}}{s} \qquad (3\text{-}21)$$

where

T_e = torsional resistance of plain concrete taken as the elastic torque, $\alpha x^2 y f'_t$ (see Eq. 2–6)

λ = efficiency coefficient of reinforcement, which varies from about $\frac{2}{3}$ to 1.0 depending on the shape of the cross section and the number of reinforcing bars

Unfortunately, Andersen's coefficient lacks rigor in derivation and is very tedious to calculate. It has not been widely accepted.

Andersen's difficulty in obtaining a simple and logical efficiency coefficient was overcome by Cowan in 1950 using a strain energy method.[3] His derivation is based strictly on St. Venant's stress and strain distribution for rectangular cross sections.

Cowan's derivation starts with a series of 45° helical reinforcement as shown in Fig. 3.11. Each helical bar makes one complete turn in a length of beam, l.

$$\mathit{l} = 2(x_1 + y_1)$$

The length of the helix itself, L, making one complete turn is:

$$L = \sqrt{2}\mathit{l}$$

Taking the horizontal spacing of the helices as s, the number of helices n is:

$$n = \frac{\mathit{l}}{s}$$

---

* It is interesting to note that Rausch also agreed to the existence of a term contributed by concrete in his 1952 book (Ref. 2).

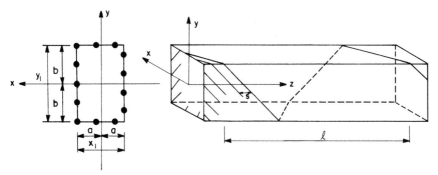

Fig. 3.11. 45° Helical reinforcement in Cowan's derivation.

To obtain the strain energy in a helical bar when the member is subjected to torsion, we return temporarily to St. Venant's theory in Chapter 1. Recalling Eqs. (1–31a) through (1–31c) for calculating the displacements:

$$u = -\theta z y \tag{3-22}$$

$$v = \theta z x \tag{3-23}$$

$$w = \theta \psi(x,y) \tag{3-24}$$

In Eq. (3–24), $\psi(x,y)$ is the warping function from Eq. (1–50):

$$\psi = xy - \frac{32a^2}{\pi^3} \sum_{n=1,3,5\ldots}^{\infty} \frac{(-1)^{(n-1)/2} \sinh \dfrac{n\pi y}{2a}}{n^3 \cosh \dfrac{n\pi b}{2a}} \sin \frac{n\pi x}{2a} \tag{3-25}$$

The maximum shear stress $\tau_{yz,\,max}$ is obtained from Eq. (1–54):

$$\tau_{yz,\,max} = kG\theta x_1 \tag{3-26}$$

where k is St. Venant's coefficient tabulated in Table 1.1. From Eqs. (1–20) and (1–21), the distortion strain components are:

$$\gamma_{yz} = \frac{\partial v}{\partial z} + \frac{\partial w}{\partial y} = \theta \left(\frac{\partial \psi}{\partial y} + x \right) \tag{3-27}$$

$$\gamma_{xz} = \frac{\partial w}{\partial x} + \frac{\partial u}{\partial z} = \theta \left(\frac{\partial \psi}{\partial x} - y \right) \tag{3-28}$$

From geometry the distortion strain component γ_{yz} will induce an elongation strain component in the 45° helical bar on the wider face, ϵ_s, as shown in Fig. 3.12:

$$\epsilon_s = \frac{\dfrac{\gamma_{yz}}{\sqrt{2}} dy}{\sqrt{2}\, dy} = \frac{\gamma_{yz}}{2} \qquad \text{for } x = \pm a \qquad (3\text{--}29)$$

Similarly, ϵ_s on the shorter face will be generated by γ_{xz}:

$$\epsilon_s = \frac{\gamma_{xz}}{2} \qquad \text{for } y = \pm b \qquad (3\text{--}30)$$

The stress in a helical reinforcement is $f_s = E_s \epsilon_s$, and the maximum stress $f_{s,\,max}$ occurs at the mid-depth of the wider face, $x = \pm a$, $y = 0$. From Eq. (3–29):

$$f_{s,\,max} = E_s \epsilon_{s,\,max} = \frac{1}{2} E_s \gamma_{yz,\,max}$$

Since $\gamma_{yz,\,max} = \tau_{yz,\,max}/G$ and using Eq. (3–26):

$$f_{s,\,max} = \frac{1}{2} k\, E_s x_1 \theta \qquad (3\text{--}31)$$

Equation (3–31) relates the maximum stress $f_{s,\,max}$ to the angle of twist θ. Referring to Rausch's space truss in Fig. 3.10, it can easily be shown

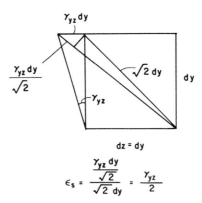

Fig. 3.12. Geometric relationship between distortion strain, γ_{yz}, and strain in the 45° bars, ϵ_s.

that when a member is subjected to torsion, the force in a 45° helical steel bar is equal to the force in a concrete compression strut. The elongation of the helical steel bar is also assumed to be equal to the shortening of a concrete compression strut. Hence the strain energies stored in a steel bar and those in a concrete strut should be equal. Using this relationship, the work of the external torque for a length of beam l should be equal to twice the strain energy stored in n helical bars of length L, each making a complete turn:

$$\frac{1}{2} T\theta l = 2 \int_0^L \frac{1}{2} n A_t f_s \epsilon_s \, dL \qquad (3\text{-}32)$$

Substituting $n = l/s$ and $f_s = E_s \epsilon_s$:

$$T = \frac{2A_t E_s}{\theta s} \int_0^L \epsilon_s^2 \, dL \qquad (3\text{-}33)$$

Using Eqs. (3–29) and (3–30):

$$T = \frac{2A_t E_s}{\theta s} 2 \left[\int_{-\sqrt{2}b}^{+\sqrt{2}\,b} \left(\frac{\gamma_{yz}}{2}\right)^2 dy + \int_{-\sqrt{2}a}^{+\sqrt{2}\,a} \left(\frac{\gamma_{xz}}{2}\right)^2 dx \right] \qquad (3\text{-}34)$$

Applying Eqs. (3–27) and (3–28) in Eq. (3–34) and noticing that $x = a$ in the first integral and $y = b$ in the second integral:

$$T = \frac{\sqrt{2}A_t E_s \theta}{s} \left[\int_{-b}^{+b} \left(\frac{\partial \psi}{\partial y} + a\right)^2 dy + \int_{-a}^{+a} \left(\frac{\partial \psi}{\partial x} - b\right)^2 dx \right] \qquad (3\text{-}35)$$

In the integrals ψ is given by Eq. (3–25). Observation of Eq. (3–35) shows that upon integration both integrals can be expressed by a parameter $a^2 b$ times a coefficient, which is a function of b/a. Hence:

$$T = \frac{\sqrt{2}A_t E_s \theta}{s} [\zeta(2a)^2(2b)] \qquad (3\text{-}36a)$$

where ζ is the coefficient obtained from integration. It is a function of b/a (or y_1/x_1) as listed in Table 3.1. Taking $x_1 = 2a$ and $y_1 = 2b$:

$$T = \zeta \frac{\sqrt{2}\, x_1 y_1 A_t}{s} E_s x_1 \theta \qquad (3\text{-}36b)$$

Table 3.1 Cowan's efficiency coefficient, λ

y_1/x_1	1.0	1.2	1.4	1.6	1.8	2.0	2.5	3.0
k	0.675	0.759	0.822	0.869	0.904	0.930	0.968	0.985
ζ	0.5636	0.6159	0.6584	0.6942	0.7250	0.7504	0.7986	0.8322
λ	0.834	0.812	0.801	0.798	0.801	0.807	0.827	0.844

Equation (3–36b) relates the external torque T to the angle of twist θ. Eliminating θ from Eqs. (3–36b) and (3–31), we obtain:

$$T = \lambda \frac{2\sqrt{2}\, x_1 y_1 \, A_t f_{s,\,max}}{s} \qquad (3\text{–}37)$$

where $\lambda = \zeta/k$, a function of y_1/x_1. The quantity λ has been called Cowan's efficiency coefficient and has been tabulated in Table 3.1 together with k and ζ. Table 3.1 shows that λ varies from 0.798 to 0.844 for y_1/x_1 less than 3. It can be taken conservatively as a constant 0.8.

Cowan also agreed with Andersen that in addition to the steel resistance concrete should also contribute a torque equal to the elastic torque, $T_e = \alpha x^2 y f'_t$. In terms of ultimate strength design, Eq. (3–37) becomes:

$$T_n = T_e + 1.6 \frac{\sqrt{2}\, x_1 y_1 \, A_t f_{sy}}{s} \qquad (3\text{–}38)$$

Equation (3–38) can be used for the design of helical reinforcement.

For beams with longitudinal bars and vertical stirrups, shown in Fig. 3.9, each direction of steel will resist a 45° component of the forces carried by the helical reinforcement. Cowan's equation becomes:

$$T_n = T_e + 1.6 \frac{x_1 y_1 \, A_t f_{sy}}{s} \qquad (3\text{–}39)$$

Equation (3–39) can be used for the design of vertical closed stirrups. The longitudinal steel area should be calculated by the principle of "equal volume" using Eq. (3–17).

It should again be emphasized that Cowan's efficiency coefficient for reinforcement is based on St. Venant's stress and strain distribution. Although this distribution is correct before cracking, its validity is doubtful in the post-cracking range. Later tests have shown that the stresses along the reinforcement are essentially uniform. They do not vary according to St. Venant's distribution. Even so, the concept of using an efficiency coefficient to improve Rausch's torsional resistance has been widely accepted.

3.2.3 PCA Tests and Skew-Bending Theory

3.2.3.1 PCA tests. According to the theories by Rausch and Cowan, the torsional strength of a rectangular member with longitudinal steel and stirrups can be expressed as:

$$T_n = T_c + \alpha_t \frac{x_1 y_1 A_t f_{sy}}{s} \qquad (3\text{-}40)$$

where

T_n = nominal torsional strength.

T_c = torque carried by concrete. It is zero in Rausch's theory and equal to the elastic torque, $\alpha x^2 y f'_t$, in Cowan's theory.

α_t = a coefficient. It is equal to 2 in Rausch's theory and 1.6 in Cowan's theory.

f_{sy} = yield strength of stirrups.

In Eq. (3-40) the parameter $x_1 y_1 A_t f_{sy}/s$ is commonly called the "reinforcement factor." If T_n is plotted against the reinforcement factor, the two theories by Rausch and Cowan can be plotted as two straight lines as shown in Fig. 3.13. The first term, T_c, appears as an intercept on the vertical axis, while α_t is the slope.

A typical series B of six beam tests made at the Portland Cement Association by the author is also plotted in Fig. 3.13. Each beam in this series has a solid cross section of 10 in. by 15 in. and is reinforced with four longitudinal corner bars plus closed stirrups. The principle of equal volume of longitudinal steel and stirrups was maintained. The material strengths are: $f_y = 50,000$ psi, $f'_c = 4,000$ psi. The only variable in the series of beam tests is the total percentage of reinforcement. Figure 3.13 shows that for this series of beams the vertical intercept $T_c = 75$ in.-kips, and the slope $\alpha_t = 1.2$. The test value of T_c is in between zero and 120 in.-k, predicted by Rausch and Cowan, respectively. The test slope of α_t, on the other hand, is considerably less than either Rausch's 2.0 or Cowan's 1.6. As a whole, the tests revealed that both Rausch's and Cowan's equations are unconservative.

Figure 3.13 also shows that the beams in Series B can be divided into three types according to the characteristics of failure:

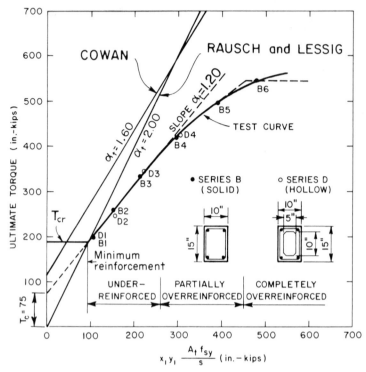

Fig. 3.13. Comparison of Rausch, Cowan, and Lessig theories with test results (Series B and D in Ref. 1).

1. Underreinforced beams: In this type of beam a moderate amount of steel is provided. Failure is caused by tensile yielding of both longitudinal bars and stirrups before the compression crushing of concrete. The failure is ductile.
2. Completely overreinforced beams: An excessive amount of steel has been provided in this type of beam. Failure is then caused by crushing of concrete before the yielding of both the longitudinal steel and the stirrups. Brittle failure is expected.
3. Partially overreinforced beams: This type of beam is reinforced with an unbalanced amount of longitudinal steel vs. stirrups. Therefore only the longitudinal bars or only the stirrups yield before crushing of concrete. (One of the two types of torsional steel does not yield at failure.) This failure could be ductile, but not as ductile as underreinforced beams.

The categorization of beams into "underreinforced" and "overreinforced" is well-known in flexural failure of reinforced concrete members. This division is also valid for torsion as described in (1) and (2). However, the third type of partially overreinforced beams is peculiar to torsion because torsional steel consists of both longitudinal steel and stirrups. These two components of torsional steel should have a proper volume ratio so as to yield simultaneously at failure. This proper ratio is known as the balanced volume ratio of longitudinal steel to stirrups, m_b. The quantity m_b should be unity according to Rausch's space truss analogy, but tests have shown that m_b may vary within a range, depending on the total percentage of reinforcement. An approximate range has been suggested[4] in conjunction with the equation in Eq. (3-42):

$$0.7 \leq m_b \frac{f_{ly}}{f_{sy}} \leq 1.5 \tag{3-41a}$$

It should be mentioned here that a theoretical derivation of the range of balanced ratio will be presented in Section 7.5.4 (Chapter 7) in terms of a variable angle of diagonal cracks in a truss model.

Figure 3.13 also shows that the curve through the test points in the partially and completely overreinforced regions is nonlinear. Only in the underreinforced region is the curve a straight line. Such underreinforced beams are both ductile and economical, while the overreinforced beams may be brittle and wasteful. In practical design, therefore, only underreinforced beams should be permitted. The total percentage of torsional steel (longitudinal steel and stirrups) dividing under- and overreinforced beams is commonly known as the balanced percentage of steel, ρ_{bt}. An approximate empirical equation for ρ_{bt} was given[4] as:

$$\rho_{bt} = \frac{2{,}400 \sqrt{f_c'}}{f_y} \tag{3-41b}$$

where f_c' and f_y are in psi and ρ_{bt} in percentage. A theoretical derivation of ρ_{bt} will also be treated in Chapter 7.

For underreinforced beams an equation was proposed based on 55 beam tests at the Portland Cement Association. Indeed, the equation can be expressed in the form of Eq. (3-40):

$$T_n = \underbrace{\frac{x^2 y}{3} (2.4\sqrt{f_c'})}_{T_c} + \underbrace{\left(0.66m \frac{f_{ly}}{f_{sy}} + 0.33 \frac{y_1}{x_1} \right)}_{\alpha_t \leq 1.5} \frac{x_1 y_1 A_t f_{sy}}{s} \tag{3-42}*$$

---

* In the original equation T_c was expressed as $(2.4/\sqrt{x})x^2 y \sqrt{f_c'}$, which includes a scale factor $\sqrt{x}$. Since the width of the test beams varies in a narrow range from 6 in. to 10 in., $\sqrt{x}$ has been simplified as 3, resulting in $T_c = (x^2 y/3)(2.4\sqrt{f_c'})$. However, for very large specimens where $x \gg 10$ in., T_c in Eq. (3-42) may be unconservative.

Note that α_t is a function of three variables:

m = ratio of volume of longitudinal steel to volume of stirrups
  $= \hat{A}_l s / A_t\ 2(x_1 + y_1)$
f_{ly}/f_{sy} = ratio of yield strengths of longitudinal steel to stirrups
y_1/x_1 = height-to-width ratio of stirrups

Normally α_t varies from 1.0 to 1.5 and is limited to an upper limit of 1.5.
  A comparison of the test value of T_c in Eq. (3–42) with the elastic theories
(Eq. 2–6), plastic theory (Eq. 2–7), and skew-bending theory (Eq. 2–12) is
illuminating. If the tensile strength of concrete, f'_t, in Eqs. (2–6) and
(2–7), and the reduced modulus of rupture 0.85 f_r, in Eq. (2–12), are taken
as $6\sqrt{f'_c}$, then:

Elastic theory:

$$T_e = (6\alpha)x^2 y\sqrt{f'_c}$$

where the coefficient 6α varies from 1.25 ($y/x = 1$) to 2.0 ($y/x = \infty$).

Plastic theory:

$$T_p = (6\alpha_p)x^2 y\sqrt{f'_c}$$

where the coefficient $6\alpha_p$ varies from 2.0 ($y/x = 1$) to 3.0 ($y/x = \infty$).

Skew-bending theory:

$$T_{np} = (2.0)\ x^2 y\sqrt{f'_c}$$

where the coefficient is a constant of 2.0.
  It can be seen that the test value of T_c in Eq. (3–42) has the same parameter
$x^2 y\sqrt{f'_c}$ as those of the above three theories. The coefficient $2.4/3 = 0.8$ is
40% of that for skew-bending theory. This percentage is greater than 40%
for elastic theory and less than 40% for plastic theory.
  What is the source of the torsional resistance of the first term, T_c? A
common view in the late fifties and early sixties held that T_c is contributed
by the concrete core within the reinforcing cage. The fact that the magnitude
of the test value, T_c, lies approximately midway between zero (according
to Rausch) and the elastic torque, T_e (according to Cowan), appears to uphold
this view.

To check this view, the PCA tests include a series D of hollow beams. The beams in this series D are identical to the solid beams of series B except that the concrete cores are removed. The test points for these hollow beams are also plotted in Fig. 3.13. It can be seen that these test points for hollow beams fall right on the curve for solid beams. In other words, other things being equal, a hollow beam (in series D) has ultimate torque equal to that of a solid beam (in series B). The concrete core, therefore, does not contribute to the ultimate strength, and the first term, T_c, cannot be contributed by the concrete core.

The source of the term T_c was revealed in the skew-bending theory developed by the author in 1968.[4] It was found that T_c is contributed by the shear resistance of the diagonal concrete struts, which was neglected in Rausch's theory. The mechanism is quite similar to that for flexural shear, where the source of the "concrete contribution" stems mainly from the shear resistance of the compression zone. The skew-bending theory to be introduced below also attempts to explain why the coefficient α_t is not a constant as suggested by Rausch and Cowan, but is a function of m, f_{ly}/f_{sy}, and y_1/x_1.

To study the mechanism of torsional failure, it is illuminating to examine the internal cracking pattern of a reinforced concrete member subjected to pure torsion. The specimens were chosen from series B with a cross section of 10 in. by 15 in. It was found that the cracks can be divided into two series as shown in Fig. 3.14 (a). One series started at the front wider face with an angle of 45° to the longitudinal axis and then progressed to the top and bottom shorter faces with an angle of between 45° and 90°. These cracks penetrated deep into the beam, reaching a depth of about 80% of the width at failure. The development of each crack was similar to that

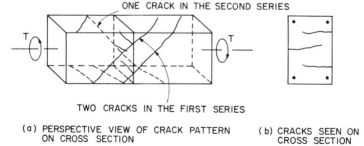

ONE CRACK IN THE SECOND SERIES

TWO CRACKS IN THE FIRST SERIES

(a) PERSPECTIVE VIEW OF CRACK PATTERN   (b) CRACKS SEEN ON
    ON CROSS SECTION   CROSS SECTION

Fig. 3.14. Internal crack pattern of a reinforced concrete member subjected to pure torsion.

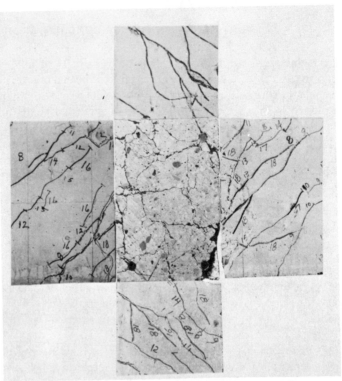

(c) CRACK PATTERN ON A SAW-CUT CROSS SECTION

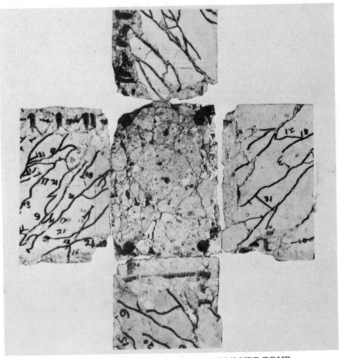

(d) SAW-CUT CROSS SECTION AT FAILURE ZONE

Fig. 3.14 continues

described in Chapter 2, Section 3 for a plain concrete beam. In other words, each crack constituted a potential failure surface.

The second series of cracks were similar to the first, except that they started at the back wider face. (Fig. 3.14a shows only one crack in this series.) When a vertical cross section was exposed by saw cut, the cracks on the cross section included both series of cracks as shown schematically in Fig. 3.14 (b). This particular cross section intersected two cracks that belonged to the first series and one crack that was in the second series. The actual photo is shown in Fig. 3.14 (c). A typical cracking picture at the failure zone is also given in Fig. 3.14 (d). The cracking was very extensive near one wider face, indicating a compression crushing of concrete there.

Since the internal cracking pattern revealed a skew-bending mode of failure, the failure surface should follow the cracking surface of one particular crack (say, in the first series of cracks). This failure surface should be similar to that shown in Fig. 2.5 for a plain concrete beam. Consequently, the same simplification of a 45° plane for plain concrete beams should also be applicable to reinforced concrete beams. This simplified failure surface is a plane that is inclined at a 45° angle to the longitudinal axis and is perpendicular to the wider face. The intersection of this failure plane and a shorter face will be oriented at an angle of 90° to the longitudinal axis. This 90° intersection on the shorter face is both simple and conservative, and is also substantiated partly by the author's tests,[1] which showed that the stresses in the shorter legs of stirrups tended to be small at failure. A perspective view of this assumed failure plane for a rectangular concrete beam is shown in Fig. 3.15.

3.2.3.2 Skew-bending theory. The assumptions used in the derivation of the skew-bending theory are:

1. Both the longitudinal steel and the stirrups yield at failure; i.e., the beam is underreinforced.
2. The tensile strength of concrete is neglected.
3. The spacing of stirrups is constant within the failure zone.
4. No external loads are present within the failure zone.
5. The effect of steel near the compression zone is neglected.
6. The area of the shear-compression zone is rectangular.

We shall now proceed to take the equilibrium of the free body about the axis of twist; i.e., the internal torque, T_{int}, must be equal to the external torque, T. The internal torsional resistance arises from three sources: (1) the axial forces of the stirrups, $A_t f_{sy}$, (2) the shear-compression force of concrete, P, and (3) the dowel forces of the longitudinal bars, Q_{lx} and Q_{ly}.

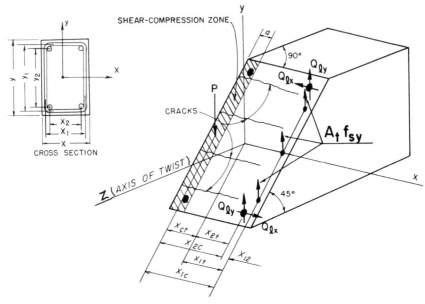

Fig. 3.15. Proposed failure surface.

The contribution of stirrup forces to the internal torsional resistance, $T_{n,s}$, is:

$$T_{n,s} = \frac{y_1}{s} A_t f_{sy} x_{1t} \tag{3-43}$$

where

y_1 = center-to-center distance of the longer leg of stirrups
s = spacing of stirrups
A_t = area of one leg of a torsional closed stirrup
f_{sy} = yield strength of stirrups
x_{1t} = distance from center of stirrups to the axis of twist (z-axis) (Fig. 3.15)

In Eq. (3-43), y_1/s is the number of stirrups intersected by a 45° crack. In connection with the symbol x_{1t}, we shall explain the meaning of the subscripts for the distance shown in Fig. 3.15. The subscript 1 represents the center of the stirrups, and the subscript t represents the axis of twist. Therefore, x_{1t} represents the distance in the shorter direction x from the

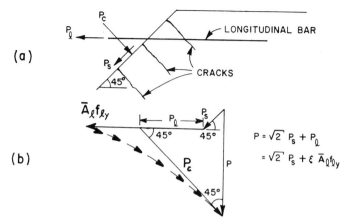

Fig. 3.16. The composition of force P.

center of the stirrups to the axis of twist. Similarly, the subscript 2 represents the center of the longitudinal corner bars, and the subscript c gives the center of the shear-compression zone.

To balance the force of the stirrups there must be a vertical force, P, at the shear-compression zone of concrete. P is the resultant of three forces, P_c, P_1, and P_s, as shown in Fig. 3.16 (a). P_c is the axial force of the concrete struts, which have been separated by the second series of cracks indicated in Fig. 3.14. P_1 is the axial force of the longitudinal steel, and P_s is the shear forces of the concrete struts. The composition of P is shown by the force polygon in Fig. 3.16 (b).

Figure 3.16 (b) shows that:

$$P = \sqrt{2}\, P_s + P_1 \qquad (3\text{--}44)$$

and P_1 in Eq. (3–44) can be expressed by:

$$P_1 = \xi \bar{A}_1 f_{1y} \qquad (3\text{--}45)$$

where

$\bar{A}_1$ = cross-sectional area of longitudinal bars within the shear-compression zone
  = ½ total longitudinal steel in case of four corner bars
f_{1y} = yield strength of longitudinal bars
ξ = efficiency coefficient of longitudinal bars, which should be less than unity

The coefficient ξ takes into account the fact that the longitudinal bars are concentrated at discrete points, instead of distributed uniformly in the shear-compression zone. The inclination of P_c would also tend to become smaller when its location is closer to the corner bars. These effects are shown schematically in Fig. 3.16 (b) by a series of P_c forces with varying directions.

Substituting Eq. (3–45) into Eq. (3–44):

$$P = \sqrt{2}\, P_s + \xi \overline{A}_l f_{ly} \tag{3–46}$$

and the contribution of P to the internal torsional resistance, $T_{n,c}$, is:

$$T_{n,c} = (\sqrt{2}\, P_s + \xi \overline{A}_l f_{ly}) x_{ct} \tag{3–47}$$

The third and last source of torsional resistance, $T_{n,q}$, stems from the dowel resistance of longitudinal bars:

$$T_{n,q} = 2\, Q_{lx} \frac{y_2}{2} + 2\, Q_{ly}\, x_{2t} \tag{3–48}$$

where Q_{lx} and Q_{ly} are the dowel force of longitudinal bars in the x and y directions.

Summing Eqs. (3–43), (3–47), and (3–48), the total torsional resistance, T_n, is:

$$
\begin{aligned}
T_n &= T_{n,c} + T_{n,s} + T_{n,q} \\
&= \sqrt{2}\, P_s x_{ct} + \xi \overline{A}_l f_{ly} x_{ct} + \frac{y_1}{s} A_t f_{sy} x_{1t} + Q_{lx} y_2 + 2 Q_{ly} x_{2t}
\end{aligned} \tag{3–49}
$$

Equation (3–49) can be simplified by considering the equilibrium of forces in the y-direction:

$$\frac{y_1}{s} A_t f_{sy} + 2 Q_{ly} = P$$

or:

$$2 Q_{ly} = \sqrt{2}\, P_s + \xi \overline{A}_l f_{ly} - \frac{y_1}{s} A_t f_{sy} \tag{3–50}$$

Substituting Eq. (3–50) into Eq. (3–49):

$$T_n = \sqrt{2}\, P_s(x_{ct} + x_{2t}) + \xi \bar{A}_1 f_{1y}(x_{ct} + x_{2t})$$

$$+ \frac{y_1 A_t f_{sy}}{s}(x_{1t} - x_{2t}) + Q_{lx} y_2 \qquad (3\text{-}51)$$

Notice that $x_{ct} + x_{2t} = x_{2c}$ and $x_{1t} - x_{2t} = x_{12}$. The quantities x_{2c} and x_{12} are independent of the axis of twist (z-axis).

$$T_n = \sqrt{2}\, P_s x_{2c} + \xi \bar{A}_1 f_{1y} x_{2c} + \frac{y_1 A_t f_{sy}}{s} x_{12} + Q_{lx} y_2 \qquad (3\text{-}52)$$

We shall now modify Eq. (3-52) into the form of Eq. (3-40). This can be done by converting the second and the fourth terms each into the form of a coefficient times the reinforcement factor, $x_1 y_1 A_t f_{sy}/s$. For the second terms we make use of the definition:

$$m = \frac{2\,\bar{A}_1\, s}{2\, A_t\, (x_1 + y_1)} \qquad (3\text{-}53)$$

$$\bar{A}_1 = \frac{m\, A_t\, (x_1 + y_1)}{s} \qquad (3\text{-}54)$$

Then:

$$\xi \bar{A}_1 f_{1y} x_{2c} = \xi m \left(\frac{f_{1y}}{f_{sy}}\right)\left(\frac{x_1 + y_1}{y_1}\right)\left(\frac{x_{2c}}{x_1}\right)\frac{x_1 y_1\, A_t\, f_{sy}}{s} \qquad (3\text{-}55)$$

For the fourth term in Eq. (3-52), we shall first evaluate Q_{lx} using two assumptions: (1) the dowel force of a longitudinal bar in the x-direction, Q_{lx}, is proportional to the area of one longitudinal bar A_l, and (2) the dowel force of a longitudinal bar in the x-direction, Q_{lx}, is proportional to the transverse displacement per unit length of the bar in the x-direction, Δ_x. Hence:

$$Q_{lx} = q\, A_l\, \Delta_x \qquad (3\text{-}56)$$

where

q = a proportional constant with the unit of stress (lb/in.^{2})
A_l = area of a longitudinal bar (in.^{2})
Δ_x = transverse displacement per unit length (in./in.)

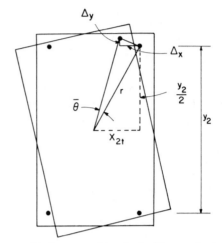

Fig. 3.17. Displacement of longitudinal bars under twisting.

From Fig. 3.17:

$$\Delta_x = \bar{\theta}\,\frac{y_2}{2} \tag{3-57}$$

where $\bar{\theta}$ = torsional angle per unit length (1/in.).

Substituting Δ_x from Eq. (3–57) and $A_l = (\frac{1}{2})\,\bar{A}_l$ from Eq. (3–54) into Eq. (3–56):

$$Q_{lx} = \left[\frac{1}{4}\left(\frac{q}{f_{sy}}\right)\bar{\theta}\left(\frac{y_2}{y_1}\right)\left(\frac{x_1+y_1}{x_1}\right)m\right]\frac{x_1 y_1\,A_t\,f_{sy}}{s} \tag{3-58}$$

Substituting Eqs. (3–55) and (3–58) into Eq. (3–52), we obtained:

$$T_n = \underbrace{\sqrt{2}\,P_s x_{2c}}_{T_c'} +$$

$$\underbrace{\left[\frac{x_2}{x_1} + \xi m\left(\frac{f_{ly}}{f_{sy}}\right)\left(1+\frac{x_1}{y_1}\right)\left(\frac{x_{2c}}{x_1}\right) + \frac{1}{4}\left(\frac{q}{f_{sy}}\right)(\bar{\theta}y_2)\left(\frac{y_2}{y_1}\right)\left(1+\frac{y_1}{x_1}\right)m\right]}_{\alpha_t}$$

$$\frac{x_1 y_1 A_t f_{sy}}{s} \tag{3-59}$$

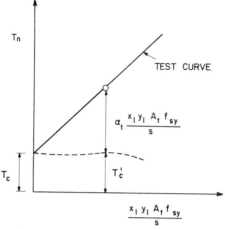

Fig. 3.18. Relationship between T_c and T_c'.

Equation (3–59) shows that the torsional resistance, T_n, can indeed be expressed by two terms, T_c' and $\alpha_t(x_1 y_1 A_t f_{sy}/s)$. Term T_c' is due to the shear resistance of the concrete struts, and α_t is shown to be a function of m, f_{ly}/f_{sy}, y_1/x_1, and y_2. When Eq. (3–59) is plotted in Fig. 3.18, the intercept T_c is seen to be directly related to T_c' and therefore is also the result of the shear resistance of concrete struts.

In conclusion, Eq. (3–59), which is derived from skew-bending theory, substantiates Eq. (3–42), which is obtained from tests.

Equation (3–42) was derived in 1968. Since then more than a dozen investigations have been carried out around the world to study pure torsion of reinforced concrete members. Based on these additional tests, a more up-to-date equation has been proposed by the author:[6]

$$T_n = \underbrace{\frac{x^2 y}{3}(2.4\sqrt{f_c'})}_{T_c} + \underbrace{\sqrt{m}\,\frac{f_{ly}}{f_{sy}}\left(1 + 0.2\,\frac{y_1}{x_1}\right)}_{\alpha_t \leq 1.6}\frac{x_1 y_1\,A_t f_{sy}}{s} \qquad (3\text{–}60)$$

The only difference between Eq. (3–42) and Eq. (3–60) is the coefficient α_t. However, α_t remains a function of m, f_{ly}/f_{sy}, and y_1/x_1.

3.2.4 Example 3.1

As shown in Fig. 3.19, a 10 in. by 20 in. rectangular beam is reinforced with four No. 7 longitudinal bars and No. 4 stirrups at 4.5 in. The clear

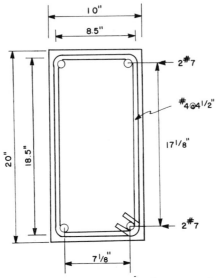

Fig. 3.19. Example 3.1—Rectangular cross section.

concrete cover is 0.5 in. Find the ultimate torsional strength using the Rausch, Cowan, and PCA methods, $f'_c = 4,000$ psi and $f_y = 60,000$ psi.

Solution

$$x_1 = 10 - 2(0.5) - 0.5 = 8.5 \text{ in.}$$
$$y_1 = 20 - 2(0.5) - 0.5 = 18.5 \text{ in.}$$
$$m = \frac{\hat{A}_1 s}{2A_t(x_1 + y_1)} = \frac{4(0.60)(4.5)}{2(0.20)(8.5 + 18.5)} = 1 \qquad \text{(equal volume)}$$

Rausch method (Eq. 3–18):

$$T_n = \frac{2\, x_1 y_1 A_t f_y}{s} = \frac{2(8.5)(18.5)(0.20)(60,000)}{4.5}$$
$$= 838,000 \text{ in.-lb (94.7 Nm)}$$

Cowan method (Eq. 3–39):

$$\frac{y}{x} = \frac{20}{10} = 2 \qquad \alpha = 0.246$$

Assume $f'_t = 5\sqrt{f'_c} = 5\sqrt{4,000} = 317$ psi

$$T_n = \alpha x^2 y f'_t + \frac{1.6\, x_1 y_1 A_t f_y}{s}$$

$$= 0.246(10)^2(20)(317) + \frac{1.6(8.5)(18.5)(0.20)(60,000)}{4.5}$$

$$= 156,000 + 671,000 = 827,000 \text{ in.-lb } (93.5 \text{ Nm})$$

PCA method (Eq. 3–42):

$$\alpha_t = 0.66\, m + 0.33\, \frac{y_1}{x_1} = 0.66(1) + 0.33\left(\frac{18.5}{8.5}\right) = 1.378 < 1.5$$

$$T_n = \frac{x^2 y}{3}(2.4\sqrt{f'_c}) + \frac{\alpha_t x_1 y_1 A_t f_y}{s}$$

$$= \frac{10^2(20)}{3}(2.4\sqrt{4,000}) + \frac{1.378(8.5)(18.5)(0.20)(60,000)}{4.5}$$

$$= 101,200 + 577,800 = 679,000 \text{ in.-lb } (76.7 \text{ Nm})$$

Comparing the three methods, it can be seen that Rausch's and Cowan's methods give close answers, but both these answers are significantly greater than that given by the PCA method and are unconservative.

3.3 POST-CRACKING TORSIONAL RIGIDITY

3.3.1 Theoretical Derivation[7]

To derive the post-cracking torsional rigidity of a reinforced concrete member, we shall start with a reinforced concrete tube of arbitrary cross-sectional shape and uniform wall thickness. Figure 3.5 shows such a reinforced concrete tube subjected to a torque, T. According to the thin-tube theory (Eq. 1–72), the shear stress τ should be:

$$\tau = \frac{T}{2Ah} \tag{3–61}$$

where

$A =$ area within the center line of hoop reinforcement
$h =$ wall thickness of the reinforced concrete tube

This shear stress τ will induce stresses and strains in the reinforcement and concrete. To evaluate these stresses and strains, the reinforced concrete tube will be idealized into Rausch's space truss, as discussed in Section 3.2.1. This space truss is shown again in Fig. 3.20, including the thickness of the tube, h.

In Rausch's analysis of the space truss, the forces in the longitudinal bars, in the hoop bars, and in the diagonal concrete struts are designated X, Y, and D, respectively. Each of these forces is constant throughout the tube. They also have the following relationship, according to Eq. (3–7):

$$X = Y = \frac{D}{\sqrt{2}} = F = \tau sh \qquad (3\text{–}62)$$

Using Eq. (3–62), the stresses σ_c, σ_l, and σ_h, in the concrete struts, in the longitudinal bars, and in the hoop bars, respectively, can be expressed as follows:

$$\sigma_c = \frac{D}{\left(\dfrac{s}{\sqrt{2}}\right) h} = \frac{\sqrt{2}\,\tau sh}{\left(\dfrac{s}{\sqrt{2}}\right) h} = 2\tau \qquad (3\text{–}63)$$

$$\sigma_l = \frac{X}{A_l} = \frac{\tau sh}{A_l} = \frac{\tau}{r_l} \qquad (3\text{–}64)$$

$$\sigma_h = \frac{Y}{A_t} = \frac{\tau sh}{A_t} = \frac{\tau}{r_h} \qquad (3\text{–}65)$$

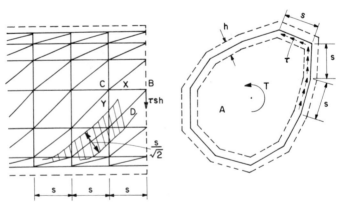

Fig. 3.20. Space truss for a tube section.

In Eq. (3–64), A_l = the area of one longitudinal bar, and $r_l = A_l/sh$ = the longitudinal reinforcement ratio with respect to the wall area. In Eq. (3–65), A_t = the area of one hoop bar, and $r_h = A_t/sh$ = the hoop reinforcement ratio with respect to the wall area.

The strains in the concrete struts and in the steel bars can be found from the stress–strain relationship and from Eqs. (3–63) through (3–65):

$$\epsilon_c = \frac{\sigma_c}{E_c} = \frac{2\tau}{E_c} \tag{3–66}$$

$$\epsilon_l = \frac{\sigma_l}{E_s} = \frac{\tau}{E_s r_l} \tag{3–67}$$

$$\epsilon_h = \frac{\sigma_h}{E_s} = \frac{\tau}{E_s r_h} \tag{3–68}$$

where ϵ_c, ϵ_l, and ϵ_h are the strains in the concrete struts, in the longitudinal bars, and in the hoop bars, respectively. E_c and E_s are the moduli of elasticity of concrete and steel.

The strains calculated by Eqs. (3–66) through (3–68) will cause a shear distortion of the tube. The shear distortion γ can be obtained from the compatibility of deformations in a basic cell of the space truss. Such basic cells are shown in Fig. 3.21. Each cell consists of one diagonal concrete strut and its surrounding steel bars, forming a square with a side length of s.

The shear distortion of cell $ABCD$ due to a compression strain ϵ_c in the concrete strut is shown in Fig. 3.21 (a). The original length of the diagonal concrete strut in a cell is designated $\overline{CB}$. After shortening, the length becomes $\overline{CE}$. The shortening $\overline{BE} = \epsilon_c(\sqrt{2}\,s)$. To maintain compatibility of the concrete strut with the steel bars, $\overline{CE}$ must rotate about point C, and $\overline{AB}$ must rotate about point A until points E and B meet at point F. For small deformation,

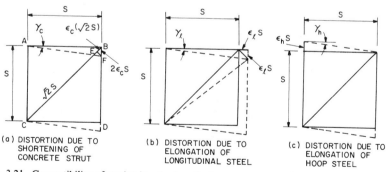

(a) DISTORTION DUE TO SHORTENING OF CONCRETE STRUT

(b) DISTORTION DUE TO ELONGATION OF LONGITUDINAL STEEL

(c) DISTORTION DUE TO ELONGATION OF HOOP STEEL

Fig. 3.21. Compatibility of strains in a basic cell of a space truss subjected to shear distortion.

$\overline{EF}$ is taken perpendicular to $\overline{CE}$ and $\overline{BF}$ perpendicular to $\overline{AB}$. From geometry, the distance $\overline{BF} = \sqrt{2}\ \overline{BE} = 2\ \epsilon_c s$. After deformation the shape of the cell beomes a parallelogram, as shown by the dotted lines. The shear distortion of the cell is then represented by the distortion angle γ_c:

$$\gamma_c = \frac{\overline{BF}}{\overline{AB}} = \frac{2\epsilon_c s}{s} = 2\epsilon_c \tag{3-69}$$

Similarly, the distortion γ_l due to the lengthening of the longitudinal bars and the distortion γ_h due to the lengthening of the hoop bars are illustrated in Fig. 3.21 (b) and (c), respectively. Accordingly,

$$\gamma_l = \frac{\epsilon_l s}{s} = \epsilon_l \tag{3-70}$$

$$\gamma_h = \frac{\epsilon_h s}{s} = \epsilon_h \tag{3-71}$$

The total shear distortion γ is:

$$\gamma = \gamma_c + \gamma_l + \gamma_h = 2\epsilon_c + \epsilon_l + \epsilon_h \tag{3-72}$$

Substituting Eqs. (3-66) through (3-68) into Eq. (3-72), we obtain:

$$\frac{\tau}{\gamma} = \frac{E_s}{\left(4n + \dfrac{1}{r_l} + \dfrac{1}{r_h}\right)}$$

where $n = E_s/E_c$. Let us define $G_{cr} = \tau/\gamma =$ post-cracking shear modulus. The above equation becomes:

$$G_{cr} = \frac{E_s}{\left(4n + \dfrac{1}{r_l} + \dfrac{1}{r_h}\right)} \tag{3-73}$$

G_{cr} describes the material property of a torsional member after cracking. It is a function of the moduli of elasticity of steel and concrete and the reinforcement ratios of longitudinal steel and hoop steel.

It is frequently more convenient and more common to express the reinforcement ratios with respect to the solid cross-sectional area of concrete A_c, instead of the area of the wall. Define:

$$\rho_l = \frac{\hat{A}_l}{A_c} = \left(\frac{uh}{A_c}\right) r_l \qquad (3\text{-}74)$$

$$\rho_h = \frac{A_h u}{A_c s} = \left(\frac{uh}{A_c}\right) r_h \qquad (3\text{-}75)$$

where

$\hat{A}_l =$ total cross-sectional area of longitudinal steel
$A_c =$ solid cross-sectional area within the outer perimeter of concrete
$u =$ perimeter of area bounded by the center line of a complete hoop bar
$A_h =$ cross-sectional area of one hoop

Substituting ρ_l and ρ_h for r_l and r_h in Eq. (3-73):

$$G_{cr} = \frac{E_s}{\left(4n + \dfrac{uh}{A_c \rho_l} + \dfrac{uh}{A_c \rho_h}\right)} \qquad (3\text{-}76)$$

The torsional geometric properties of a thin tube of homogeneous materials were studied in Chapter 1. According to Bredt's Eq. (1-80), the post-cracking torsional constant C_{cr} of a reinforced concrete member with uniform wall thickness can be expressed as:

$$C_{cr} = \frac{4 A_1^2 h}{u} \qquad (3\text{-}77)$$

where A_1 is defined as the area bounded by the center line of the hoop reinforcement. It should be noted that A_1 was defined in Eq. (1-80) as the area bounded by the center line of the wall thickness for tubes of homogeneous materials. In the case of reinforced concrete members, however, it seems more logical to define A_1 by the dimension of the reinforcement, rather than that of the concrete. This is so because in the post-cracking stage the concrete has cracked, and the torsional resistance is contributed mainly by the tension reinforcement in conjunction with the concrete in compression.

Combining Eqs. (3-76) and (3-77) gives the post-cracking torsional rigidity, $G_{cr} C_{cr}$:

$$G_{cr} C_{cr} = \frac{4 E_s A_1^2 A_c}{u^2 \left(\dfrac{4 n A_c}{uh} + \dfrac{1}{\rho_l} + \dfrac{1}{\rho_h}\right)} \qquad (3\text{-}78)$$

The denominator in Eq. (3–78) consists of three terms. They represent, in sequence, the contributions to the post-cracking torsional rigidity of the concrete struts, the longitudinal steel, and the hoop steel.

Equation (3–78) is applicable to arbitrary cross sections. It can easily be reduced to the special cases for circular and rectangular sections. For circular sections, $A_c = \pi d^2/4$, $A_1 = \pi d_1^2/4$, and $u = \pi d_1$.

$$G_{cr}C_{cr} = \frac{E_s \pi d_1^2 \, d^2}{\left(\dfrac{nd^2}{d_1 h} + \dfrac{1}{\rho_l} + \dfrac{1}{\rho_h}\right)} \tag{3–79}$$

where d and d_1 are, respectively, the diameters of the concrete cross section and the diameter of the circle formed by the center line of a hoop bar. For rectangular sections, $A_c = xy$, $A_1 = x_1 y_1$, and $u = 2(x_1 + y_1)$.

$$G_{cr}C_{cr} = \frac{E_s \, x_1^2 \, y_1^2 \, xy}{(x_1 + y_1)^2 \left[\dfrac{2n \, xy}{(x_1 + y_1)h} + \dfrac{1}{\rho_l} + \dfrac{1}{\rho_h}\right]} \tag{3–80}$$

In Eq. (3–80), x and y are the shorter and longer dimensions of the rectangular concrete section. x_1 and y_1 are the shorter and longer dimensions of a rectangular stirrup.

Equations (3–78) through (3–80) appear to be quite complex. It would be more straightforward in practice to calculate G_{cr} and C_{cr} separately by Eqs. (3–76) and (3–77) and then multiply them to obtain the torsional rigidity, $G_{cr}C_{cr}$.

3.3.2 Modifications of Theoretical Equations

3.3.2.1 Effective Wall Thickness (Rectangular Section).

In the foregoing theoretical derivation of post-cracking torsional rigidity we have assumed a reinforced concrete tube with a uniform wall thickness, h. An obvious question arises. What is the effective wall thickness, h_e, that is applicable to members with solid cross sections and thick hollow sections? This effective wall thickness has been derived for rectangular sections by comparing Eq. (3–80) with the PCA test results. Substituting the experimental values of $G_{cr}C_{cr}$ into Eq. (3–80), we can solve for h and get the effective wall thickness, h_e. The nondimensional ratio h_e/x is then plotted against the total reinforcement ratio, $\rho_l + \rho_h$, in Fig. 3.22. It can be seen that h_e/x is proportional to $\rho_l + \rho_h$, and the proportional constant is 1.4. Hence:

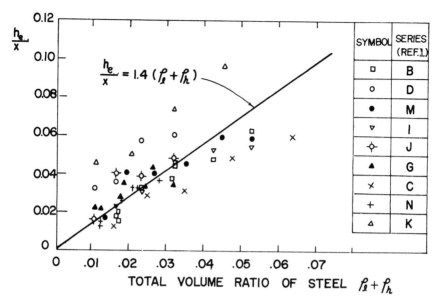

Fig. 3.22. Effective wall thickness as a function of steel ratio (Ref. 7).

$$h_e = 1.4(\rho_l + \rho_h)x \qquad (3\text{-}81)$$

The term h_e is usually quite small. It is an empirical quantity that fits the test results and should not be construed as the actual required wall thickness at ultimate strength.

3.3.2.2 Vertical Intercept.

A typical torque-twist curve is shown in Fig. 3.23. Before cracking, the slope of the curve represents St. Venant's torsional rigidity, G_cC. After cracking, the curve starts out as a straight line and then gradually curves toward horizontal when the maximum torque is approached. The slope of the straight portion represents the post-cracking torsional rigidity calculated by Eq. (3–80) using the effective wall thickness, h_e, from Eq. (3–81). The extrapolation of the straight portion will intersect the vertical axis, giving a vertical intercept. This vertical intercept was found to have the same parameters as $T_c = (x^2y/3)(2.4\sqrt{f_c'})$ and was defined as ηT_c, where η is a coefficient. The post-cracking torque-twist relationship can therefore be expressed as:

$$T = \eta T_c + G_{cr}C_{cr}\theta \qquad (3\text{-}82)$$

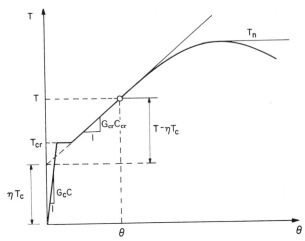

Fig. 3.23. Typical torque-twist curve of reinforced concrete beams.

The coefficient η is evaluated from three series of PCA tests with identical dimensions and materials, except the wall thickness. The first series has a solid cross section, i.e., $h/x = 0.5$ where h is the wall thickness and x is the smaller dimension of a rectangular cross section. The second and the third series have h/x values of 0.25 and 0.15, respectively. The torque-twist curves of three beams with identical reinforcement but with different wall thickness are plotted in Fig. 3.24. It can be seen that the post-cracking torsional rigidity is the same, but the vertical intercept increases with the wall thickness. Consequently, η is plotted against h/x in Fig. 3.25. For the three series of beams the test points can be approximated by a straight line that can be expressed by:

$$\eta = 0.57 + 2.86 \frac{h}{x} \qquad (3\text{–}83)$$

For a solid section, $\eta = 2$.

The fact that η is a function of the wall thickness is quite interesting. In the study of ultimate strength of a reinforced concrete torsional member, tests have shown that the concrete core has no effect on the ultimate strength. However, Eq. (3–83) appears to indicate that the concrete core does have an effect on the post-cracking torsional behavior via the vertical intercept.

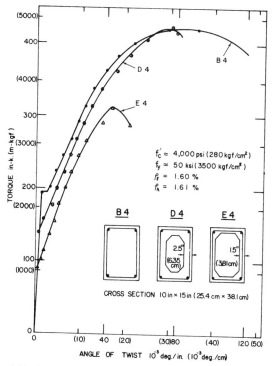

Fig. 3.24. Torque-twist curves for beams B4, D4, and E4 (Ref. 7).

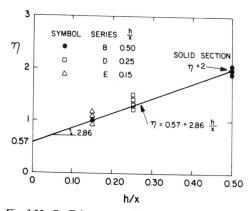

Fig. 3.25. Coefficient η as a function of wall thickness.

111

3.3.3 Summary of Torsional Rigidities

Before cracking $(T \leqslant T_{cr})$:

$$G_c C = \frac{T}{\theta} \qquad (3\text{--}84)$$

where

G_c = shear modulus of concrete = $E_c/2(1 + \nu)$
C = St. Venant's torsional constant (Section 1.2.4)

After cracking $(T > T_{cr})$:

$$G_{cr} C_{cr} = \frac{T - \eta T_c}{\theta} \qquad (3\text{--}85)$$

where:

$$G_{cr} = \frac{E_s}{\left(4n + \dfrac{u h_e}{A_c \rho_l} + \dfrac{u h_e}{A_c \, \rho_h}\right)} \qquad (3\text{--}86)$$

$$C_{cr} = \frac{4 \, A_1^2 \, h_e}{u} \qquad (3\text{--}87)$$

and:

$$h_e = 1.4 \, (\rho_l + \rho_h) x \qquad (3\text{--}88)$$

$$\eta = 0.57 + 2.86 \frac{h}{x} \qquad (3\text{--}89)$$

The cracking torque T_{cr} may be calculated by the skew-bending theory or by the thin-tube theory:

Skew-bending theory (applicable to solid as well as hollow sections):

$$T_{cr} = 6(x^2 + 10)y \sqrt[3]{f_c'} \left(\frac{4h}{x}\right) \quad \text{where } h \leqslant \frac{x}{4} \; \left(\text{use } h = \frac{x}{4} \text{ when } h > \frac{x}{4}\right)$$

Thin-tube theory $\left(\text{applicable only to thin hollow sections where h} \leq \dfrac{x}{4}\right)$:

$$T_{cr} = 2\,x_1 y_1 h f_t' \qquad \text{where } f_t' \text{ may be taken as } 5\sqrt{f_c'}$$

In the case of nonuniform wall thickness, the minimum wall thickness can be conservatively used for h in the above two equations.

3.3.4 Simplification of Torsional Rigidity

The post-cracking torsional rigidity expressed by Eqs. (3–85) through (3–89) can be greatly simplified using two assumptions:

1. Take $\eta = 0$; i.e., the vertical intercept is zero, and the torque-twist curve passes through the origin. Equation (3–85) becomes:

$$G_{cr}C_{cr} = \frac{T}{\theta} \tag{3–90}$$

2. Neglect the contribution of the concrete struts to the torsional rigidity; i.e., neglect the first term in the denominator of Eq. (3–86):

$$G_{cr} = \frac{E_s}{h_e\left(\dfrac{u}{\hat{A}_1} + \dfrac{s}{A_t}\right)} \tag{3–91}$$

When Eq. (3–91) is combined with Eq. (3–87):

$$G_{cr}C_{cr} = \frac{4E_s A_1^2}{u\left(\dfrac{u}{\hat{A}_1} + \dfrac{s}{A_t}\right)} \tag{3–92}$$

Notice that h_e has been canceled out in the multiplication, so that Eq. (3–88) is no longer necessary.

The simplified $G_{cr}C_{cr}$, Eq. (3–92), is compared with the more elaborate $G_{cr}C_{cr}$, expressed by Eqs. (3–85) through (3–89), in Fig. 3.26. It is interesting to note that the first assumption lowers the straight line, whereas the second assumption increases the slope of the straight line. These two assumptions tend to cancel each other in the post-cracking region. For this reason, the simplified $G_{cr}C_{cr}$ can often be used with acceptable accuracy in the post-

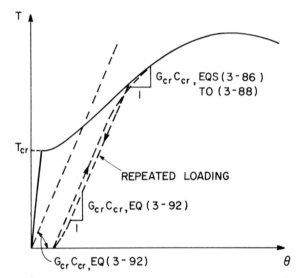

Fig. 3.26. Simplified torsional rigidity $G_{cr}C_{cr}$ calculated by Eq. (3–92).

cracking stage. It has been used by the author in the post-cracking analysis of horizontally curved beams.[8] It has also been suggested that this simplified post-cracking torsional rigidity is reasonable to represent the post-cracking elastic behavior under repeated loading (see Fig. 3.26).

3.3.5 Example 3.2

A hollow rectangular reinforced concrete beam with cross section shown in Fig. 3.27 is subjected to pure torsion. The longitudinal steel consists of four No. 7 corner bars, and the closed stirrups are No. 4 at 4.5-in. spacing, $f'_c = 4,000$ psi and $f_y = 60,000$ psi. Find the angle of twist under a torque of 350 in.-k using (1) the accurate method of Eqs. (3–85) through (3–89), and (2) the simplified method of Eq. (3–92).

Solution

$$T_{cr} = 6(x^2 + 10)y \sqrt[3]{f'_c} \left(\frac{4h}{x}\right)$$

$$= 6(10^2 + 10)20 \sqrt[3]{4,000} \left(\frac{4(2.5)}{10}\right)$$

$$= 209,500 \text{ in.-lb} < 350,000 \text{ in.-lb}$$

Use post-cracking properties.

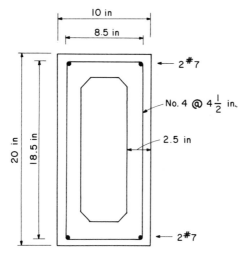

Fig. 3.27. Example 3.2—Box section.

(1) Accurate method:

$$\rho_l = \frac{\hat{A}}{A_c} = \frac{4(0.6)}{200} = 0.012$$

$$\rho_h = \frac{A_t u}{A_c s} = \frac{0.2(54)}{200(4.5)} = 0.012$$

$$h_e = 1.4(\rho_l + \rho_h)x = 1.4(0.012 + 0.012)10 = 0.336 \text{ in.}$$

$$\frac{u h_e}{A_c} = \frac{54(0.336)}{200} = 0.0907$$

$$n = \frac{E_s}{E_c} = \frac{29,000}{57\sqrt{4,000}} = 8.04$$

$$G_{cr} = \frac{E_s}{\left(4n + \dfrac{u h_e}{A_c \rho_l} + \dfrac{u h_e}{A_c \rho_n}\right)} = \frac{29,000}{\left[4(8.04) + \dfrac{0.0907}{0.012} + \dfrac{0.0907}{0.012}\right]}$$

$$= 613 \text{ ksi}$$

$$C_{cr} = \frac{4 A_1^2 h_e}{u} = \frac{4(8.5 \times 18.5)^2\, 0.336}{54} = 615 \text{ in.}^4$$

$$G_{cr}C_{cr} = 613(615) = 377,000 \text{ k-in.}^2$$

$$\eta = 0.57 + 2.86\frac{h}{x} = 0.57 + 2.86\frac{2.5}{10} = 1.285$$

$$T_c = \frac{x^2 y}{3}(2.4\sqrt{f_c'}) = \frac{10^2(20)}{3} 2.4\sqrt{4,000} = 101,200 \text{ in.-lb}$$

$$\theta = \frac{T - \eta T_c}{G_{cr}C_{cr}} = \frac{350 - 1.285(101.2)}{377,000}$$

$$= 0.583 \times 10^{-3} \text{ rad/in.}$$

(2) Simplified method:

$$A_1 = 8.5(18.5) = 157.25 \text{ in.}^2$$
$$u = 2(x_1 + y_1) = 2(8.5 + 18.5) = 54 \text{ in.}$$
$$G_{cr}C_{cr} = \frac{4E_s A_1^2}{u\left(\dfrac{u}{A_1} + \dfrac{s}{A_t}\right)} = \frac{4(29,000)(157.25)^2}{54\left(\dfrac{54}{2.4} + \dfrac{4.5}{0.2}\right)}$$
$$= 1,180,000 \text{ k-in.}^2$$
$$\theta = \frac{T}{G_{cr}C_{cr}} = \frac{350}{1,180,000} = 0.297 \times 10^{-3} \text{ rad/in.}$$

Note that θ calculated by the simplified method is only about one-half of that calculated by the accurate method. However, it is probably close to the post-cracking elastic rotation under repeated loading.

SUMMARY

In this chapter we have studied the behavior of reinforced concrete members subjected to torsion, emphasizing particularly the ultimate strength and the post-cracking torsional stiffness. This discussion should provide an in-depth understanding of the torsion phenomenon in reinforced concrete, particularly after cracking.

Three theories, by Rausch, Cowan, and the author, were presented here with a historical perspective. These theories led to the formulation of the ACI torsion design criteria, which will be discussed in Chapter 4.

The theory for post-cracking torsional stiffness of reinforced concrete members has not yet found its way into the ACI Code because the torsion provisions in the ACI Code are intended only for the ultimate strength design of a reinforced concrete member when the torsional moment is known. This code does not provide guidance for the analysis of torsional moments in a structure. It should be pointed out, however, that formulas for post-cracking torsional stiffness have been incorporated in Section 8.7 of the 1978 CEB-FIP Model Code.[9]

REFERENCES

1. Hsu, T. T. C., "Torsion of Structural Concrete—Behavior of Reinforced Concrete Rectangular Members," *Torsion of Structural Concrete,* SP-18, American Concrete Institute, Detroit, 1968, pp. 261–306.
2. Rausch, E., "Design of Reinforced Concrete in Torsion" (Berechnung des Eisenbetons gegen Verdrehung), Ph.D. thesis, Technische Hochschule, Berlin, 1929, 53 pp. (in German). A second edition was published in 1938 and a third in 1952. The third edition has the title "Drillung (Torsion), Schub and Scheren in Stahlbetonbau," Deutcher Ingenieur-Verlag GmbH, Dusseldorf, 168 pp.
3. Cowan, H. J., "Elastic Theory for Torsional Strength of Rectangular Reinforced Concrete Beams," *Magazine of Concrete Research* (London), Vol. 2, No. 4, July 1950, pp. 3–8.
4. Hsu, T. T. C., "Ultimate Torque of Reinforced Rectangular Beams," *Journal of the Structural Division, ASCE,* Vol. 94, ST 2, February 1968, pp. 485–510.
5. Andersen, P., "Experiments with Concrete in Torsion," *Transactions, ASCE,* Vol. 100, 1935, pp. 949–983. Also, *Proceedings, ASCE,* Vol. 60, May 1934, pp. 641–652.
6. Hsu, T. T. C., Discussion of "Pure Torsion in Rectangular Sections—A Re-examination," by McMullen, A. E. and Rangan, B. V., *Journal of the American Concrete Institute,* Proc., Vol. 76, No. 6, June 1979, pp. 741–746.
7. Hsu, T. T. C., "Post-Cracking Torsional Rigidity of Reinforced Concrete Sections," *Journal of the American Concrete Institute,* Proc., Vol. 70, No. 5, May 1973, pp. 352–360.
8. Hsu, T. T. C., M. Inan, and L. Fonticiella, "Behavior of Reinforced Concrete Horizontally Curved Beams," *Journal of the American Concrete Institute,* Proc., Vol. 75, No. 4, April 1978, pp. 112–123.
9. CEB-FIP, "Model Code for Concrete Structures," CEB-FIP International Recommendations, 3rd ed. Comite Euro-International du Beton (CEB), 1978.

Plate 4: American Hospital Association Buildings, Chicago. The continuous transfer girder with 5 ft by 9 ft cross section at the second floor level was the first to be designed by the ACI torsion criteria (Ref. 2), 1967.

4
ACI Torsion Design Criteria

4.1 INTRODUCTIONS

4.1.1 Brief History

Codification of torsion provisions for reinforced concrete began in earnest in the 1950s. A survey by Fisher and Zia[1] revealed that by 1960 there were eight countries in the world that had "full" specifications and five countries that had "partial" specifications. A "full" specification implies that the torsion provisions give both the permissible torsional stresses for concrete and the formulas for the design of torsional reinforcement. For "partial" specifications, only the former are given. It is interesting to note that five of the eight "full" specifications—namely, those from Egypt, Germany, Hungary, Poland, and Russia—utilized Rausch's Eqs. (3–12), (3–17), and (3–20a) for the design of torsional reinforcement. In contrast, the torsion provision of the Australian Code of 1958 was based on Cowan's theory, Eqs. (3–38) and (3–39). Cowan's theory also influenced the codes of several countries in the 1960s.

Although Rausch's and Cowan's theories were adopted by many codes in the 1950s and 1960s, the validity of these theories was not proved by systematic testing. In 1958, ACI Committee 438, Torsion, was created to study the torsion problem and to recommend a suitable provision for the 1963 ACI Building Code. However, owing to the lack of knowledge on the torsional behavior of reinforced concrete members, the committee decided that it could not recommend any detailed design provision for the 1963 Code. Hence only the following clause was included: "In edge or spandrel beams the stirrups provided shall be closed and at least one longitudinal bar shall be placed in each corner of the beam section, the bar to be at least the diameter of the stirrups or ½ in., whichever is greater."

Under the active promotion and prodding of ACI Committee 438, extensive experimental research was carried out in the 1960s. Based on these tests, ACI torsion criteria were formulated in 1969. These criteria were embodied in the "Tentative Recommendations for the Design of Reinforced Concrete Members to Resist Torsion."[2] The background and practical application of

119

these criteria were given in Ref. 3. With minor modifications these recommendations were incorporated into the 1971 ACI Building Code.[4]

The torsion design provisions of the 1971 Code were continued in the 1977 and the 1983 codes but in a somewhat different format. In this chapter we shall study those design criteria using both the 1971 and the 1977 (or 1983) code format. In addition, a new torsional limit design method was included in the 1977 and 1983 codes for the design of spandrel beams. This limit design method will be treated in Chapter 9.

ACI design criteria for torsion follow very closely the ACI design philosophy for flexural shear. In this way, the interaction between shear and torsion follows the same logic, and the design provisions for shear and torsion are included in the same Chapter 11 of the ACI Code. In view of this close connection between shear and torsion, we shall first review the ACI method for flexural shear design.

4.1.2 Review of Flexural Shear Design

Design procedures for flexural shear are based on the ACI philosophy that the shear strength of members with web reinforcement consists of two parts. The first part, V_c, contributed by concrete, is equal to the shear strength of the beam without web reinforcement. The second part, V_s, is contributed by the web reinforcement, using the well-known truss analogy. Based on this philosophy the nominal shear strength, V_n, is:

$$V_n = V_c + V_s \qquad (4\text{--}1)$$

where:

$$V_c = \left(1.9\sqrt{f_c'} + 2{,}500\,\frac{\rho_w V_u d}{M_u} \right) b_w d \qquad (4\text{--}2)$$

or:

$$V_c = 2\sqrt{f_c'}\, b_w d \qquad \text{(simplified)} \qquad (4\text{--}3)$$

and:

$$V_s = A_v f_y \frac{d}{s} \qquad (4\text{--}4)$$

In design, the applied factored shear, V_u, should satisfy the condition:

$$V_u \leqslant \phi(V_c + V_s) \tag{4-5}$$

where ϕ is the strength reduction factor.

According to this method, a three-step procedure of shear design can be formulated.

Step 1. Calculate the applied factored shear, V_u, from structural analysis and the shear contributed by concrete, V_c, by Eqs. (4-2) or (4-3).

If $V_u/\phi \leqslant V_c$, no shear reinforcement is required.
If $V_u/\phi > V_c$, provide web reinforcement.

Step 2. Calculate web reinforcement.

$$V_s = (V_u/\phi) - V_c \tag{4-6}$$

and:

$$A_v = \frac{V_s s}{d f_y} \tag{4-7}$$

Step 3. Provide minimum web reinforcement where necessary to ensure ductility.

$$A_{v,\,\text{min}} = \frac{50\ b_w s}{f_y} \tag{4-8}$$

The maximum web reinforcement is limited indirectly by:

$$V_s \leqslant 8\sqrt{f_c'}\ b_w d \tag{4-9}$$

The maximum spacing is limited to $d/2$ when $V_s \leqslant 4\sqrt{f_c'}\ b_w d$ and to $d/4$ when $V_s > 4\sqrt{f_c'}\ b_w d$.

4.2 NOMINAL TORSIONAL MOMENT STRENGTHS

4.2.1 Rectangular Sections

The fundamental equation for torsional strength of reinforced concrete members is based on the PCA tests described in Section 3.2.3. Assuming $m = 1$, $f_{ly} = f_{sy} = f_y$; Eq. (3-42) is simplified to become:

$$T_n = \frac{x^2 y}{3} \underbrace{(2.4\sqrt{f_c'}\,)}_{T_c} + \underbrace{\alpha_t \frac{x_1 y_1 A_t f_y}{s}}_{T_s} \qquad (4\text{--}10)$$

where

$\alpha_t = 0.66 + 0.33\ y_1/x_1 \leqslant 1.5$

T_n = nominal torsional moment strength

T_c = nominal torsional moment strength provided by concrete

T_s = nominal torsional moment strength provided by torsion reinforcement

For members without web reinforcement, we use the torsional strength for plain concrete members, Eq. (2–12), derived from skew-bending theory:

$$T_{np} = \frac{x^2 y}{3}(0.85 f_r) \qquad (4\text{--}11)$$

When $0.85 f_r$ is taken very conservatively as $2.4\sqrt{f_c'}$, the right-hand side of Eq. (4–11) is identical to the first term, T_c, in Eq. (4–10). Therefore, T_c can also be viewed very conservatively as the torsional moment strength of a member without web reinforcement. In this way the ACI philosophy of shear design can also be applied to torsion. Equation (4–10) states that the torsional capacity of a member with torsional reinforcement is made up of two parts. One part of the torsional strength, T_c, contributed by concrete, is taken to be the torsional strength of a member without torsional reinforcement. The other part of the torsional strength, T_s, is assumed to be contributed by torsional reinforcement. In practical design, the applied factored torque, T_u, should satisfy the following condition:

$$T_u \leqslant \phi(T_c + T_s) \qquad (4\text{--}12)$$

where ϕ is the strength reduction factor.

Based on this design criterion, the design procedures for torsion are identical to those for shear:

Step 1. Calculate the applied factored torque, T_u, and the torque contributed by concrete, T_c, by:

$$T_c = \frac{x^2 y}{3} 2.4\sqrt{f_c'} \qquad (4\text{--}13)$$

If $T_u/\phi \leqslant T_c$, no torsional reinforcement is required.

If $T_u/\phi > T_c$, provide torsional reinforcement.

Step 2. Calculate torsional reinforcement:

$$T_s = (T_u/\phi) - T_c \qquad (4\text{-}14)$$

$$A_t = \frac{T_s s}{\alpha_t x_1 y_1 f_y} \qquad (4\text{-}15)$$

where $\alpha_t = 0.66 + 0.33\ y_1/x_1 \leqslant 1.5$. For longitudinal steel the principle of equal volume should be used, since m has been assumed to be unity in Eq. (4-10).

$$\hat{A}_l = A_t \frac{2(x_1 + y_1)}{s} \qquad (4\text{-}16)$$

where $\hat{A}_l$ = the total area of longitudinal steel.

Step 3. Provide minimum torsional reinforcement when necessary to ensure ductility. The maximum permissible torque and the maximum stirrup spacing should also be limited. These three limitations will be derived in Sections 4.4.2, 4.3.3, and 4.4.3.1, respectively.

EXAMPLE 4.1

A 20 in. by 36 in. (508 mm by 914 mm) rectangular section as shown in Fig. 4.1 is subjected to a factored design torque of 200 k-ft (271 kN-m). Design the torsional reinforcement using $f'_c = 4,000$ psi (27.6 MN/m²) and $f_y = 60,000$ psi (414 MN/m²). The clear concrete cover is 1.5 in. (3.81 cm).

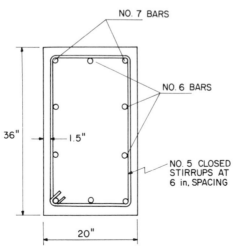

Fig. 4.1. Example 4.1—Rectangular section.

Solution

$$T_c = \frac{x^2 y}{3} 2.4\sqrt{f'_c} = \frac{(20)^2(36)}{3} 2.4\sqrt{4{,}000} = 728{,}600 \text{ in.-lb}$$

$$= 60.7 \text{ k-ft } (82.3 \text{ kN-m})$$

$$\frac{T_u}{\phi} = \frac{200}{0.85} = 235.3 \text{ k-ft } (319.0 \text{ kN-m}) > T_c$$

Torsional reinforcement is required.

$$T_s = \frac{T_u}{\phi} - T_c = 235.3 - 60.7 = 174.6 \text{ k-ft } (236.7 \text{ kN-m})$$

Assume No. 5 stirrups.

$$x_1 = 20 - 2(1.5) - 0.625 = 16.375 \text{ in. } (41.6 \text{ cm})$$
$$y_1 = 36 - 2(1.5) - 0.625 = 32.375 \text{ in. } (82.2 \text{ cm})$$
$$\alpha_t = 0.66 + 0.33\, y_1/x_1 = 1.31 < 1.5 \quad \text{O.K.}$$
$$\frac{A_t}{s} = \frac{T_s}{\alpha_t x_1 y_1 f_y} = \frac{174.6(12{,}000)}{1.31(16.375)(32.375)(60{,}000)}$$
$$= 0.0503 \text{ in.}^2/\text{in. } (0.128 \text{ cm}^2/\text{cm})$$

For No. 5 stirrups:

$$A_t = 0.31 \text{ in.}^2 (2.0 \text{ cm}^2)$$

$$s = \frac{0.31}{0.0503} = 6.16 \text{ in. } (15.6 \text{ cm})$$

Use No. 5 stirrups at 6 in. (15.2 cm) spacing.

$$\hat{A}_l = A_t \frac{2(x_1 + y_1)}{s} = 0.31 \frac{(2)(16.375 + 32.375)}{6.16} = 4.91 \text{ in.}^2 (31.68 \text{ cm}^2)$$

Use four No. 7 + six No. 6, $\hat{A}_l = 5.04$ in.^{2} (32.5 cm^2) > 4.91 in.^{2}

The four No. 7 corner bars and the six No. 6 longitudinal bars are distributed around the cross section as shown in Fig. 4.1. The spacing of the longitudinal bars satisfies the maximum spacing of 12 in. The 6-in. spacing of stirrups satisfies the maximum spacing of $(x_1 + y_1)/4 = 12.2$ in. or 12 in., whichever is less. These maximum spacing requirements will be discussed in Section 4.4.3.1.

The minimum torsional reinforcement will be derived later in Section 4.4.2, Eq. (4–67). We shall illustrate the application of this equation here for pure torsion. Taking $V_u = 0$:

$$\hat{A}_l = \left[\frac{400\,xs}{f_y} \left(\frac{T_u}{T_u + \dfrac{V_u}{3C_t}} \right) - 2A_t \right] \frac{x_1 + y_1}{s}$$

$$= \left[\frac{400(20)(6)}{60,000}\,(1) - 2(0.31) \right] \frac{(16.375 + 32.375)}{6}$$

$$= 1.46 \text{ in.}^2 \ (9.42 \text{ cm}^2) < 5.04 \text{ in.}^2 \qquad \text{O.K.}$$

Similarly, the maximum permissible torque can be checked by Eq. (4–50) in Section 4.3.3. Taking $V_u = 0$:

$$T_{n,\,\text{max}} = \frac{4\sqrt{f_c'}\ \Sigma x^2 y}{\sqrt{1 + \left(\dfrac{0.4 V_u}{C_t T_u} \right)^2}} = 4\sqrt{4,000}\ (20)^2(36)$$

$$= 3,643,000 \text{ in.-lb} = 303.6 \text{ k-ft} \ (411.9 \text{ kN-m}) > 235.3 \text{ k-ft} \qquad \text{O.K.}$$

4.2.2 Flanged Sections (T, L, I)

In Chapter 2, Section 2.4, a summation assumption was utilized for the calculation of torsional strength of plain concrete members with flanged sections. Using this assumption, the torsional strength of a flanged section is taken as the sum of the strengths of its rectangular components. This assumption was also found to be conservative for members with longitudinal steel and web reinforcement.[5-7] Hence:

$$T_n = \Sigma \left[\frac{x^2 y}{3} 2.4\sqrt{f_c'} + \alpha_t \frac{x_1 y_1 A_t f_y}{s} \right)$$

$$= \underbrace{\frac{\Sigma x^2 y}{3} 2.4\sqrt{f_c'}}_{T_c} + \underbrace{\Sigma \alpha_t \frac{x_1 y_1 A_t f_y}{s}}_{T_s} \qquad (4\text{–}17)$$

where the symbol Σ means the sum of all rectangular components.

The calculation of the quantity $\Sigma x^2 y$ in the first term of Eq. (4–17) influences the arrangement of the closed stirrups as shown in Fig. 4.2. In dividing the component rectangles, the ACI Code allows us to maximize $\Sigma x^2 y$. In

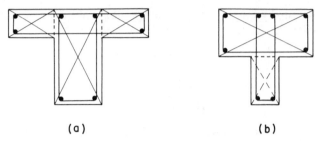

Fig. 4.2. Selection of rectangular components in flanged sections to calculate $\Sigma x^2 y$.

the common case of Fig. 4.2 (a), the primary closed stirrups are installed in the stem. The $\Sigma x^2 y$ value should be taken as the $x^2 y$ value of the web extending through the overall depth of the section plus the $x^2 y$ values of the outstanding flanges. In the less common case of Fig. 4.2 (b), however, it would be more advantageous to install the primary closed stirrups in the upper wider rectangular portion. The $\Sigma x^2 y$ value should then be taken as the sum of the $x^2 y$ values of the upper wider component rectangle and the narrow vertical outstanding flange. In both cases, the outstanding flange width of the cross section should not exceed three times the thickness of the flange.

Figure 4.3 (a) shows the common case of a rectangular spandrel beam with floor slab on one side. In this case, closed stirrups are provided only in the stem of the beam. In calculating the web reinforcement, the Σ symbol in the second term of Eq. (4–17) consists of only one rectangular component. The three-step design procedures for rectangular sections using Eqs. (4–13) through (4–16) should therefore be applicable, except that T_c in Eq. (4–13)

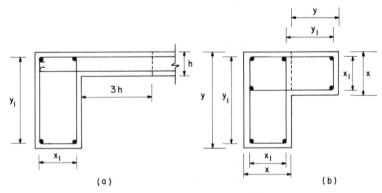

Fig. 4.3. Design of torsional stirrups in flanged sections.

must include the Σ symbol, accounting for both the beam and the flange, i.e.:

$$T_c = \frac{\Sigma x^2 y}{3} 2.4\sqrt{f_c'} \qquad (4\text{-}18)$$

Figure 4.3 (b) shows the case of an isolated L-beam where stirrups may be provided in both the rectangular components. To design stirrups for *each* component it may be assumed that T_s, the torsional resistance contributed by stirrups, is distributed among the component rectangles in proportion to the parameter $x^2 y$.

$$T_{s1} = T_s \frac{x^2 y}{\Sigma x^2 y} \qquad (4\text{-}19)$$

where T_{s1} refers to T_s for a particular rectangular component being designed. The design of torsional reinforcement for *each* component rectangle should then follow the design procedures for rectangular sections, Eqs. (4-14) through (4-16), except that T_s should be substituted by T_{s1} in Eq. (4-19). The x, y, x_1, and y_1 values for the two rectangular components are illustrated in Fig. 4.3 (b). Similar treatment can be applied to the T-beams in Fig. 4.2 (a) and (b).

EXAMPLE 4.2

An ell-shape cross section as shown in Fig. 4.4 is subjected to a factored design torque of 200 k-ft (271 kN-m). Design the torsional reinforcement.

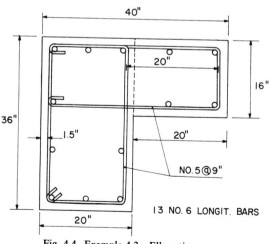

Fig. 4.4. Example 4.2—Ell section.

in both the web and the flange using $f'_c = 4,000$ psi (27.6 MN/m^2) and $f_y = 60,000$ psi (414 MN/m^2). The clear concrete cover is 1.5 in. (3.81 cm).

Solution
The ell section is divided into the web and the flange, as shown by the dotted line in Fig. 4.4, to maximize the parameter $\Sigma x^2 y$.

Web: $x^2 y = 20^2(36) = 14,400$ in.^{3}
Flange: $x^2 y = 16^2(20) = 5,120$ in.^{3}
Total: $\Sigma x^2 y = 14,400 + 5,120 = 19,520$ in.^{3} (319,900 cm^3)

From Eq. (4–18):

$$T_c = \frac{\Sigma x^2 y}{3} 2.4\sqrt{f'_c} = \frac{19,520}{3} 2.4\sqrt{4,000} = 987,600 \text{ in.-lb}$$
$$= 82.3 \text{ k-ft (111.6 kN-m)}$$
$$\frac{T_u}{\phi} = \frac{200}{0.85} = 235.3 \text{ k-ft (319.0 kN-m)} > T_c$$

Torsional reinforcement will be placed in the web and the flange:

$$T_s = \frac{T_u}{\phi} - T_c = 235.3 - 82.3 = 153.0 \text{ k-ft (207.5 kN-m)}$$

For the web:
From Eq. (4–19):

$$T_{s, web} = T_s \frac{(x^2 y)_{web}}{\Sigma x^2 y} = 153.0 \frac{14,400}{19,520} = 112.9 \text{ k-ft (153.1 kN-m)}$$

Assume No. 5 stirrups:

$$x_1 = 20 - 2(1.5) - 0.625 = 16.375 \text{ in. (41.6 cm)}$$
$$y_1 = 36 - 2(1.5) - 0.625 = 32.375 \text{ in. (82.2 cm)}$$
$$\alpha_t = 0.66 + 0.33 \, y_1/x_1 = 1.31 < 1.5 \quad \text{O.K.}$$
$$\frac{A_t}{s} = \frac{T_{s, web}}{\alpha_t x_1 y_1 f_y} = \frac{112.9(12,000)}{1.31(16.375)(32.375)(60,000)}$$
$$= 0.0325 \text{ in.}^2/\text{in. (0.8826 cm}^2/\text{cm)}$$

For No. 5 stirrups:

$$A_t = 0.31 \text{ in.}^2 \ (2.0 \text{ cm}^2)$$

$$s = \frac{0.31}{0.0325} = 9.54 \text{ in. (24.2 cm)} \qquad \text{Use 9 in. (22.9 cm)}$$

$$\hat{A}_{l,w} = A_t \frac{2(x_1 + y_1)}{s} = 0.31 \frac{2(16.375 + 32.375)}{9.54} = 3.17 \text{ in.}^2 \ (20.4 \text{ cm}^2)$$

For the flange:

$$T_{s,\text{ flange}} = T_s \frac{(x^2 y)_{\text{flange}}}{\Sigma x^2 y} = 153.0 \frac{5,120}{19,520}$$

$$= 40.1 \text{ k-ft (54.4 kN-m)}$$

$$x_1 = 16 - 2(1.5) - 0.625 = 10.375 \text{ in. (26.4 cm)}$$

$$y_1 = 20 \text{ in. (50.8 cm)}$$

$$\alpha_t = 0.66 + 0.33 \, y_1/x_1 = 1.30 < 1.5 \qquad \text{O.K.}$$

$$\frac{A_t}{s} = \frac{T_{s,\text{ flange}}}{\alpha_t x_1 y_1 f_y} = \frac{40.1(12,000)}{1.30(10.375)(20)(60,000)}$$

$$= 0.0297 \text{ in.}^2/\text{in. } (0.0755 \text{ cm}^2/\text{cm})$$

For No. 5 stirrups:

$$A_t = 0.31 \text{ in.}^2$$

$$s = \frac{0.31}{0.0297} = 10.4 \text{ in. (265 cm)} \qquad \text{Use 9 in. (22.9 cm)}$$

$$\hat{A}_{l,f} = A_t \frac{2(x_1 + y_1)}{s} = 0.31 \frac{2(10.375 + 20)}{10.4} = 1.81 \text{ in.}^2 \ (12.64 \text{ cm}^2)$$

Total area of longitudinal bars:

$$\hat{A}_l = \hat{A}_{l,w} + \hat{A}_{l,f} = 3.17 + 1.81 = 4.98 \text{ in.}^2 \ (33.1 \text{ cm}^2)$$

Use 13 No. 6 bars:

$$\hat{A}_l = 13(0.44) = 5.72 \text{ in.}^2 \ (36.9 \text{ cm}^2) > 4.98 \text{ in.}^2 \qquad \text{O.K.}$$

The 13 longitudinal bars are distributed as shown in Fig. 4.4 with all the spacings less than 12 in. The stirrups are No. 5 at 9-in. spacings for both the web and the flange. The 9-in. spacing is less than the maximum spacing of $(x_1 + y_1)/4 = 12.2$ in. for the web, but is greater than $(x_1 + y_1)/4 = 7.6$ in. for the flange. Such a slight violation of the ACI spacing requirement can be tolerated in flanged cross sections.

The minimum torsional reinforcement will be derived later. (See Eq. 4–67 in Section 4.4.2.) This equation will be illustrated here for pure torsion. Taking $V_u = 0$:

For the web:

$$\hat{A}_{1,w} = \left[\frac{400(20)(9)}{60,000}(1) - 2(0.31)\right]\frac{16,375 + 32.375}{9}$$
$$= 3.14 \text{ in.}^2 \ (20.3 \text{ cm}^2) < 3.17 \text{ in.}^2 \qquad \text{O.K.}$$

For the flange:

$$\hat{A}_{1,f} = \left[\frac{400(20)(9)}{60,000}(1) - 2(0.31)\right]\frac{(10.375 + 20)}{9}$$
$$= 1.96 \text{ in.}^2 \ (12.6 \text{ cm}^2) > 1.81 \text{ in.}^2 \qquad \text{say O.K.}$$

This small difference can be tolerated. This calculation shows that this beam is reinforced with approximately minimum reinforcement for pure torsion.

The maximum permissible torque can be checked by Eq. (4–50) in Section 4.3.3. Taking $V_u = 0$:

$$T_{n,\max} = 4\sqrt{4,000} \ (19,520) = 4,938,000 \text{ in.-lb}$$
$$= 411.5 \text{ k-ft} \ (558 \text{ kN-m}) > 235.3 \text{ k-ft} \qquad \text{O.K.}$$

4.2.3   Box Sections

It was shown in Fig. 3.13 that, all things being equal, the ultimate strength of a box section with wall thickness $h \geq x/4$ is equal to that of a solid section. In other words, Eq. (4–10) for solid sections is also applicable to box sections when $h \geq x/4$. For wall thickness $h < x/4$, however, the ultimate torque is less than that of a solid section, and Eq. (4–10) should be modified by a reduction factor R. This reduction factor, taken as $4h/x$, is applied to the first term T_c of Eq. (4–10), resulting in:

$$T_n = \left(\frac{4h}{x}\right) \underbrace{\frac{x^2 y}{3} 2.4\sqrt{f'_c}}_{T_c} + \underbrace{\alpha_t \frac{x_1 y_1 A_t f_y}{s}}_{T_s} \qquad (4\text{--}20)$$

where $\dfrac{x}{10} \leqslant h \leqslant \dfrac{x}{4}$.

There are two justifications for using the reduction factor $R = 4h/x$. First, using the reduction factor, the first term T_c in Eq. (4–20) is a simple approximation of the torsional strength T_n of an unreinforced thin tube of constant thickness. According to Eq. (1–72):

$$T = 2hA\tau = 2h(x - h)(y - h)\tau \qquad (4\text{--}21)$$

Taking $\tau = 2.4\sqrt{f'_c}$, T_n can be written as:

$$T_n = \underbrace{6\frac{h}{x}\left(1 - \frac{h}{x}\right)\left(1 - \frac{h}{y}\right)}_{R}\frac{x^2 y}{3} 2.4\sqrt{f'_c} \qquad (4\text{--}22)$$

The R shown in Eq. (4–22) is plotted as a function of h/x in Fig. 4.5, together with $R = 4h/x$ in Eq. (4–20). It can be seen that $R = 4h/x$ is a good approximation of R derived from thin-tube theory when $h \leq x/4$.

Second, Eq. (4–20) was checked directly by torsion tests of box beams with longitudinal steel and stirrups as shown in Fig. 4.6. The tests are taken from PCA (Ref. 1 of Chapter 3), as well as from Stuttgart[8] and Zurich.[9] The ratios of test strength to the strength calculated by Eq. (4–20) are plotted as a function of the ratio h/x. It can be seen that Eq. (4–20) is valid for h down to $0.15x$. In view of the lack of tests and the possibility of local wall failure when $h < 0.1x$, the ACI Code limits the applicability of Eq. (4–20) to $h \geq x/10$.

Figure 4.6 shows three PCA beams that have torsional strengths 20% less than calculated. These rectangular box beams have only four longitudinal bars at the corners, and some of the stirrups did not yield at failure. This premature failure appears to be caused by the lack of longitudinal bars within the thin wall. It is, therefore, suggested that the spacing of the longitudinal bars in a box section should not exceed two times the thickness of the wall, $2h$.

In view of the applicability of Eq. (4–20), the design procedures for box sections are the same as those for rectangular solid sections in Eqs. (4–13)

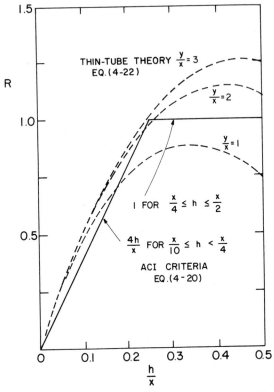

Fig. 4.5. Comparison of reduction factor, R, for thin-tube theory and for ACI criteria.

through (4–16), except that T_c in Eq. (4–13) should be multiplied by the reduction factor $4h/x$, i.e.:

$$T_c = \left(\frac{4h}{x}\right)\frac{x^2y}{3}2.4\sqrt{f_c'} \qquad (4\text{–}23)$$

where $x/10 \leqslant h \leqslant x/4$.

EXAMPLE 4.3

A 50 in $\times$ 72 in. box cross section, shown in Fig. 4.7, is subjected to a factored design torque of 1,100 k-ft (1,492 kN-m). The wall thicknesses of the top and bottom flanges are 8 in. The vertical webs have a larger thickness of 10 in. (vertical webs are called up to resist shear in addition to torsion, but combined shear and torsion is not illustrated in this example). Design

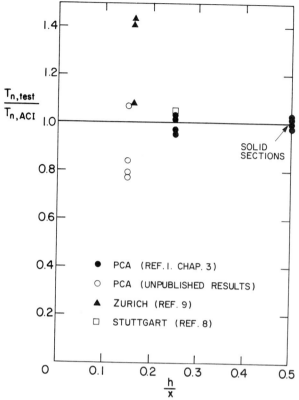

Fig. 4.6. Comparison of hollow beam tests with ACI criteria, Eq. (4–20).

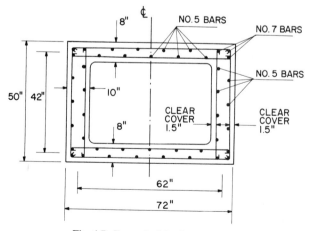

Fig. 4.7. Example 4.3—Box section.

the torsional reinforcement using $f'_c = 5,000$ psi (34.5 MN/m²) and $f_y = 60,000$ psi (414 MN/m²). The clear concrete cover is 1.5 in. (3.81 cm).

Solution

$$\frac{x}{4} = \frac{50}{4} = 12.5 \text{ in.}$$

$$\frac{x}{10} = \frac{50}{10} = 5.0 \text{ in.}$$

$$5.0 \text{ in.} < h = 8 < 12.5 \text{ in.} \qquad \text{O.K.}$$

$$\frac{4h}{x} = \frac{4(8)}{50} = 0.64$$

$$T_c = \left(\frac{4h}{x}\right)\frac{x^2 y}{3}(2.4\sqrt{f'_c}) = 0.64 \frac{50^2(72)}{3}(2.4\sqrt{5,000})$$

$$= 6,517,000 \text{ in.-lb} = 543.1 \text{ k-ft} \ (736.4 \text{ kN-m})$$

$$\frac{T_u}{\phi} = \frac{1,100}{0.85} = 1,294 \text{ k-ft} \ (1,755 \text{ kN-m}) > T_c$$

$$T_s = \frac{T_u}{\phi} - T_c = 1,294 - 543.1 = 750.9 \text{ k-ft} \ (1,018 \text{ kN-m})$$

For transverse reinforcement use two bars symmetrically in each wall:

$$x_1 = 50 - 8 = 42 \text{ in.}$$
$$y_1 = 72 - 10 = 62 \text{ in.}$$
$$\alpha_t = 0.66 + 0.33 \, y_1/x_1 = 1.15 < 1.5 \qquad \text{O.K.}$$
$$\frac{A_t}{s} = \frac{T_s}{\alpha_t x_1 y_1 f_y} = \frac{750.9(12,000)}{1.15(42)(62)(60,000)}$$
$$= 0.0501 \text{ in.}^2/\text{in.} \ (0.127 \text{ cm}^2/\text{cm})$$

For 2 No. 4 bars:

$$A_t = 2(0.20) = 0.40 \text{ in.}^2$$
$$s = \frac{0.40}{0.0501} = 7.98 \text{ in.} \qquad \text{Use } s = 8 \text{ in.} \ (20.3 \text{ cm})$$

The stirrups are No. 4 at 8-in. spacing. The 8-in. stirrups spacing is less than the maximum spacing of $(x_1 + y_1)/4 = 26$ in., $2h = 16$ in., or 12 in., whichever is less.

It should be noted that the reduction factor $4h/x$ has been calculated from the thickness of the flanges, $h = 8$ in. This is correct for the flanges, but is conservative for the web, which has a large thickness of 10 in. If the torsional reinforcement for the flanges and the webs can be installed independently, the torsional reinforcement in the web can be reduced by using a larger reduction factor for the web:

$$\frac{4h}{x} = \frac{4(10)}{50} = 0.80$$

$$T_c = 0.80 \frac{50^2(72)}{3} (2.4\sqrt{5,000}) = 8,146,000 \text{ in.-lb}$$

$$= 678.8 \text{ k-ft (920.5 kN-m)}$$

$$T_s = \frac{T_u}{\phi} - T_c = 1,294 - 678.8 = 615.2 \text{ k-ft (834.2 kN-m)}$$

$$\frac{A_t}{s} = \frac{615.2(12,000)}{1.15(42)(62)(60,000)} = 0.0411 \text{ in.}^2/\text{in. (0.1044 cm}^2/\text{cm)}$$

$$s = \frac{0.40}{0.0411} = 9.73 \text{ in.} \qquad \text{Use } s = 9.5 \text{ in. (24.1 cm)}$$

It can be seen that the stirrups in the webs can be designed as two No. 4 at 9.5-in. spacing, if desired.

The longitudinal bars can be calculated as follows:

$$\hat{A}_l = A_t \frac{2(x_1 + y_1)}{s} = 0.40 \frac{2(42 + 62)}{8} = 10.4 \text{ in.}^2 \text{ (67.1 cm}^2\text{)}$$

The minimum torsional reinforcement can be checked by calculating the minimum longitudinal steel area given later in Eq. (4–67). For a hollow box section, Eq. (4–67) has to be modified by the reduction factor $4h/x$. To be conservative the larger reduction factor of 0.8 for the webs will be used. Taking $V_u = 0$:

$$\hat{A}_l = \left[\left(\frac{4h}{x}\right) \left(\frac{400xs}{f_y}\right) \left(\frac{T_u}{T_u + \dfrac{V_u}{3C_t}}\right) - 2A_t \right] \frac{x_1 + y_1}{s}$$

$$= \left[(0.80) \frac{400(50)(8)}{60,000} (1) - 2(0.40) \right] \frac{(42 + 62)}{8}$$

$$= 17.3 \text{ in.}^2 \text{ (111.8 cm}^2\text{)} > 10.4 \text{ in.}^2$$

Use 12 No. 7 corner bars plus 32 No. 5 bars in the walls.

$$\hat{A}_l = 12(0.60) + 32(0.31) = 17.1 \text{ in.}^2 \; (110.5 \text{ cm}^2) \approx 17.3 \text{ in.}^2$$

These 44 longitudinal bars are distributed in the cross section as shown in Fig. 4.7. All the bar spacings are close to but less than the maximum spacing of 12 in. It should be emphasized that the minimum reinforcement governs only in this case of pure torsion. In the case of combined shear and torsion, the minimum reinforcement would require much less steel area and usually does not govern.

The maximum permissible torque can be checked by Eq. (4–50). For hollow sections, however, the parameter $\Sigma x^2 y$ in Eq. (4–50), and in the constant C_t (see Eq. 4–39), must be multiplied by the reduction factor $4h/x$. Taking $V_u = 0$:

$$T_{n,\max} = 4\sqrt{f_c'} \; \Sigma x^2 y \left(\frac{4h}{x}\right) = 4\sqrt{5,000} \; (50)^2(72)(0.64)$$

$$= 32,580,000 \text{ in.-lb} = 2,715 \text{ k-ft} \; (3,682 \text{ kN-m})$$

$$> 1,294 \text{ k-ft} \quad \text{O.K.}$$

4.3 INTERACTION OF TORSION AND SHEAR

For beams without web reinforcement, the interaction of torsion, shear, and flexure has been studied at the University of Washington[10] and the University of Texas.[11-13] It was found that the strength in combined torsion and shear is not significantly affected by the magnitude of the moment–shear ratio. The interaction between shear and torsion can be expressed by a nondimensional interaction diagram as shown in Fig. 4.8. The vertical axis represents the ratio V_n/V_{no} and the horizontal axis T_n/T_{no}. The definitions of these quantities are:

V_n = shear strength in combined torsion and shear
V_{no} = shear strength with zero torsion
T_n = torsional strength in combined torsion and shear
T_{no} = torsional strength in pure torsion

In Fig. 4.8 the point on the vertical axis where $V_n/V_{no} = 1$ represents a beam subjected to flexural shear without torsion. The point on the horizontal axis where $T_n/T_{no} = 1$ gives a beam in pure torsion. All other beams that are subjected to combined torsion and shear lie in between these two points.

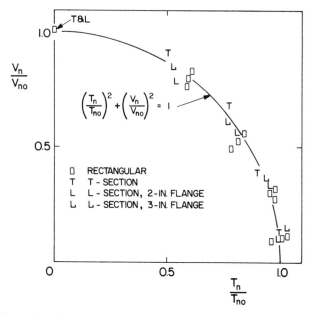

Fig. 4.8. Interaction of torsion and shear for beams without web reinforcement (Ref. 10).

The test trend shows clearly that the interaction of torsion and shear can be approximated by an elliptical equation as follows:

$$\left(\frac{T_n}{T_{no}}\right)^2 + \left(\frac{V_n}{V_{no}}\right)^2 = 1 \qquad (4\text{-}24)$$

4.3.1 Cracking Stresses

The interaction of torsional cracking and shear cracking for beams without stirrups can also be expressed by an elliptical equation similar to Eq. (4-24) because the diagonal cracking is followed very closely by failure. Therefore, we can also write:

$$\left(\frac{T_{cr}}{T_{cro}}\right)^2 + \left(\frac{V_{cr}}{V_{cro}}\right)^2 = 1 \qquad (4\text{-}25)$$

where

T_{cr} = diagonal cracking torque under torsion and shear
T_{cro} = diagonal cracking torque under pure torsion
V_{cr} = diagonal cracking shear under torsion and shear
V_{cro} = diagonal cracking shear with zero torsion

Equation (4–25) was found to be also applicable to beams with web reinforcement.[14-17]

Equation (4–25) is expressed in terms of torsional moment T and shear force V. Since T and V do not have the same units, it is not convenient to use them in the following derivation of equations for torsion–shear interaction. For this reason we shall convert shear force and torsion moment into shear stress and torsional stress, which have the same unit of psi, using the following definitions:

$$\text{nominal shear stress } v = \frac{V}{b_w d} \qquad (4\text{–}26)$$

$$\text{nominal torsional stress } \tau = \frac{3T}{\Sigma x^2 y} \qquad (4\text{–}27)$$

All the subscripts that are used to clarify V and T, such as u (factored), n (nominal strength), c (concrete contributions), and cr (cracking), remain valid for the stresses v and τ.

Express Eq. (4–25) in terms of stresses:

$$\left(\frac{\tau_{cr}}{\tau_{cr0}}\right)^2 + \left(\frac{v_{cr}}{v_{cr0}}\right)^2 = 1 \qquad (4\text{–}28)$$

where

$\tau_{cr} = $ torsional stress at diagonal cracking under torsion and shear
$\tau_{cr0} = $ torsional stress at diagonal cracking in pure torsion
$v_{cr} = $ shear stress at diagonal cracking under torsion and shear
$v_{cr0} = $ shear stress at diagonal cracking with zero torsion

Equation (4–28) is plotted as the elliptical curve DE in Fig. 4.9, assuming $\tau_{cr0} = 6\sqrt{f_c'}$ and $v_{cr0} = 2\sqrt{f_c'}$. Rearranging Eq. (4–28):

$$\left(\frac{\tau_{cr}}{\tau_{cr0}}\right)^2 \left[1 + \left(\frac{\tau_{cr0}}{v_{cr0}}\right)^2 \left(\frac{v_{cr}}{\tau_{cr}}\right)^2\right] = 1$$

$$\tau_{cr} = \frac{\tau_{cr0}}{\sqrt{1 + \left(\frac{\tau_{cr0}}{v_{cr0}}\right)^2 \left(\frac{v_{cr}}{\tau_{cr}}\right)^2}} \qquad (4\text{–}29)$$

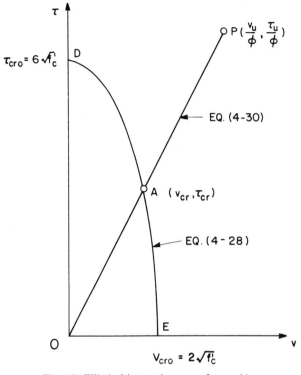

Fig. 4.9. Elliptical interaction curve for cracking.

We then assume that the ratio of shear stress to torsional stress remains a constant during the history of loading:

$$\frac{v_{cr}}{\tau_{cr}} = \frac{v_u}{\tau_u} \qquad (4\text{-}30)$$

Equation (4–30) is represented by the straight line OP in Fig. 4.9. Substituting Eq. (4–30), $\tau_{cr0} = 6\sqrt{f'_c}$ and $v_{cr0} = 2\sqrt{f'_c}$ into Eq. (4–29) results in:

$$\tau_{cr} = \frac{6\sqrt{f'_c}}{\sqrt{1 + (3\,v_u/\tau_u)^2}} \qquad (4\text{-}31)$$

In a similar manner, we can derive a conjugate equation for v_{cr}:

$$v_{cr} = \frac{2\sqrt{f'_c}}{\sqrt{1 + (\tau_u/3v_u)^2}} \qquad (4\text{-}32)$$

Equations (4–31) and (4–32) are a pair of conjugate cracking stresses representing point A in Fig. 4.9, which is the intersection of the elliptical curve, Eq. (4–28), and the straight line OP, Eq. (4–30).

4.3.2 Contribution of Concrete

In combined torsion and shear the interaction of torsional and shear stresses contributed by concrete, τ_c and v_c, is not as simple as the interaction of cracking stresses, τ_{cr} and v_{cr}. Unlike the cracking stresses, it is very difficult to obtain the interaction relationship between τ_c and v_c directly from tests. In the absence of direct guidance from tests, we shall model our derivation for τ_c and v_c after the previous derivation for cracking stresses. This derivation requires two assumptions:

Assumption 1. In combined torsion and shear the interaction relationship between τ_c and v_c can be represented by an elliptical curve, i.e.:

$$\left(\frac{\tau_c}{\tau_{c0}}\right)^2 + \left(\frac{v_c}{v_{c0}}\right)^2 = 1 \tag{4–33}$$

where

> τ_c = torsional stress contributed by concrete under torsion and shear
> τ_{c0} = torsional stress contributed by concrete under pure torsion $(2.4\sqrt{f'_c})$
> v_c = shear stress contributed by concrete under torsion and shear
> v_{c0} = shear stress contributed by concrete under zero torsion $(2\sqrt{f'_c})$

Rearranging Eq. (4–33):

$$\left(\frac{\tau_c}{\tau_{c0}}\right)^2 \left[1 + \left(\frac{\tau_{c0}}{v_{c0}}\right)^2 \left(\frac{v_c}{\tau_c}\right)^2\right] = 1$$

$$\tau_c = \frac{\tau_{c0}}{\sqrt{1 + \left(\frac{\tau_{c0}}{v_{c0}}\right)^2 \left(\frac{v_c}{\tau_c}\right)^2}} \tag{4–34}$$

Assumption 2. The ratio of shear stress to torsional stress remains constant during the history of loading:

$$\frac{v_c}{\tau_c} = \frac{v_u}{\tau_u} \tag{4–35}$$

Substituting Eq. (4–35), $\tau_{c0} = 2.4\sqrt{f'_c}$ and $v_{c0} = 2\sqrt{f'_c}$ into Eq. (4–34) gives:

$$\tau_c = \frac{2.4\sqrt{f'_c}}{\sqrt{1 + \left(1.2\,\dfrac{v_u}{\tau_u}\right)^2}} \qquad (4\text{–}36)$$

Similarly, the conjugate equation for v_c is:

$$v_c = \frac{2\sqrt{f'_c}}{\sqrt{1 + \left(\dfrac{\tau_u}{1.2\,v_u}\right)^2}} \qquad (4\text{–}37)$$

Equations (4–33) and (4–35) are plotted in Fig. 4.10 as the elliptical curve FE and a straight line OP, respectively. The point of intersection B gives the coordinates τ_c and v_c, expressed by a pair of conjugate Eqs. (4–36) and (4–37). These two equations were used in the 1971 ACI Code.

In the 1977 and 1983 ACI codes, Eqs. (4–36) and (4–37) are expressed

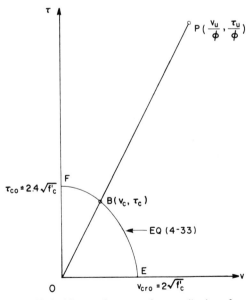

Fig. 4.10. Elliptical interaction curve for contribution of concrete.

in terms of torsional moment T and shear force V. From the definition of τ_u and v_u in Eqs. (4–26) and (4–27), we can write:

$$\frac{\tau_u}{v_u} = \frac{T_u}{V_u} \left(3 \frac{b_w d}{\Sigma x^2 y} \right)$$ (4–38)

The 1977 and 1983 ACI codes define:

$$C_t = \frac{b_w d}{\Sigma x^2 y}$$ (4–39)

so that Eq. (4–38) becomes:

$$\frac{\tau_u}{v_u} = \frac{T_u}{V_u} (3C_t)$$ (4–40)

Substituting Eqs. (4–40), (4–26), and (4–27) into Eqs. (4–36) and (4–37) results in:

$$T_c = \frac{0.8\sqrt{f'_c}\,\Sigma x^2 y}{\sqrt{1 + \left(\dfrac{0.4 V_u}{C_t T_u} \right)^2}}$$ (4–41)

$$V_c = \frac{2\sqrt{f'_c}\,b_w d}{\sqrt{1 + \left(\dfrac{C_t T_u}{0.4 V_u} \right)^2}}$$ (4–42)

4.3.3 Maximum Permissible Stresses

In Chapter 3 we studied the torsional balanced reinforcement ratio, ρ_{bt}, which divides the underreinforced beams and the overreinforced beams. Since only underreinforced beams are acceptable in design practice, the balanced reinforcement serves as an upper limit for the torsional reinforcement ratio. Figure 4.11 shows a typical curve for T_n vs. torsional reinforcement ratio, ρ_t. The balanced point, ρ_{bt}, is also identified. When $\rho_t < \rho_{bt}$, a beam is underreinforced, and it fails in a ductile manner. When $\rho_t > \rho_{bt}$, a beam is overreinforced, and the failure mode may be brittle.

The upper limit of web reinforcement for shear without torsion has been specified indirectly by a maximum shear strength of $V_{n0,\max} = 10\sqrt{f'_c}\,b_w d$ (or a maximum shear stress of $10\sqrt{f'_c}$). This approach has also been adopted

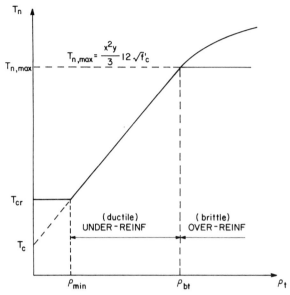

Fig. 4.11. Typical curve relating torsional strength T_n to total torsional reinforcement ratio ρ_t.

in torsion. Figure 4.11 shows that ρ_{bt} corresponds to a torsional moment, $T_{n\,0,\max}$. From tests, this $T_{n\,0,\max}$ has been taken very conservatively to be:

$$T_{n\,0,\max} = \frac{\Sigma x^2 y}{3} 12\sqrt{f'_c} \qquad (4\text{-}43)$$

or in terms of torsional stress:

$$\tau_{n\,0,\max} = 12\sqrt{f'_c} \qquad (4\text{-}44)$$

In combined torsion and shear, two assumptions are also made in the derivation of $\tau_{n,\max}$ and $v_{n,\max}$:

Assumption 1. The interaction relationship between $\tau_{n,\max}$ and $v_{n,\max}$ can be represented by an elliptical curve, i.e.:

$$\left(\frac{\tau_{n,\max}}{\tau_{n\,0,\max}}\right)^2 + \left(\frac{v_{n,\max}}{v_{n\,0,\max}}\right)^2 = 1 \qquad (4\text{-}45)$$

Rearranging gives:

$$T_{n,\,max} = \frac{12\sqrt{f'_c}}{\sqrt{1 + \left(\dfrac{T_{n0,\,max}}{v_{n0,\,max}}\right)^2 \left(\dfrac{v_{n,\,max}}{T_{n,\,max}}\right)^2}} \tag{4-46}$$

Assumption 2. The ratio of shear stress to torsional stress remains constant during the history of loading:

$$\frac{v_{n,\,max}}{T_{n,\,max}} = \frac{v_u}{T_u} \tag{4-47}$$

Substituting Eq. (4-47) and $T_{n0,\,max}/v_{n0,\,max} = 12\sqrt{f'_c}/10\sqrt{f'_c} = 1.2$ into Eq. (4-46):

$$T_{n,\,max} = \frac{12\sqrt{f'_c}}{\sqrt{1 + \left(\dfrac{1.2 v_u}{T_u}\right)^2}} \tag{4-48}$$

Similarly, the conjugate equation for $v_{n,\,max}$ is:

$$v_{n,\,max} = \frac{10\sqrt{f'_c}}{\sqrt{1 + \left(\dfrac{T_u}{1.2\,v_u}\right)^2}} \tag{4-49}$$

Equation (4-45) is plotted in Fig. 4.12 as the elliptical curve *GH*, and Eq. (4-47) is the straight line *OC*. The point of intersection *C* is represented by a pair of equations, Eqs. (4-48) and (4-49). To give a clear picture of the region where we can design shear and torsional reinforcement, the curve for cracking *DE* and the curve for concrete contribution *FE* are transferred from Figs. 4.9 and 4.10 to Fig. 4.12. The allowable region to design web reinforcement should lie in the area *FEHG*. If the applied design stresses v_u/ϕ and T_u/ϕ lie within the area *FED*, a minimum amount of torsional reinforcement must be provided. This minimum reinforcement will be derived in Section 4.4.2.

Equations (4-48) and (4-49) are two conjugate equations that serve the same purpose of checking maximum permissible stress (or maximum reinforcement ratio). In practice, we select one convenient equation for checking. Thus, only Eq. (4-48) appears in the 1971 ACI Code. In terms of torsional moment, Eq. (4-48) becomes:

$$T_{n,\,max} = \frac{4\sqrt{f'_c}\,\Sigma x^2 y}{\sqrt{1 + \left(\dfrac{0.4 V_u}{C_t T_u}\right)^2}} \tag{4-50}$$

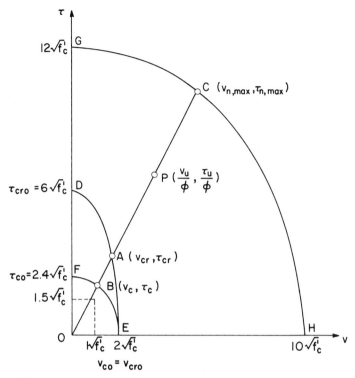

Fig. 4.12. Interaction curve for maximum shear and torsional stresses.

Comparison of Eqs. (4–50) and (4–41) reveals that $T_{n,\,max} = 5T_c$. In design:

$$T_s + T_c \leqslant T_{n,\,max} = 5T_c$$

Therefore, the 1977 and the 1983 ACI Codes specify that:

$$T_s \leqslant 4T_c \qquad (4\text{–}51)$$

4.4 DESIGN OF TORSIONAL REINFORCEMENT

4.4.1 Design Equations and Arrangements of Reinforcement

The design of shear and torsional reinforcement in combined loadings follows the familiar three-step procedure. The first two steps are:

Step 1. Calculate the applied factored torque and shear, T_u and V_u, as well as the contribution of concrete, T_c and V_c, by Eqs. (4–41) and (4–42).

If $T_u/\phi \le T_c$ (or $V_u/\phi \le V_c$), no web reinforcement is required.
If $T_u/\phi > T_c$ (or $V_u/\phi > V_c$), provide web reinforcement.

Step 2. Design web reinforcement and the torsional longitudinal reinforcement. The stirrups required by shear and torsion are:

$$A_v = \frac{V_s s}{d f_y} \qquad \text{where } V_s = (V_u/\phi) - V_c \qquad (4\text{-}52)$$

$$A_t = \frac{T_s s}{\alpha_t x_1 y_1 f_y} \qquad \text{where } T_s = (T_u/\phi) - T_c \qquad (4\text{-}53)$$

The term A_v in Eq. (4–52) is the total area of vertical stirrups, while A_t in Eq. (4–53) is the area of one leg of a closed stirrup. The torsional longitudinal steel is determined using the equal volume principle, assuming that the yield strengths of longitudinal steel and stirrups are equal:

$$\hat{A}_l = A_t \frac{2(x_1 + y_1)}{s} \qquad (4\text{-}54)$$

This longitudinal steel should be added to that required by flexure.

The arrangement of web and longitudinal reinforcement in Step 2 requires further discussion. Figure 4.13 shows two ways to arrange web reinforcement. For very large girders, as in Fig. 4.13 (a), it would be economical to provide torsional and shear reinforcement separately. Torsional web reinforcement consists of closed stirrups along the periphery, while the shear web reinforcement is in the form of vertical bars distributed along the width of the girder. Such an arrangement has been used in the torsional girders of the American Hospital Association building in Chicago. These girders have large cross sections of 9 ft by 5 ft.

For beams of smaller size, as shown in Fig. 4.13 (b), it is often more

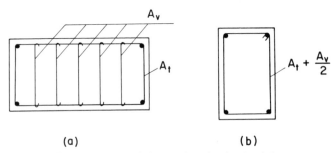

(a)   (b)

Fig. 4.13. Arrangement of shear and torsional web reinforcement.

convenient to combine shear and torsional web reinforcement in the form of closed stirrups only. In this case one can calculate $A_t + (A_v/2)$ and provide this amount of closed stirrups.

For torsional longitudinal bars, ACI Code requires that they should be uniformly distributed around the perimeter of the cross section with a spacing not exceeding 12 in. At least one longitudinal bar should be placed in each corner of the closed stirrups. Those torsional longitudinal bars that are located in the flexural tension zone and flexural compression zone may be combined with the flexural steel. A typical arrangement of longitudinal bars is shown in Fig. 4.14.

The addition of torsional and flexural longitudinal reinforcement in the flexural compression zone is obviously too conservative. This is so because it is illogical to add torsional steel that is in tension to the flexural steel that is in compression. The method of adding torsional longitudinal steel to the flexural longitudinal steel (regardless of whether the latter is in tension or in compression) is adopted purely for simplicity. In the author's opinion a design refinement may be made in the flexural compression zone by using the greater of the flexural compression steel or the torsional steel that lies in the flexural compression zone. In terms of the typical example in Fig. 4.14, we may take A'_s or $\hat{A}_l/3$, whichever is greater, as the design steel area in the flexural compression zone.

In beams subjected to bending in addition to significant torsion and shear (represented by points in the region $FEHG$ in Fig. 4.12), the interaction of bending with shear and torsion for such beams is accounted for by adding the flexural longitudinal steel to the torsional longitudinal steel. The validity of this method was substantiated in Refs. 18 and 3. On the other hand, for beams subjected to small shear and torsion (points lying in region OEF in Fig. 4.12), it is possible to use beams without web reinforcement. The torsional

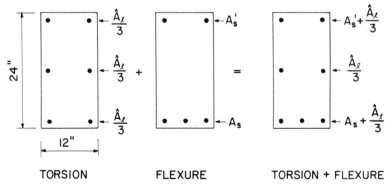

TORSION   FLEXURE   TORSION + FLEXURE

Fig. 4.14. Arrangement of torsional and flexural longitudinal steel.

strength of such beams without stirrups would be $T_n = T_c = (x^2y/3) \, 2.4\sqrt{f_c'}$ from Eq. (4–10). When such a beam is subjected to bending, shear, and torsion, the effect of bending moment on the torsional strength is neglected according to the ACI Code. Is this measure safe? The answer to this question is positive because the torsional stress of $\tau_c = 2.4\sqrt{f_c'}$ is much less than the torsional cracking stress of $\tau_{cr} = 6\sqrt{f_c'}$. This conservatism in adopting $\tau_c = 2.4\sqrt{f_c'}$ for the failure of beams without web reinforcement more than counteracts the unconservatism in neglecting the effect of bending moment on torsional strength.

4.4.2 Minimum Torsional Reinforcement

A minimum torsional reinforcement is required to ensure the ductility of the beam when it cracks. For a beam subjected to pure torsion, a minimum amount of torsional reinforcement must be provided such that:

$$T_n = T_{cr} \qquad (4\text{–}55)$$

Substituting T_n from Eq. (4–10) and $T_{cr} = (x^2y/3)(6\sqrt{f_c'})$ into Eq. (4–55) and solving for A_t:

$$A_{t, \min} = \frac{xs}{f_y} \left[\left(\frac{1.2}{\alpha_t} \right) \left(\frac{x}{x_1} \right) \left(\frac{y}{y_1} \right) \sqrt{f_c'} \right]$$

Assuming $\alpha_t = 1.2$, $x/x_1 = y/y_1 = 1.2$, and $f_c' = 5,000$ psi:

$$A_{t, \min} = \frac{100xs}{f_y} \qquad (4\text{–}56)$$

This equation is derived from pure torsion and does not take into account the interaction of torsion and shear. A more refined minimum torsional reinforcement, which considers the interaction of shear and torsion, is derived below.

Figure 4.15 shows two interaction curves, one for the cracking stresses τ_{cr} and v_{cr} and another for the contribution of concrete τ_c and v_c. These two curves divide the diagram into three regions. Beams in region (1) require no web reinforcement, whereas those in region (3) require shear and torsional reinforcement as calculated by Eqs. (4–52) through (4–54). In region (2), however, a beam designed by Eqs. (4–52) through (4–54) will fail in a brittle manner because the shear and torsional steel so designed cannot develop the cracking strength under shear and torsion. To prevent this undesirable

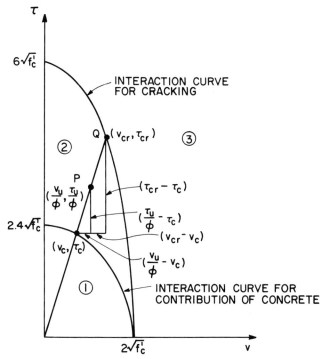

Fig. 4.15. Region where minimum web reinforcement is required to prevent brittle failure.

failure mode, we must provide sufficient reinforcement such that the beam will develop an ultimate strength equal to the cracking strength. For example, a beam subjected to the applied design stresses τ_u/ϕ and v_u/ϕ is represented by point P in region (2). To prevent brittle failure this beam must be designed for point Q to develop the cracking strength. In other words, we must provide shear and torsion reinforcement according to the stresses $(\tau_{cr} - \tau_c)$ and $(v_{cr} - v_c)$, instead of $[(\tau_u/\phi) - \tau_c]$ and $[(v_u/\phi) - v_c]$.

Torsional web reinforcement required to resist the torsional stress $(\tau_{cr} - \tau_c)$ is:

$$A_t = \frac{(\tau_{cr} - \tau_c)\, x^2ys}{3\alpha_t x_1 y_1 f_y}$$

$$= \frac{xs}{f_y}\left[(\tau_{cr} - \tau_c)\frac{1}{3\alpha_t}\left(\frac{x}{x_1}\right)\left(\frac{y}{y_1}\right)\right] \qquad (4\text{--}57)$$

Assuming $\alpha_t = 1.2$ and $x/x_1 = y/y_1 = 1.2$, Eq. (4–57) can be simplified:

$$A_t = \frac{xs}{f_y} [0.40(\tau_{cr} - \tau_c)] \qquad (4\text{-}58)$$

where τ_{cr} and τ_c are given in Eqs. (4–31) and (4–36). $(\tau_{cr} - \tau_c)$ can be expressed as a function of τ_u/v_u, if f'_c is assumed.

Shear web reinforcement required to resist the shear stress $(v_{cr} - v_c)$ is:

$$A_v = \frac{b_w s}{f_y} (v_{cr} - v_c) \qquad (4\text{-}59)$$

where v_{cr} and v_c are calculated from Eqs. (4–32) and (4–37). $(v_{cr} - v_c)$ is also a function of τ_u/v_u, if f'_c is assumed.

Notice that the parameter $b_w s/f_y$ in Eq. (4–59) is a little different from the parameter xs/f_y in Eq. (4–58). For the common case of a beam in which the height is greater than the width, $b_w = x$ as shown in Fig. 4.16 (a). The parameters in Eqs. (4–58) and (4–59) are the same. However, in the case of a shallow beam shown in Fig. 4.16 (b), b_w is not equal to x. Even so, we shall assume that $b_w = x$ for simplicity. Adding Eqs. (4–58) and (4–59) gives:

$$2A_t + A_v = \frac{xs}{f_y} [0.80(\tau_{cr} - \tau_c) + (v_{cr} - v_c)] \qquad (4\text{-}60)$$

Remembering that the stresses τ_{cr}, v_{cr}, τ_c, and v_c must be calculated from equations that are not simple, the expression in Eq. (4–60) is obviously too complicated for practical use. To simplify this equation we notice that the right-hand side of Eq. (4–60) is a function of τ_u/v_u, representing the interaction of torsion and shear. Let us introduce a more convenient parameter, $\tau_u/(\tau_u + v_u)$, that can also represent the interaction of torsion and shear.

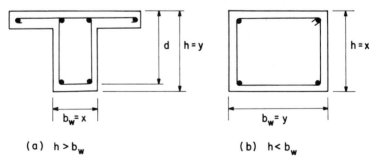

Fig. 4.16. Definitions of b_w and x.

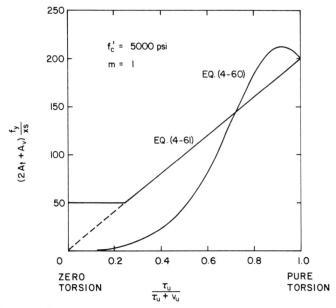

Fig. 4.17. Minimum web reinforcement according to Eq. (4–60) and Eq. (4–61).

For pure torsion $\tau_u/(\tau_u + v_u) = 1$, and for zero torsion $\tau_u/(\tau_u + v_u) = 0$.

We can now plot $(2A_t + A_v)f_y/xs$ as a function of $\tau_u/(\tau_u + v_u)$ in Fig. 4.17 according to Eq. (4–60), assuming $f'_c = 5,000$ psi. Equation (4–60) gives an S-shaped curve that can be roughly approximated by a slanted straight line:

$$2A_t + A_v = \frac{200xs}{f_y}\left(\frac{\tau_u}{\tau_u + v_u}\right) \qquad (4\text{--}61)$$

This slanted straight line is cut off by a horizontal straight line with the equation:

$$2A_t + A_v = \frac{50\,b_w s}{f_y} \qquad (4\text{--}62)$$

Equation (4–62) is designed to match the requirement for minimum shear steel. Equations (4–61) and (4–62) have been considered in the ACI Committee 438 closure of the "Tentative Recommendations."[19]

If we provide all the minimum web reinforcement in the form of closed stirrups, then we can substitute $2A_t + A_v$ in Eq. (4–61) by $2A_t$, resulting in:

$$2A_t = \frac{200xs}{f_y} \left(\frac{\tau_u}{\tau_u + \nu_u} \right) \tag{4–63}$$

The minimum longitudinal steel is obtained from the principle of equal volume of longitudinal bars and stirrups (i.e., the volume ratio of longitudinal steel to stirrups, $m = 1$). Substituting $2A_t = \hat{A}_l s/(x_1 + y_1)$ from Eq. (3–19) into Eq. (4–63) gives:

$$\hat{A}_l \frac{s}{(x_1 + y_1)} = \frac{200xs}{f_y} \left(\frac{\tau_u}{\tau_u + \nu_u} \right) \tag{4–64}$$

The total amount of minimum reinforcement, including both the stirrups and the longitudinal steel, is obtained by adding Eqs. (4–63) and (4–64):

$$2A_t + \hat{A}_l \frac{s}{(x_1 + y_1)} = \frac{400xs}{f_y} \left(\frac{\tau_u}{\tau_u + \nu_u} \right) \tag{4–65}$$

Solving for $\hat{A}_l$ gives:

$$\hat{A}_l = \left[\frac{400xs}{f_y} \left(\frac{\tau_u}{\tau_u + \nu_u} \right) - 2A_t \right] \frac{(x_1 + y_1)}{s} \tag{4–66}$$

Equation (4–66) was incorporated into the 1971 ACI Building Code and appeared in the 1977 and 1983 ACI codes as follows:

$$\hat{A}_l = \left[\frac{400xs}{f_y} \left(\frac{T_u}{T_u + \dfrac{V_u}{3C_t}} \right) - 2A_t \right] \frac{(x_1 + y_1)}{s} \tag{4–67}$$

In Eq. (4–67) the term $2A_t$ need not be taken less than $50b_w s/f_y$ because this is the minimum amount specified for web reinforcement in Eq. (4–62).

The requirement of a total amount of minimum reinforcement in Eq. (4–65) implies that a reduction of stirrups can be compensated by an increase in longitudinal steel as long as the total steel volume remains unchanged. In other words, ACI Code permits us to design minimum reinforcement using an m value greater than unity. This measure is very economical because, pound for pound, stirrups are much more expensive and much more difficult to install than longitudinal steel. The validity of this measure has been substantiated in Ref. 1 of Chapter 3 for the critical case of pure torsion.

If Eq. (4–65) is divided on both sides of the equation by $2A_t$, and noticing that $m = \hat{A}_l s / 2A_t (x_1 + y_1)$, then Eq. (4–65) can be written as:

$$m = \frac{400}{\left(\dfrac{2A_t f_y}{xs}\right)} \left(\frac{\tau_u}{\tau_u + v_u}\right) - 1 \qquad (4\text{–}68)$$

The relation among the three parameters, $(2A_t f_y / xs)$, $\tau_u/(\tau_u + v_u)$, and m, in Eq. (4–68) is shown by a curved surface in Fig. 4.18. The curved surface is valid only when $2A_t f_y / xs$ is less than 200. When $\tau_u/(\tau_u + v_u)$ is close to unity, the value of m increases rapidly with decreasing $2A_t f_y / xs$. When $2A_t f_y / xs = 50$ and $\tau_u/(\tau_u + v_u) = 1$ (pure torsion), the curved surface gives a maximum m value of 7.

Incidentally, when $2A_t f_y / xs$ is greater than 200, Eq. (4–68) is no longer valid, because design is governed by strength and not by ductility. Equation (4–54) should then govern based on $m = 1$.

Equation (4–67) has been derived from a solid rectangular cross section. For a hollow box section, it can be shown that Eq. (4–67) is applicable if

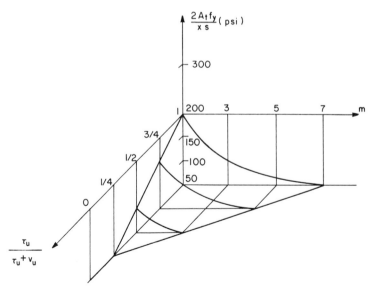

Fig. 4.18. Relationship among $2A_t f_y / xs$, $\tau_u/(\tau_u + v_u)$, and m in ACI criterion for minimum web reinforcement.

the first term, $(400xs/f_y)[T_u/(T_u + V_u/3C_t)]$, is multiplied by the reduction factor $4h/x$ for box sections (see Section 4.2.3 and Example 4.3).

4.4.3 Practical Limitations and Details

4.4.3.1 Spacing of Stirrups, s. The purpose of limiting the spacing of stirrups is twofold: first, to prevent a drastic drop of shear or torsional strengths when the spacing is too large; second, to control crack widths. The spacing of stirrups specified in the ACI Code for shear and torsion is based on these two requirements.

In evaluating the flexural shear strength of a reinforced concrete beam, the contribution of steel, V_s, is computed by dA_vf_y/s. This parameter includes a ratio, A_v/s. Similarly, in calculating the torsional strength, all the available theories utilize the reinforcement factor, $x_1y_1A_tf_y/s$. This parameter includes a ratio, A_t/s. The use of A_v/s in shear and A_t/s in torsion, in units of area per unit length, implies that the stirrup area is continuously distributed along the length of the beam. If a series of vertical bars, representing either the shear or the torsion stirrups, is intersected by a 45° crack, the area of bars theoretically being cut is $(A_v/s)y_1$ or $(y_1/s)A_v$. In actuality, however, the stirrups are concentrated at discrete points along the beam. The 45° crack will therefore intersect only a discrete number of bars having an area nA_v, where n is the number of bars actually intersected. Area nA_v is less than or at most equal to the theoretical $(y_1/s)A_v$.

Figure 4.19 shows three possible cases of a 45° crack intersecting a series of vertical bars uniformly distributed with spacing s. In all three cases, three bars have been cut such that $n = 3$. However, (y_1/s) may vary from 3 to

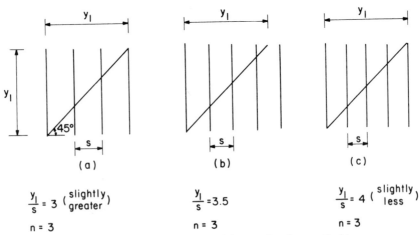

Fig. 4.19. Three cases of a 45° crack intersecting three vertical bars.

4. In case (a), (y_1/s) is slightly greater than but approaching 3. Therefore we can write in a general way:

$$n = \frac{y_1}{s} \quad \text{or} \quad \frac{n}{\left(\dfrac{y_1}{s}\right)} = 1 \qquad (4\text{-}69)$$

In case (b), (y_1/s) is taken as 3.5. A more general expression is:

$$n = \left(\frac{y_1}{s}\right) - 0.5 \quad \text{or} \quad \frac{n}{\left(\dfrac{y_1}{s}\right)} = 1 - \frac{0.5}{\left(\dfrac{y_1}{s}\right)} \qquad (4\text{-}70)$$

In case (c), (y_1/s) is slightly less than 4, but approaching 4. For this case:

$$n = \left(\frac{y_1}{s}\right) - 1 \quad \text{or} \quad \frac{n}{\left(\dfrac{y_1}{s}\right)} = 1 - \frac{1}{\left(\dfrac{y_1}{s}\right)} \qquad (4\text{-}71)$$

Equations (4–69) through (4–71) are plotted in Fig. 4.20, which shows $n/(y_1/s)$ as a function of y_1/s. Equation (4–69) gives the horizontal line (a), which

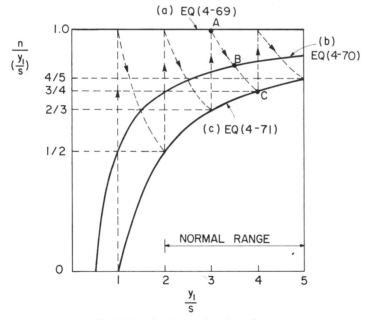

Fig. 4.20. $n/(y_1/s)$ as a function of y_1/s.

is the upper limit. Equation (4–71) is plotted as curve (c), the lower limit. The most practical curve should be curve (b) from Eq. (4–70), which lies midway between the upper and lower limits. The three points A, B, and C represent, respectively, the three cases (a), (b), and (c) in Fig. 4.19. The dotted curve, which has a saw-tooth shape and passes through points A, B, and C between $y_1/s = 3$ and 4, gives a continuous variation of $n/(y_1/s)$ between $y_1/s = 1$ and 5.

Let us now examine curve (b) in Fig. 4.20. The curve is quite flat when $y_1/s > 2$. However, when $y_1/s < 2$, the ordinate $n/(y_1/s)$ decreases very rapidly with decreasing y_1/s. This means that a drastic loss of shear or torsional strengths will occur when spacing of stirrups is increased beyond $y_1/2$. This general trend is supported by tests in shear and in torsion.

Another factor also appears to reduce the torsional strength when the spacing of stirrups exceeds $y_1/2$. For large stirrup spacing, the number of diagonal cracks in a beam decreases, and the cracks tend to arrange themselves so as to avoid intersecting the stirrups. Therefore, the stirrups may not yield at failure, even though the beam is underreinforced according to the parameter A_v/s.

In view of the drastic loss of strength when $s > y_1/2$, the maximum spacing of shear stirrups is specified in the ACI Code as $d/2$, where d is the effective depth of cross section and is approximately equal to y_1. For torsional stirrups, a maximum spacing of $y_1/2$ has been proposed in the "Tentative Recommendations."[2]

The maximum spacing of $y_1/2$ for torsional stirrups considers only the depth of the cross section, but not the width. To consider the width, an additional limitation of $4x_1/3$ has been suggested in the "Tentative Recommendations." This limitation appears to be excessively conservative for cross sections with a high depth-to-width ratio. Combining the effect of y_1 and x_1, Committee 438 adopted a maximum stirrup spacing for torsion stirrups as follows:*[21]

$$s_{max} = \frac{x_1 + y_1}{4} \qquad (4\text{–}72)$$

This limitation is supported by tests shown in Fig. 4.21. In Fig. 4.21 all the test specimens satisfying this limitation have developed the yield strength of the stirrups. Figure 4.21 also includes the old limitations in the "Tentative Recommendations" for comparison. In view of the simplicity and the validity

---

* Suggested initially by M. Collins and P. Lampert, March 1970.

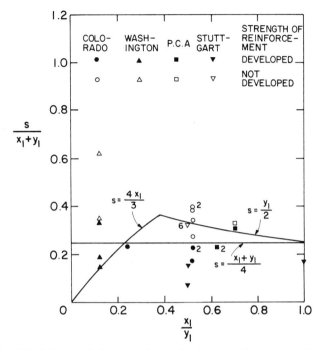

Fig. 4.21. Influence of stirrup spacing on development of strength of stirrups.

of Eq. (4–72), it was first incorporated into the 1971 ACI Code, and continued into the 1977 and 1983 ACI codes.

A more general expression of maximum spacing for arbitrary bulky sections without re-entrant corners is:

$$s_{max} = \frac{u}{8} \qquad (4\text{–}73)$$

where u is the periphery of a stirrup. Equation (4–73) reduces to Eq. (4–72) in the case of rectangular sections.

The 1977 and 1983 ACI codes also specify a maximum stirrup spacing of 12 in. The purpose of this limitation is to control crack width in large-size girders. This 12-in. maximum spacing is also applicable to longitudinal bars.

4.4.3.2 Other ACI Code Limitations.
There are three other limitations in the 1977 and 1983 ACI codes that should be mentioned:

1. Section 11.6.7.4 requires that the design yield strength of torsion rein-

forcement shall not exceed 60,000 psi. This limitation ensures the yielding of torsional steel before failure, since all the torsional design provisions are based on the yielding of steel. This limitation is identical to that required for shear reinforcement and serves the same purpose.

2. Section 11.6.7.6 specifies that torsional reinforcement shall be provided at least a distance $(d + b)$ beyond the point theoretically required. This specification is more stringent than the corresponding Section 12.11.3 for flexure that requires only a distance d or $12d_b$, whichever is greater. This is done because torsional cracks develop in a helical shape inclined at approximately 45° to the axis of the beam.

For beams subjected to significant torsion, good practice suggests that closed stirrups should be placed throughout the beam's length, observing at least the requirements of minimum reinforcement and maximum stirrup spacing. In addition, it is suggested that at least one-fourth of the negative steel at the column face shall be continued throughout the beam.

3. Section 11.6.1 allows us to neglect torsion if the factored torsional moment T_u is less than $\phi(0.5\sqrt{f'_c}\ x^2y)$, or, in terms of stresses, if the torsional stress τ_u is less than $\phi 1.5\sqrt{f'_c}$. This section is very powerful because it exempts many structural elements that are subjected to small torsion from the complexity of torsion design. Take a typical floor of a building, for example. All the interior beams that are subjected to small torsion may be exempted from torsion analysis on account of this provision. Such interior beams usually constitute a great majority of all the beams in the building. Therefore, only the spandrel beams and edge beams normally require torsional analysis.

4.4.3.3 Additional Details.

Additional details for torsional steel have been discussed by Mitchell and Collins.[20,21] For longitudinal corner bars,

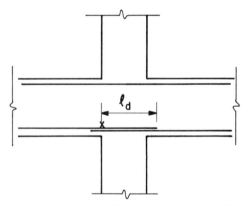

Fig. 4.22. Anchorage of longitudinal bottom bars at interior support.

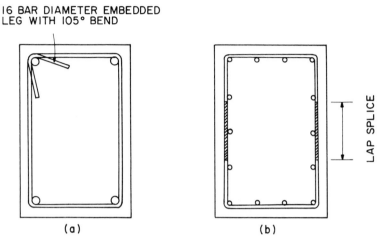

16 BAR DIAMETER EMBEDDED
LEG WITH 105° BEND

(a)   (b)

Fig. 4.23. Detailing for torsional web reinforcement.

they suggested that the diameter should be greater than $\frac{1}{16}$ of the stirrup spacing so as to prevent the so-called push-off failure of the corner bars. The anchorage of the longitudinal bottom bars into the interior support must have an anchorage length of at least l_d (see Fig. 4.22). This is so because these longitudinal bottom bars may be subjected to tension due to large torsion.

The closed stirrups for torsion are recommended to have hooks with at least a 105° bend and with a 16-bar-diameter length of embedded leg as shown in Fig. 4.23 (a). This is more stringent than the standard stirrup hooks allowed by Section 7.1 (b) and (c) of the 1977 ACI Code.

Owing to the difficulties in installing one-piece, closed stirrups, a type of closed stirrups, each made of a pair of U-shaped "hair pins" has been suggested, as shown in Fig. 4.23 (b). For such closed stirrups, great care must be exercised with regard to the lap splices. Although Section 12.14.5 of the 1977 ACI Code specifies a splice length of $1.7l_d$ (l_d = development length), tests given in Ref. 20 showed that a splice length of 40 bar diameters was insufficient.

4.5 DESIGN EXAMPLE 4.4

An 18 in. by 28 in. beam with a clear span of 30 ft from column faces supports a 12 in. by 28 in. cantilever at midspan [see Fig. 4.24 (a) and (b)]. A concentrated load P is applied at the end of the cantilever at a distance 2 ft from the center line of the beam. This superimposed load is

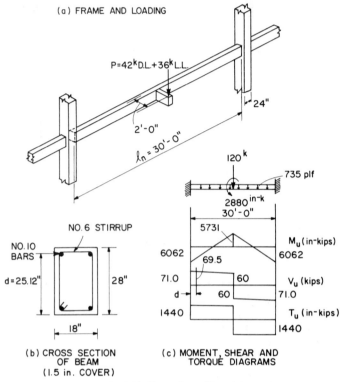

(a) FRAME AND LOADING

$P = 42^k D.L. + 36^k L.L.$

24"

2'-0"

$\ell_n = 30'-0"$

120^k

735 plf

2880^{in-k}
30'-0"

NO. 6 STIRRUP

NO. 10 BARS

d = 25.12"

28"

18"

5731

6062

69.5

71.0

d → ⊢ 60

1440

M_u (in-kips)

6062

60

V_u (kips)

71.0

T_u (in-kips)

1440

(b) CROSS SECTION
OF BEAM
(1.5 in. COVER)

(c) MOMENT, SHEAR AND
TORQUE DIAGRAMS

Fig. 4.24. Example problem 4.4.

made up of a service dead load of 42 kips and a service live load of 36 kips. Design the reinforcement for the beam according to the 1977 or 1983 ACI Building Code. The material strengths are $f'_c = 4,000$ psi and $f_y = 60,000$ psi.

Solution

Flexural Moments, Shears, and Torques:
Beam dead weight:

$$w_d = 1.4 \, (150) \, \frac{18(28)}{144} = 735 \text{ plf}$$

Superimposed concentrated load:

$$
\begin{array}{lll}
\text{Dead load} & 1.4 \, (42) = & 58.8 \text{ k} \\
\text{Live load} & 1.7 \, (36) = & \underline{61.2 \text{ k}} \\
\text{Factored load } P_u & = & 120.0 \text{ k}
\end{array}
$$

Bending moment at face of column (assumed fixed ends):

$$M_u = \frac{P_u \ell}{8} + \frac{w_d \ell^2}{12} = \frac{120(30)(12)}{8} + \frac{0.735(30)^2(12)}{12} = 6{,}062 \text{ in.-k}$$

Bending moment at midspan:

$$M_u = \frac{P_u \ell}{8} + \frac{w_d \ell^2}{24} = \frac{120(30)(12)}{8} + \frac{0.735(30)^2(12)}{24} = 5{,}731 \text{ in.-k}$$

Shear at face of column:

$$V_u = \frac{P_u}{2} + \frac{w_d \ell}{2} = \frac{120}{2} + \frac{0.735(30)}{2} = 71.0 \text{ k}$$

Shear at a distance d from face of column (assume $d = 25.12$ in.; see Fig. 4.24b):

$$V_u = \frac{P_u}{2} + w_d \left(\frac{\ell}{2} - d \right) = \frac{120}{2} + 0.735 \left(15 - \frac{25.12}{12} \right) = 69.5 \text{ k}$$

Shear at midspan:

$$V_u = \frac{P_u}{2} = 60.0 \text{ k}$$

Torsional moment throughout beam:

$$T_u = \frac{120(2)(12)}{2} = 1{,}440 \text{ in.-k}$$

The flexural moment, shear, and torque diagrams are given in Fig. 4.24 (c). The flexural moment diagram should be slightly curved owing to the small uniform dead weight of the beam.

Shear and Torsional Web Reinforcement:

$$\Sigma x^2 y = 18^2 (28) = 9{,}072 \text{ in.}^3$$
$$b_w d = 18 \,(25.12) = 452 \text{ in.}^2$$

Eq. (4–39): $$C_t = \frac{b_w d}{\Sigma x^2 y} = \frac{452}{9{,}072} = 0.0499 \frac{1}{\text{in.}}$$

Section 4.4.3.2(3):

$$\phi(0.5\sqrt{f_c'}\Sigma x^2 y) = 0.85\,(0.5\sqrt{4,000})(9,072)$$
$$= 243,849 \text{ in.-lb} = 244 \text{ in.-k} < 1,440 \text{ in.-k}$$

Torsion cannot be neglected in design.

At a distance d from the face of the column, $V_u = 69.5$ k and $T_u = 1,440$ in.-k:

Eq. (4-50): $T_{n,\,max} = \dfrac{4\sqrt{f_c'}\,\Sigma x^2 y}{\sqrt{1 + \left(\dfrac{0.4V_u}{C_t T_u}\right)^2}} = \dfrac{4\sqrt{4,000}\,(9,072)}{\sqrt{1 + \left[\dfrac{0.4(69.5)}{0.0499(1,440)}\right]^2}}$

$$= 2,140,447 \text{ in.-lb} = 2,140 \text{ in.-k} > 1,440 \text{ in.-k}/\phi \quad \text{O.K.}$$

The size of the beam is sufficient.

Eq. (4-41): $T_c = \dfrac{0.8\sqrt{f_c'}\,\Sigma x^2 y}{\sqrt{1 + \left(\dfrac{0.4V_u}{C_t T_u}\right)^2}} = \dfrac{T_{n,\,max}}{5} = \dfrac{2,140}{5} = 428$ in.-k

Eq. (4-42): $V_c = \dfrac{2\sqrt{f_c'}\,b_w d}{\sqrt{1 + \left(\dfrac{C_t T_u}{0.4V_u}\right)^2}} = \dfrac{2\sqrt{4,000}(452)}{\sqrt{1 + \left[\dfrac{0.0499(1,440)}{0.4(69.5)}\right]^2}}$

$$= 20,629 \text{ lb} = 20.6 \text{ k}$$

Eq. (4-53): $T_s = \dfrac{T_u}{\phi} - T_c = \dfrac{1,440}{0.85} - 428 = 1,266$ in.-k

Use No. 6 bars (concrete cover 1.5 in.).

$$x_1 = 18 - 2(1.5 + 0.375) = 14.25 \text{ in.}$$
$$y_1 = 28 - 2(1.5 + 0.375) = 24.25 \text{ in.}$$
$$\alpha_t = 0.66 + 0.33\,\frac{y_1}{x_1} = 0.66 + 0.33\,\frac{24.25}{14.25} = 1.22 < 1.5$$
$$\frac{A_t}{s} = \frac{T_s}{\alpha_t x_1 y_1 f_y} = \frac{1,266}{1.22(14.25)(24.25)(60)} = 0.0500 \text{ in.}^2/\text{in.}$$

Eq. (4–52):
$$V_s = \frac{V_u}{\phi} - V_c = \frac{69.5}{0.85} - 20.6 = 61.2 \text{ k}$$

$$\frac{A_v}{s} = \frac{V_s}{df_y} = \frac{61.2}{25.12(60)} = 0.0406 \text{ in.}^2/\text{in.}$$

Total web reinforcement:

$$\frac{A_t}{s} + \frac{1}{2}\frac{A_v}{s} = 0.0500 + \frac{0.0406}{2} = 0.0703 \text{ in.}^2/\text{in.}$$

$$s = \frac{0.44}{0.0703} = 6.26 \text{ in.} < 12 \text{ in.}$$

Eq. (4–72):
$$s_{max} = \frac{x_1 + y_1}{4} = \frac{14.25 + 24.25}{4} = 9.625 \text{ in.} > 6.26 \text{ in.} \qquad \text{O.K.}$$

$$\frac{d}{2} = \frac{25.12}{2} = 12.56 \text{ in.} > 6.26 \text{ in.} \qquad \text{O.K.}$$

Eq. (4–62):
$$\frac{50b_w}{f_y} = \frac{50(18)}{60,000} = 0.0150 \text{ in.}^2/\text{in.} < 2(0.0703 \text{ in.}^2/\text{in.}) \qquad \text{O.K.}$$

Use No. 6 closed stirrups at 6-in. spacing throughout the beam.

Torsional Longitudinal Reinforcement:

Eq. (4–54):
$$\hat{A}_l = \frac{A_t}{s} 2(x_1 + y_1) = 0.0500(2)(14.25 + 24.25) = 3.85 \text{ in.}^2$$

Eq. (4–67):

$$\hat{A}_l = \left[\frac{400xs}{f_y} \left(\frac{T_u}{T_u + \frac{V_u}{3C_t}} \right) - 2A_t \right] \frac{x_1 + y_1}{s}$$

$$= \left[\frac{400(18)(6.26)}{60,000} \left(\frac{1,440}{1,400 + \frac{69.5}{3(0.0499)}} \right) - 2(0.44) \right] \frac{14.25 + 24.25}{6.26}$$

$$= -1.92 \text{ in.}^2 < 3.85 \text{ in.}^2 \qquad \text{O.K.}$$

Since the spacing of longitudinal bars is limited to 12 in., it is necessary to provide longitudinal bars at mid-height of the cross section. Assume three layers of steel with equal distribution, each layer having an area of:

$$\frac{\hat{A}_l}{3} = \frac{3.85}{3} = 1.28 \text{ in.}^2$$

Using two No. 7 bars at mid-height will give an area of $2(0.60) = 1.20$ in.$^2 \approx$ 1.28 in.^{2} Then the top and bottom longitudinal bar areas are:

$$\hat{A}_{l,\text{top}} = \hat{A}_{l,\text{bot}} = \frac{1}{2}(3.85 - 1.20) = 1.33 \text{ in.}^2$$

Flexural Longitudinal Reinforcement:
At face of column, $M_u = 6{,}062$ in.-k.
Assume No. 10 bars:

$$d = 28 - 1.5 - 0.75 - \frac{1.27}{2} = 25.12 \text{ in.}$$

$$M_{\text{max}} = 937bd^2 = 937(18)(25.12)^2 = 10{,}651 \text{ in.-k}$$
$$> 6{,}062 \text{ in.-k} \qquad \text{Singly reinforced}$$

Assume $a = 4.8$ in.:

$$A_s = \frac{M_u}{\phi f_y \left(d - \dfrac{a}{2}\right)} = \frac{6{,}062}{0.9(60)\left(25.12 - \dfrac{4.8}{2}\right)} = 4.94 \text{ in.}^2$$

$$a = \frac{A_s f_y}{0.85 f'_c b} = \frac{4.94(60)}{0.85(4)(18)} = 4.84 \text{ in.} \approx 4.8 \text{ in.} \qquad \text{O.K.}$$

Arrangement of Longitudinal Bars (see Fig. 4.25):

At face of column:

Top bars:

$$A_s + \hat{A}_{l,\text{top}} = 4.94 + 1.33 = 6.27 \text{ in.}^2$$
$$\text{Use 5 No. 10.} \qquad 5(1.27) = 6.35 \text{ in.}^2 > 6.27 \text{ in.}^2 \qquad \text{O.K.}$$

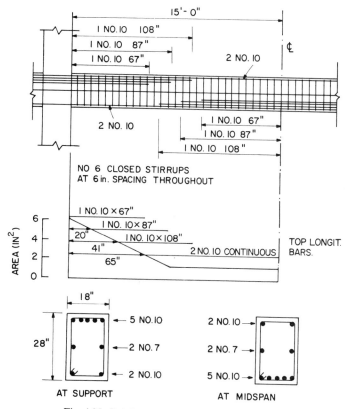

Fig. 4.25. Reinforcement details (Example 4.4).

Bottom bars:

$$\hat{A}_{1,\text{bot}} = 1.33 \text{ in.}^2$$

Use 2 No. 10. $2(1.27) = 2.54 \text{ in.}^2 > 1.33 \text{ in.}^2$ O.K.

The top and bottom longitudinal bars required at the other selected sections are summarized in Table 4.1. The sections are chosen at 2.5-ft intervals, except near the column face. The areas of the top longitudinal bars required at these sections are plotted in Fig. 4.25. From this diagram the cut-off points for the top bars can be determined.

Table 4.1 Summary of reinforcement (Example 4.4)

ITEM	UNITS	AT COLUMN FACE	AT DISTANCE d FROM COLUMN FACE	DISTANCE FROM CENTER OF SPAN (FT)				
				10.0	7.5	5.0	2.5	0
Top Longitudinal Bars								
$-M_u$	in.-kips	6,062	4,297	1,912	—	—	—	—
T_s	in.-kips	1,266	1,266	1,265	1,264	1,262	1,261	1,259
A_s	in.^{2}	4.94	3.40	1.45	—	—	—	—
$A_{l,\text{top}}$	in.^{2}	1.33	1.33	1.33	1.33	1.32	1.32	1.32
$A_s + A_{l,\text{top}}$	in.^{2}	6.27	4.73	2.78	1.33	1.32	1.32	1.32
Use		5 No. 10		2 No. 10		2 No. 10	2 No. 10	2 No. 10
Bottom Longitudinal Bars								
$+M_u$	in.-kips	—	—	—	83	2,021	3,903	5,731
T_s	in.-kips	1,266	1,266	1,265	1,264	1,262	1,261	1,259
A_s	in.^{2}	—	—	—	0.06	1.54	3.06	4.64
$A_{l,\text{bot}}$	in.^{2}	1.33	1.33	1.33	1.33	1.32	1.32	1.32
$A_s + A_{l,\text{bot}}$	in.^{2}	1.33	1.33	1.33	1.39	2.86	4.38	5.96
Use		2 No. 10		2 No. 10		2 No. 10	2 No. 10	5 No. 10

Note: The values underlined have been calculated in the text.

Cut-Off Points for Negative Bars (see Fig. 4.25):

$$d = 25.12 \text{ in.} \approx 25 \text{ in.} \quad \text{governs}$$
$$12d_b = 12(1.27) = 15.2 \text{ in.}$$
$$\ell_d = 1.4 \frac{0.04 A_s f_y}{\sqrt{f'_c}} = 1.4 \frac{0.04(1.27)(60,000)}{\sqrt{4,000}} = 67 \text{ in.}$$

First cut-off point:

Steel area requirement $= 20 + 25 = 45$ in.
Development length requirement $= 67$ in.   governs

Second cut-off point:

$$41 + 25 = 66 \text{ in.}$$
$$20 + 67 = 87 \text{ in.} \quad \text{governs}$$

Third cut-off point:

$$65 + 25 = 90 \text{ in.}$$
$$41 + 67 = 108 \text{ in.} \quad \text{governs}$$

The same cut-off lengths will be used for positive bars. This simplification will be on the conservative side.

Checking Stirrups at Midspan Joint:
In Section 9.3.6 an equation is given to check the stirrup area at the joint:

Eq. (9–53):

$$V_n = \frac{b_f + d_s}{s} A_v f_y = \frac{12 + 25.13}{6} (2 \times 0.44)(60) = 327 \text{ k}$$
$$V_u/\phi = 120/0.85 = 141 \text{ k} < 327 \text{ k} \quad \text{O.K.}$$

Example for Spandrel Beam. It should be mentioned that another example problem of a spandrel beam supporting closely spaced joists is given as Design Example 9.1 in Chapter 9.

Design Aid. To aid the design of torsion in reinforced concrete members, the Portland Cement Association has published a design handbook specifically for this purpose.[22]

REFERENCES

1. Fisher, G. P. and P. Zia, "Review of Code Requirements for Torsion Design," *Journal of the American Concrete Institute,* Proc., Vol. 61, No. 1, January 1964, pp. 1–44.

2. ACI Committee 438, "Tentative Recommendations for the Design of Reinforced Concrete Members to Resist Torsion," *Journal of the American Concrete Institute,* Proc., Vol. 66, No. 1, January 1969, pp. 1–8.

3. Hsu, T. T. C. and E. L. Kemp, "Background and Practical Application of Tentative Design Criteria for Torsion," *Journal of the American Concrete Institute,* Proc., Vol. 66, No. 1, January 1969, pp. 12–23.

4a. ACI Standard 318–71, *Building Code Requirements for Reinforced Concrete (ACI 318–71),* American Concrete Institute, Detroit, 1971, 78 pp.

4b. ACI Standard 318–77, *Building Code Requirements for Reinforced Concrete (ACI 318–77),* American Concrete Institute, Detroit, 1977, 102 pp.

4c. ACI Standard 318–83, *Building Code Requirements for Reinforced Concrete (ACI 318–83),* American Concrete Institute, Detroit, 1983.

5. Hsu, T. T. C., Unpublished results of tests at Portland Cement Association, 1968.

6. Liao, H. M. and P. M. Ferguson, "Combined Torsion in Reinforced Concrete L-Beams with Stirrups," *Journal of the American Concrete Institute,* Proc., Vol. 66, No. 17, December 1969, pp. 986–993.

7. Kirk, D. W. and S. D. Lash, "T-Beams Subjected to Combined Bending and Torsion," *Journal of the American Concrete Institute,* Proc., Vol. 68, No. 2, February 1971, pp. 150–159.

8. Leonhardt, F. and G. Schelling, "Torsionsversuche an Stahlbetonbalken," Heft 239, *Deutscher Ausschuss Fur Stahlbeton,* Berlin, 1974, 122 pp.

9. Lampert, P. and B. Thurlimann, "Torsion Tests on Reinforced Concrete" (Torsionsversuche an Stahlbetonbalken), Bericht, Nr. 6506–2, Institute fur Baustatik ETH, Zurich, June 1968 (in German); translated by Dept. of Civil Engineering, University of Miami, Coral Gables, Florida.

10. Mattock, A. H., C. J. Birkeland, and M. E. Hamilton, "Strength of Reinforced Concrete Beams without Web Reinforcement in Combined Torsion, Shear and Bending," *Trend in Engineering,* University of Washington, Vol. 19, No. 4, October 1967, pp. 8–12.

11. Farmer, L. E. and P. M. Ferguson, "T-Beams under Combined Bending, Shear and Torsion," *Journal of the American Concrete Institute,* Proc., Vol. 64, No. 11, November 1967, pp. 757–766.

12. Ersoy, U. and P. M. Ferguson, "Behavior and Strength of Concrete L-Beams under Combined Torsion and Shear," *Journal of the American Concrete Institute,* Proc., Vol. 64, No. 12, December 1967.

13. Victor, D. J. and P. M. Ferguson, "Beams under Distributed Load Creating Moment, Shear and Torsion," *Journal of the American Concrete Institute,* Proc., Vol. 65, No. 4, April 1968.

14. Pandit, G. S. and J. Warwaruk, "Torsional Strength and Behavior of Concrete Beams in Combined Loading." Research Report, University of Alberta, Edmonton, Canada, July 1965.

15. McMullen, A. E. and J. Warwaruk, "The Torsional Strength of Rectangular Reinforced Concrete Beams Subjected to Combined Loading." Report No. 2, Dept. of Civil Engineering, University of Alberta, Edmonton, Canada, July 1967.

16. Collins, M. P., P. F. Walsh, F. E. Archer, and A. S. Hall, "Reinforced Concrete Beams

Subjected to Combined Torsion, Bending and Shear." UNICIV Report, No. R-14, University of New South Wales, October 1965.

17. Behera, U. and P. M. Ferguson, "Torsion, Shear and Bending in Stirruped L-Beams," *Journal of the Structural Division, ASCE,* Proc., Vol. 96, ST 7, July 1970, pp. 1271–1286.

18. Mattock, A. H., "How to Design for Torsion," *Torsion of Structural Concrete,* SP-18, American Concrete Institute, Detroit, 1968, pp. 469–495.

19. ACI Committee 438, Torsion, Closure to "Tentative Recommendations for the Design of Reinforced Concrete Members to Resist Torsion," *Journal of the American Concrete Institute,* Proc., Vol. 66, No. 7, July 1969, pp. 586–588.

20. Mitchell, D. and M. P. Collins, "Detailing for Torsion," *Journal of the American Concrete Institute,* Proc., Vol. 73, No. 9, September 1976, pp. 506–511.

21. Mitchell, D., P. Lampert, and M. P. Collins, "The Effect of Stirrup Spacing and Longitudinal Restraint on the Behavior of Reinforced Concrete Beams Subjected to Pure Torsion," Publication 71-22, Department of Civil Engineering, University of Toronto, October 1971, 75 pp.

22. PCA Handbook, *Design of Concrete Beams for Torsion, A Design Aid,* Portland Cement Association, Skokie, Illinois, 1983.

Plate 5: Prestressed concrete curved beam used for the aerial guideway at Disney World, Orlando, Florida. Large torsional moment is induced by the curvature of the beam, by centrifugal forces and by the wind load. (Courtesy ABAM Engineers, Inc. Tacoma, Washington)

5
Prestressed Concrete

5.1 MEMBERS WITHOUT WEB REINFORCEMENT

It is well known that prestress increases the cracking strength of a concrete member subjected to a bending moment. This is so because the prestress force creates a compressive stress that counterbalances the tensile stress due to the bending moment. Cracking can occur only when the tensile stress due to bending exceeds the compressive stress due to prestress by an amount equal to the modulus of rupture.

Prestress will also increase the cracking strength of a concrete member subjected to torsion or shear. In these cases, the prestress creates a compressive stress that, in combination with the shear stress created by the torsional moment or shear force, results in a shear–compression biaxial state of stress. This biaxial stress condition delays the cracking of the concrete.

To evaluate the effectiveness of prestress in raising the torsional cracking strength of a concrete member, we shall first study the failure criteria of concrete under biaxial stresses. From the failure criteria it is possible to derive a simple prestress factor, which is defined as the strength ratio of a prestressed member to a nonprestressed member. It will also be shown that this prestress factor can be derived from the elastic and plastic theories as well as the skew-bending theory.

For a prestressed member without web reinforcement, torsional cracking will be followed shortly by failure. Consequently, the cracking torsional strength of a prestressed member without web reinforcement can be taken as the ultimate strength. The prestress factor, therefore, is applicable to both the cracking torsional strength and the ultimate torsional strength.

5.1.1 Failure Criteria of Concrete Under Biaxial Stresses

A rectangular beam subjected to torsion T and prestress σ is shown in Fig. 5.1 (a). An element A is taken from the surface of the beam at mid-height and is enlarged in Fig. 5.1 (b). This element is subjected to a torsional shear stress τ due to torsion T on all four faces and a compressive stress

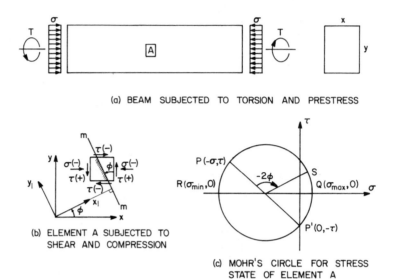

(a) BEAM SUBJECTED TO TORSION AND PRESTRESS

(b) ELEMENT A SUBJECTED TO SHEAR AND COMPRESSION

(c) MOHR'S CIRCLE FOR STRESS STATE OF ELEMENT A

Fig. 5.1. Stress condition on an element in a beam subjected to torsion and prestress.

σ due to prestress on the vertical faces. The stress state of this element can be expressed by a Mohr's circle in a $\sigma-\tau$ coordinate system, as can be seen in Fig. 5.1 (c).

In Fig. 5.1 (c) the point P is located at the coordinates $(-\sigma, \tau)$, which represent the stress state on a vertical face of element A. The point P' $(0, -\tau)$, on the other hand, represents the stress state of a horizontal face, where the normal stress is zero and the shear stress is negative with respect to the rotating x_1-y_1 axes (to be explained later). Connecting points P and P', we can construct a Mohr circle using the straight line PP' as a diameter. According to the principle of stress transformation, the stresses on an arbitrary surface $m-m$, which has an angle ϕ measured counterclockwise from the vertical surface, can be represented by the coordinates of point S, which has a subtending angle -2ϕ, from the point P. The negative sign of the angle 2ϕ in Fig. 5.1 (c) indicates that this angle is rotating in an opposite direction to the angle ϕ in Fig. 5.1 (b).

In understanding the Mohr circle it is necessary to define the sign convention of the stresses. The sign convention for the stresses on the vertical and horizontal faces of element A is determined according to the rotating coordinate axes x_1-y_1, which start from the position of the $x-y$ axes. A positive stress is defined as a stress acting on a position face and pointing toward the *positive* direction of a rotating axis. A position face is defined as a face with its normal pointing outward. This outward normal is always taken as the positive x_1-axis of the rotating x_1–y_1 coordinate. Using this sign convention, both of the shear stresses shown on the two vertical faces are positive. The downward shear stress on the left vertical face is positive because the y_1-axis is also pointed downward when the x_1–y_1 coordinate axes are rotated 180 degrees from their starting position.

A negative stress is defined as a stress acting on a positive face and pointing toward the *negative* direction of a rotating axis. Therefore, both of the shear stresses shown on the two horizontal faces are negative. The rightward-pointing shear stress on the top face is negative because the y_1-axis should be pointing to the left, when the x_1–y_1 coordinate axes are rotated 90 degrees from the starting position. Similarly, the two normal compressive stresses on the vertical surfaces are negative because these stresses are pointing toward the negative direction of the rotating x_1-axis.

Failure of element A will occur when the biaxial stresses acting on the element reach a critical value. The most widely accepted failure criterion for concrete is Mohr's failure theory, which states that failure occurs due to slipping on a definite plane within the material. At failure the shear and normal stresses on this plane, τ and σ, are related by a certain functional relationship:

$$\tau = F(\sigma) \tag{5-1}$$

This relationship, which is a characteristic of the material, is shown in Fig. 5.2. It can be obtained by testing concrete to failure under various combinations of biaxial stresses. Three types of tests are most commonly used, namely,

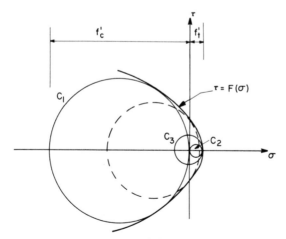

Fig. 5.2. Mohr's failure envelope.

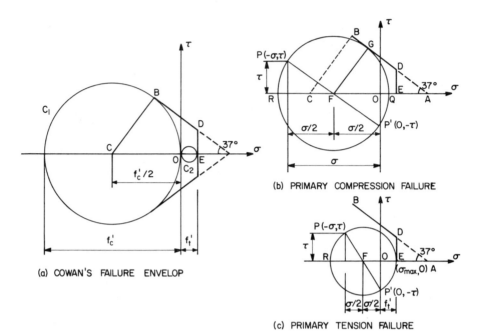

(a) COWAN'S FAILURE ENVELOP

(b) PRIMARY COMPRESSION FAILURE

(c) PRIMARY TENSION FAILURE

Fig. 5.3. Cowan's dual failure criteria.

the uniaxial compression, the uniaxial tension, and the pure shear tests. The Mohr's circles for these three stress states are shown as C_1, C_2, and C_3, respectively. Another circle, shown by the dotted curve, represents an arbitrary biaxial loading condition. All these failure circles must be tangential to the curve $\tau = F(\sigma)$. Hence, this curve is also known as Mohr's failure envelope.

Attempts have been made by many investigators to determine Mohr's failure envelope and to express it mathematically. However, this curved relationship is difficult to establish and difficult to use. Consequently, several simplifications have been proposed. In the author's opinion, Cowan's dual failure criteria[1] for concrete are the simplest to use and are very concise in concept. These criteria are illustrated in Fig. 5.3.

Figure 5.3 (a) shows that Mohr's curved failure envelope has been simplified into two straight lines BD and DE (considering the symmetry of the failure envelope). The inclined line BD is derived from Mohr circle C_1 for uniaxial compression. It is assumed to have an angle to the horizontal of 37° and is tangential to circle C_1 at point B. Such a criterion is known as the internal friction theory with an internal friction angle of 37°. It will govern when the concrete fails primarily in compression. In contrast, the vertical line

DE is tangential to Mohr circle C_2 for uniaxial tension. Such a criterion is known as the maximum stress failure theory and will govern when concrete fails primarily in tension.

The case of primary compression failure of element *A* is shown in Fig. 5.3 (b). This circle has a center at *F* and is tangential to the failure envelope *BDE* at point *G*. The length of the straight line *FG* can be calculated as follows:

$$CB = OC = 0.5f'_c \quad \text{(see Fig. 5.3a)}$$

$$AC = \frac{CB}{\sin 37°} = 0.831f'_c$$

$$AO = AC - OC = 0.831f'_c - 0.5f'_c = 0.331f'_c$$

$$AF = AO + \frac{\sigma}{2} = 0.331f'_c + 0.5\sigma$$

From similar triangles *AFG* and *ACB:*

$$FG = AF\frac{CB}{AC} = (0.331f'_c + 0.5\sigma)\frac{0.5f'_c}{0.831f'_c} = 0.199f'_c + 0.301\sigma$$

Since $FG = FP = \sqrt{\left(\frac{\sigma}{2}\right)^2 + \tau^2}$, then:

$$\sqrt{\left(\frac{\sigma}{2}\right)^2 + \tau^2} = 0.199f'_c + 0.301\sigma$$

$$\frac{\tau}{f'_c} = \sqrt{0.0396 + 0.120\frac{\sigma}{f'_c} - 0.1594\left(\frac{\sigma}{f'_c}\right)^2} \tag{5-2}$$

Equation (5–2) expresses the torsional shear stress τ at failure as a function of the prestress σ in a nondimensional form. It is applicable only to the case of primary compression failure.

The case of primary tension failure of element *A* is shown in Fig. 5.3 (c). The circle has a center at *F* and is tangential to the failure envelope *BDE* at point *E*. The length *OE* is equal to the uniaxial tensile strength, f'_t. This maximum stress failure criterion can be stated as follows: Failure occurs when the maximum principal tensile stress, σ_{max}, reaches the uniaxial tensile strength of concrete, f'_t (i.e., $\sigma_{max} = f'_t$).

$$\sigma_{\max} = OE = FE - FO = FP - FO$$

$$= \sqrt{\left(\frac{\sigma}{2}\right)^2 + \tau^2} - \frac{\sigma}{2}$$

Since $\sigma_{\max} = f_t'$:

$$\sqrt{\left(\frac{\sigma}{2}\right)^2 + \tau^2} - \frac{\sigma}{2} = f_t'$$

$$\tau = f_t' \sqrt{1 + \frac{\sigma}{f_t'}} \tag{5-3}$$

Again, the torsional shear stress τ at failure is expressed as a function of prestress σ in Eq. (5–3). This equation is applicable only to the case of

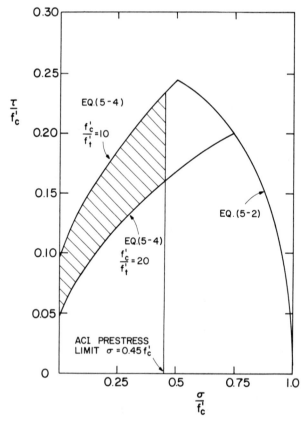

Fig. 5.4. Failure shear stress as a function of prestress (Cowan's dual failure criteria).

primary tension. Equation (5–3) can also be written in a nondimensional form:

$$\frac{\tau}{f'_c} = \frac{1}{\left(\frac{f'_c}{f'_t}\right)} \sqrt{1 + \left(\frac{f'_c}{f'_t}\right)\frac{\sigma}{f'_c}} \tag{5–4}$$

In this nondimensional form, τ/f'_c is a function of the strength ratio, f'_c/f'_t.

Equations (5–2) and (5–4) are plotted in Fig. 5.4 using the nondimensional ratios τ/f'_c and σ/f'_c as coordinates. Since the strength ratio, f'_c/f'_t, varies normally in the range from 10 to 20, curves are given for these two approximate boundaries. For $f'_c/f'_t = 10$ it can be seen that τ/f'_c increases with prestress according to Eq. (5–4) up to $\sigma/f'_c = 0.5$, and decreases thereafter according to Eq. (5–2). When σ/f'_c approaches unity, τ/f'_c becomes zero, and a uniaxial compression state of stress is reached. This general trend is also true for the case $f'_c/f'_t = 20$, except that the maximum torsional shear stress occur when the prestress σ is $0.75 f'_c$. Summarizing the above observations, it can be concluded that the failure shear stress will always increase with prestress up to the ACI prestress limit of $\sigma = 0.45 f'_c$. Within this ACI limit, Eq. (5–4) based on maximum tensile stress theory will always govern.

5.1.2 Elastic and Plastic Theories

A concentrically prestressed concrete beam without web reinforcement and subjected to torsion was shown in Fig. 5.1 (a). The failure of this beam has been assumed to occur when the maximum principal tensile stress on the most critical element A reaches the uniaxial tensile strength of concrete. In other words, Eq. (5–3), which is based on the maximum stress theory, will govern. The evaluation of the torsional shear stress τ on element A can be based on the elastic theory of Eq. (1–62) or on the plastic theory of Eq. (1–100).

Elastic Theory. The elastic theory has been used by Cowan[2] and Humphreys.[3] From Eqs. (1–62) and (5–3)

$$T = \alpha x^2 y \, \tau_{\max} = \alpha x^2 y f'_t \sqrt{1 + \frac{\sigma}{f'_t}} \tag{5–5}$$

where $\tau_{\max}$ is the torsional shear stress on element A. Since $\alpha x^2 y f'_t$ in Eq. (5–5) is the elastic torque without prestress, T_e, we can write:

$$T = T_e \underbrace{\sqrt{1 + \frac{\sigma}{f_t'}}}_{\gamma} \qquad (5\text{--}6)$$

Equation (5–6) states that the failure torque of a prestressed beam T is equal to the failure torque of a nonprestressed beam T_e times a square root factor. This square root factor is known as the prestress factor, γ.

Plastic theory. The plastic theory was proposed by Nylander.[4] From Eqs. (1–100) and (5–3):

$$T = \alpha_p x^2 y \tau = \alpha_p x^2 y f_t' \sqrt{1 + \frac{\sigma}{f_t'}} \qquad (5\text{--}7)$$

Since $\alpha_p x^2 y f_t'$ is the plastic torque without prestress, T_p, Eq. (5–7) can be written as:

$$T = T_p \underbrace{\sqrt{1 + \frac{\sigma}{f_t'}}}_{\gamma} \qquad (5\text{--}8)$$

Again, the failure torque of a prestressed beam, T, can be expressed as the failure torque of a nonprestressed beam, T_p, times the prestress factor, γ.

The prestress factor, γ, in Eqs. (5–6) and (5–8) is based on the uniaxial tensile strength of concrete, f_t'. The quantity f_t' can be converted into the uniaxial compressive strength f_c' by assuming $f_c'/f_t' = 10$. Then:

$$\gamma = \sqrt{1 + 10\frac{\sigma}{f_c'}} \qquad (5\text{--}9)$$

This expression of the prestress factor can also be derived from the skew-bending theory.

5.1.3 Skew-Bending Theory[5]

In Section 2.2.3 we derived the torsional strength of a plain concrete beam based on the skew-bending mechanism of failure. This failure mechanism is also observed in a prestressed concrete beam without web reinforcement. In the skew-bending theory, failure is due to bending along a skew failure

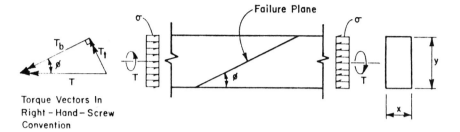

Fig. 5.5. Skew-bending failure.

surface. This failure surface can be assumed as a plane inclined at an angle ϕ to the longitudinal axis of the beam, as shown in Fig. 5.5. In prestressed beams, the angle ϕ was observed to decrease with increasing prestress.

The same logic of derivation used in Section 2.2.3 for plain concrete beams will be employed here, except that the prestress will be taken into account. Referring to Fig. 5.5, the applied torque T on a beam can be divided into two components acting on the failure plane. These are the bending component, T_b, and the twisting component, T_t. The bending component is responsible for the bending-type of failure and induces a tensile stress, f_1, on the tensile face of the failure plane:

$$f_1 = \frac{T_b}{\frac{x^2y}{6}\csc\phi} = \frac{T\cos\phi}{\frac{x^2y}{6}\csc\phi} = \frac{T\sin 2\phi}{\frac{x^2y}{3}} \tag{5-10}$$

Similarly, the uniform prestress induces a compressive stress, f_2, on the failure plane:

$$f_2 = -\sigma\sin^2\phi \tag{5-11}$$

Summing the two stresses obtained from Eqs. (5-10) and (5-11) gives:

$$\frac{T}{\frac{x^2y}{3}\csc 2\phi} - \sigma\sin^2\phi = f_1 + f_2 \tag{5-12}$$

Failure occurs when $f_1 + f_2$ reaches $0.85f_r$, where the factor 0.85 accounts for the effect of the twisting component, T_t. Then Eq. (5-12) becomes:

$$T_n = \frac{x^2 y}{3}(0.85 f_r) \csc 2\phi \left(1 + \frac{\sigma}{0.85 f_r}\sin^2\phi\right) \qquad (5\text{-}13)$$

To find the angle ϕ that corresponds to the minimum torsional resistance, we differentiate T_n in Eq. (5–13) and equate it to zero:

$$\frac{dT_n}{d\phi} = -2\cot 2\phi \csc 2\phi + \frac{\sigma}{0.85 f_r}\frac{\sec^2\phi}{2} = 0 \qquad (5\text{-}14)$$

Simplifying Eq. (5–14) gives:

$$\csc\phi = \sqrt{2 + \frac{\sigma}{0.85 f_r}} \qquad (5\text{-}15)$$

It can be seen from Eq. (5–15) that $\phi = 45°$ when $\sigma = 0$. This is the same answer obtained in Eq. (2–10) for plain concrete members. Equation (5–15) also shows that ϕ decreases when the prestress σ is increased.

The minimum torsional resistance can be obtained by substituting ϕ from Eq. (5–15) into Eq. (5–13). To do this we shall find the trigonometric relationship of ϕ from Fig. 5.6:

$$\sin\phi = \frac{1}{\sqrt{2 + \dfrac{\sigma}{0.85 f_r}}} \qquad (5\text{-}16)$$

$$\cos\theta = \frac{\sqrt{1 + \dfrac{\sigma}{0.85 f_r}}}{\sqrt{2 + \dfrac{\sigma}{0.85 f_r}}} \qquad (5\text{-}17)$$

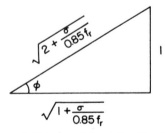

Fig. 5.6. Trigonometric relationship of ϕ.

$$\csc 2\phi = \frac{1}{2\sin\phi\cos\phi} = \frac{2 + \dfrac{\sigma}{0.85 f_r}}{2\sqrt{1 + \dfrac{\sigma}{0.85 f_r}}} \tag{5-18}$$

$$\sin^2\phi = \frac{1}{2 + \dfrac{\sigma}{0.85 f_r}} \tag{5-19}$$

Substituting $\csc 2\phi$ and $\sin^2\phi$ into Eq. (5–13) gives:

$$T_n = \frac{x^2 y}{3}(0.85 f_r)\sqrt{1 + \frac{\sigma}{0.85 f_r}} \tag{5-20}$$

Note that $(x^2 y/3)(0.85 f_r)$ is the torsional strength of a plain concrete beam, T_{np}, from Eq. (2–12). Equation (5–20) can be written as:

$$T_n = T_{np}\underbrace{\sqrt{1 + \frac{\sigma}{0.85 f_r}}}_{\gamma} \tag{5-21}$$

It can be seen from Eq. (5–21) that the failure torque of a prestressed beam is equal to the failure torque of a nonprestressed beam times a prestress factor, which is:

$$\gamma = \sqrt{1 + \frac{\sigma}{0.85 f_r}} \tag{5-22}$$

This prestress factor is expressed in terms of the modulus of rupture, f_r. If f_r is converted into the compression strength f'_c by $f_r = f'_c/8.5$, then:

$$\gamma = \sqrt{1 + 10\frac{\sigma}{f'_c}} \tag{5-23}$$

This expression of the prestress factor is exactly the same as those derived from the elastic or plastic theories in Eq. (5–9).

The prestress factor in Eq. (5–23) has been compared to tests in Fig. 5.7. In this figure, γ is expressed as follows:

$$\gamma = \frac{T_{n,\text{test}}}{T_{np}} = \frac{T_{n,\text{test}}}{6(x^2 + 10)y(f'_c)^{1/3}} \tag{5-24}$$

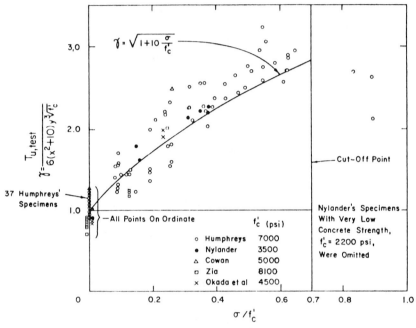

Fig. 5.7. Verification of prestress factor γ by tests.

where $T_{n,\text{test}}$ is the test value of a prestressed beam and T_{np} is expressed in terms of f'_c from Eq. (2–14). Figure 5.7 shows that the prestress factor is quite valid and also illustrates the effectiveness of the prestress in increasing the torsional strength of a beam without web reinforcement. For the ACI maximum prestress of $\sigma = 0.45\,f'_c$, the prestress factor $\gamma = 2.36$.

5.2 MEMBERS WITH WEB REINFORCEMENT (GENERALIZED ACI DESIGN METHOD)

5.2.1 Nominal Torsional Strength

For the torsion design of prestressed concrete members with web reinforcement, we shall introduce a method that is a generalization of the ACI design method. The basic equation for pure torsional strength of prestressed concrete members with web reinforcement has been developed from test results. Figure 5.8 shows two series of tests conducted by the author,[6] one for prestressed concrete members and another for nonprestressed concrete members. For these two series of beams the two curves intersect the ordinate and are mutually parallel. This means that the nominal torsional strength of either a pre-

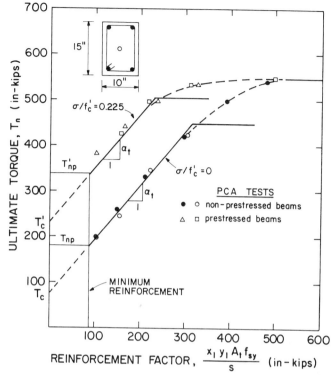

Fig. 5.8. Results of PCA pure torsion tests.

stressed or a nonprestressed beam can be expressed as the sum of a strength contributed by concrete and a strength contributed by web reinforcement. The effect of prestress is simply to increase the contribution of concrete while the contribution of the reinforcement remains unchanged. Thus, the ultimate torques of prestressed concrete beams can be expressed as:

$$T_n = T'_c + \alpha_t \frac{x_1 y_1 A_t f_{sy}}{s} \qquad (5\text{-}25)$$

where $\alpha_t = 0.66 + 0.33\ y/x \leq 1.5$. T'_c is the vertical intercept in Fig. 5.8 and represents the contribution of concrete to the torsional strength.

To evaluate the contribution of concrete, T'_c, we observe another interesting fact from Fig. 5.8. The prestressed beams appear to require approximately the same minimum torsional reinforcement as the nonprestressed beams. Let us assume that the minimum torsional reinforcement is the same for both

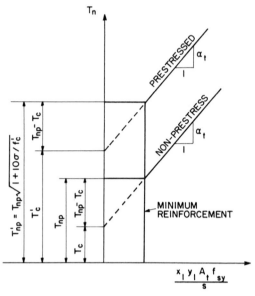

Fig. 5.9. Relationship between contributions of concrete T_c' and T_c for prestressed and nonprestressed reinforced beams.

prestressed and nonprestressed beams; then the contribution of concrete for prestressed beams T_c' can be related to that for nonprestressed beams T_c as shown in Fig. 5.9:

$$T_c' = T_{np} \sqrt{1 + 10\frac{\sigma}{f_c'}} - (T_{np} - T_c) \tag{5-26}$$

Since $T_c = (x^2y/3)(2.4\sqrt{f_c'})$ and $T_{np} = (x^2y/3)(6\sqrt{f_c'}) = 2.5\ T_c$:

$$T_c' = T_c \underbrace{\left(2.5\sqrt{1 + 10\frac{\sigma}{f_c'}} - 1.5 \right)}_{\gamma_1} \tag{5-27}$$

It can be seen that the contribution of concrete for prestressed concrete, T_c', is equal to the contribution of nonprestressed concrete, T_c, times a prestress factor, γ_1.

Substituting T_c' into Eq. (5–25), we obtain the basic equation for the torsional strength of prestressed concrete beams:

$$T_n = \underbrace{\frac{x^2 y}{3} 2.4\sqrt{f'_c}}_{T_c} \underbrace{\left(2.5\sqrt{1 + 10\frac{\sigma}{f'_c}} - 1.5\right)}_{\gamma_1} + \underbrace{\alpha_t \frac{x_1 y_1 A_t f_y}{s}}_{T_s} \qquad (5\text{-}28)$$

where $\alpha_t = 0.66 + 0.33\, y_1/x_1 \leq 1.5$. Equation (5–28) for prestressed concrete is the same as Eq. (4–10) for nonprestressed concrete, except that the contribution of concrete, T_c, should be multiplied by the prestress factor, γ_1. If $\sigma = 0$, $\gamma_1 = 1$, and the two equations become identical.

The interaction of torsion and shear will be treated in the following two sections for the contribution of concrete (T_c and V_c) and the maximum permissible torque ($T_{n,\,\text{max}}$).

5.2.2 Contribution of Concrete

The contribution of concrete in prestressed members subjected to torsion and shear can be derived in the same manner as in Section 4.3.2 for nonprestressed concrete, except that the effect of prestress on both torsion and shear strength must be taken into account. The same two assumptions are made regarding the elliptical interaction relationship and the constant stress ratio during loading. From Eqs. (4–34) and (4–35) we obtain:

$$\tau_c = \frac{\tau_{c0}}{\sqrt{1 + \left(\dfrac{\tau_{c0}}{v_{c0}}\dfrac{v_u}{\tau_u}\right)^2}} \qquad (5\text{-}29)$$

The conjugate equation is:

$$v_c = \frac{v_{c0}}{\sqrt{1 + \left(\dfrac{v_{c0}}{\tau_{c0}}\dfrac{\tau_u}{v_u}\right)^2}} \qquad (5\text{-}30)$$

In terms of shear forces and torques:

$$T_c = \frac{T_{c0}}{\sqrt{1 + \left(\dfrac{V_{c0}}{T_{c0}}\dfrac{T_u}{V_u}\right)^2}} \qquad (5\text{-}31)$$

$$V_c = \frac{V_{c0}}{\sqrt{1 + \left(\dfrac{T_{c0}}{V_{c0}}\dfrac{V_u}{T_u}\right)^2}} \qquad (5\text{-}32)$$

In Eqs. (5–31) and (5–32), T_{c0} is given by the first term in Eq. (5–28):

$$T_{c0} = \frac{x^2 y}{3} 2.4\sqrt{f_c'} \left(2.5\sqrt{1 + 10\frac{\sigma}{f_c'}} - 1.5 \right) \tag{5-33}$$

and V_{c0} can be obtained from Eqs. (11–11) through (11–13) of the 1977 ACI Code:

$$V_{ci} = 0.6\sqrt{f_c'}\ b_w d + V_d + \frac{V_i M_{cr}}{M_{\max}} \tag{5-34}$$

$$\geq 1.7\sqrt{f_c'}\ b_w d$$

where:

$$M_{cr} = \left(\frac{I}{y_t}\right)(6\sqrt{f_c'} + f_{pe} - f_d) \tag{5-35}$$

and:

$$V_{cw} = (3.5\sqrt{f_c'} + 0.3f_{pc})b_w d + V_p \tag{5-36}$$

V_{c0} should be taken as the lesser of V_{ci} or V_{cw}. For the special case of nonprestressed concrete members, V_{c0} may be taken as $2\sqrt{f_c'}\, b_w d$.

Equations (5–31) and (5–32) for prestressed concrete are generalization of Eqs. (4–41) and (4–42) for nonprestressed concrete. The former equations become identical to the latter when the prestress is equal to zero.

5.2.3 Maximum Permissible Torque

In proportioning members to resist torsion, consideration should be given to preventing the overreinforcing of the beams so that tensile yielding of reinforcement will precede the compression crushing of concrete. It was shown in Section 4.3.3 that this purpose is achieved by imposing a maximum permissible torque. In reinforced concrete subjected to pure torsion, the maximum permissible torque has been taken as $T_{n0,\max} = (\Sigma x^2 y/3)(12\sqrt{f_c'})$. For prestressed concrete, $T_{n0,\max}$ should be modified to account for the effect of prestress:

$$T_{n0,\max} = \gamma \frac{\Sigma x^2 y}{3}(C\sqrt{f_c'}) \tag{5-37}$$

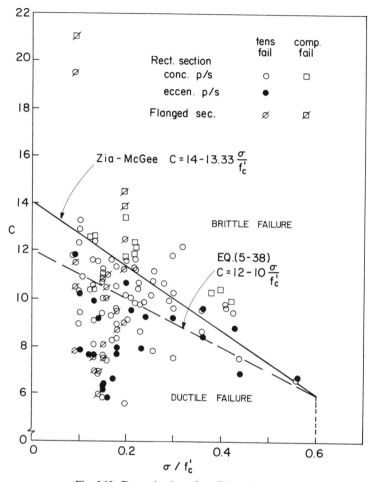

Fig. 5.10. Determination of coefficient C by tests.

where $\gamma = \sqrt{1 + 10\,\sigma/f'_c}$ is the prestress factor from Eq. (5–23), and C is an empirical coefficient. This coefficient was originally determined by Zia and McGee[7] to be $C = 14 - 13.33(\sigma/f'_c)$ as shown in Fig. 5.10. For nonprestressed concrete, $C = 14$ as the prestress σ becomes zero. This value of 14 does not coincide with the conservative value of 12 adopted by the ACI Code. To be consistent with the ACI Code, a conservative coefficient is recommended[8] as follows:

$$C = 12 - 10\frac{\sigma}{f'_c} \qquad (5\text{–}38)$$

Substituting Eqs. (5–38) and (5–23) into (Eq. 5–37) gives:

$$T_{n0,\max} = \underbrace{\frac{\Sigma x^2 y}{3}(12\sqrt{f_c'})}_{\substack{T_{n0,\max} \\ \text{nonprestressed} \\ \text{concrete}}}\underbrace{\left(1 - 0.833\frac{\sigma}{f_c'}\right)\left(\sqrt{1 + 10\frac{\sigma}{f_c'}}\right)}_{\gamma_2} \qquad (5\text{–}39)$$

Equation (5–39) shows that the maximum permissible torque for prestressed concrete is equal to that for a nonprestressed concrete times a prestress factor, γ_2:

$$\gamma_2 = \frac{C\gamma}{12} = \left(1 - 0.833\frac{\sigma}{f_c'}\right)\sqrt{1 + 10\frac{\sigma}{f_c'}} \qquad (5\text{–}40)$$

This prestress factor is plotted in Fig. 5.11 as a function of σ/f_c'.

For prestressed members subjected to torsion and shear, the derivation of maximum permissible torque is exactly the same as in Section 4.3.3 for

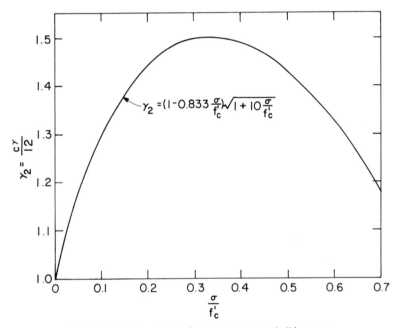

Fig. 5.11. Prestress factor for maximum permissible torque.

nonprestressed members except that the maximum torsional stress $\tau_{n\,0,\max}$ should be multiplied by the prestress factor γ_2. Equation (4–48) then becomes:

$$\tau_{n,\max} = \frac{12\gamma_2\sqrt{f_c'}}{\sqrt{1 + \left(\dfrac{1.2\gamma_2 v_u}{\tau_u}\right)^2}} \tag{5-41}$$

and in terms of torsional moment:

$$T_{n,\max} = \frac{4\gamma_2\sqrt{f_c'}\,\Sigma x^2 y}{\sqrt{1 + \left(\dfrac{0.4\gamma_2 V_u}{C_t T_u}\right)^2}} \tag{5-42}$$

where γ_2 is given in Eq. (5–40) and C_t in Eq. (4–39). Equation (5–42) for prestressed concrete will degenerate into Eq. (4–50) for nonprestressed concrete when prestress $\sigma = 0$ and $\gamma_2 = 1$. Since γ_2 is always greater than unity (see Fig. 5.11), it would be conservative to use Eq. (4–50) for both prestressed and nonprestressed concrete. This, of course, would considerably simplify the equation for maximum permissible torque.

5.2.4   Minimum Torsional Reinforcement

The development of minimum torsional reinforcement for nonprestressed concrete was discussed in Section 4.4.2, resulting in Eq. (4–67) for longitudinal steel and the limitation of Eq. (4–62) for web reinforcement. Equations (4–67) and (4–62) are rewritten here for convenience:

$$\hat{A}_t = \left[\frac{400xs}{f_y}\left(\frac{T_u}{T_u + \dfrac{V_u}{3c_t}}\right) - 2A_t\right]\frac{x_1 + y_1}{s} \tag{5-43}$$

$$2A_t + A_v = \frac{50b_w s}{f_y} \tag{5-44}$$

For prestressed concrete, it was found[9] that Eq. (5–43) for longitudinal steel remains valid, and that only a slight modification is necessary for Eq. (5–44) to account for the effect of prestress on the web reinforcement. The modified equation is:

$$2A_t + A_v = \frac{50b_w s}{f_y}\left(1 + 12\frac{\sigma}{f_c'}\right) \tag{5-45}$$

The factor $(1 + 12 \, \sigma/f'_c)$ accounts for the prestress, and Eq. (5–45) degenerates into Eq. (5–44) when the prestress σ becomes zero.

To prove the validity of Eq. (5–43) for prestressed concrete, we shall first rearrange Eq. (5–43) and let the stress ratio $R = T_u/(T_u + V_u/3C_t)$:

$$400R = \underbrace{\frac{\hat{A}_l f_y}{x(x_1 + y_1)}}_{(rf_y)_{\text{long}}} + \underbrace{\frac{2A_t f_y}{xs}}_{(rf_y)_{\text{web}}} \tag{5–46}$$

Notice that the last term, $2A_t f_y/xs$, is the web reinforcement index $(rf_y)_{\text{web}}$, and the term $\hat{A}_l f_y/x(x_1 + y_1)$ can be defined as the longitudinal reinforcement index, $(rf_y)_{\text{long}}$. For prestressed concrete $\hat{A}_l f_y$ should be interpreted as the total available longitudinal steel forces to resist torsion, including both the mild steel and the prestressed steel. Then Eq. (5–46) can be written as:

$$\frac{(rf_y)_{\text{long}}}{R} + \frac{(rf_y)_{\text{web}}}{R} = 400 \tag{5–47}$$

In Eq. (5–47) the two terms on the left-hand side which are divided by the stress ratio R may be referred to as the modified longitudinal reinforcement index and the modified web reinforcement index. Equation (5–47) is plotted in Fig. 5.12 as the straight line AB, connecting the abscissa and the ordinate at the values of 400.

In Fig. 5.12 another 45° straight line, OD, is plotted from Eq. (4–16), representing equal volumes of longitudinal steel and web steel:

$$\hat{A}_l = A_t \frac{2(x_1 + y_1)}{s} \tag{5–48}$$

In the ACI Code design method the longitudinal reinforcement is designed according to Eq. (5–48), represented by the solid straight line DC, when the modified web reinforcement index is greater than 200. When the modified web reinforcement index is less than 200, however, the longitudinal steel must be designed by Eq. (5–47), represented by the solid straight line CA. The increase of longitudinal steel to compensate for the decrease of web reinforcement at low web reinforcement index is necessary to ensure the ductility of a member.

In Fig. 5.12 we have also plotted all the suitable test data available in the literature.[10-15] The hollow and solid points represent ductile and brittle

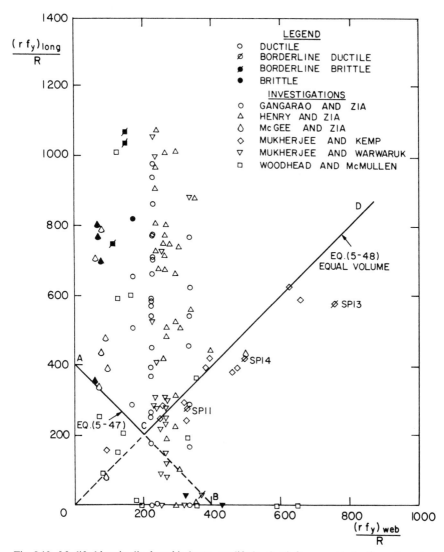

Fig. 5.12. Modified longitudinal steel index vs. modified web reinforcement index for rectangular sections.

failure, respectively. Figure 5.12 shows that seven points representing brittle failure occur above the line *AC*. To exclude these undesirable points, we plot the web reinforcement indices as a function of the prestress ratios σ/f'_c in Fig. 5.13. It can be seen that Eq. (5–45) has excluded all the seven test members exhibiting brittle failure. To further confirm the applicability

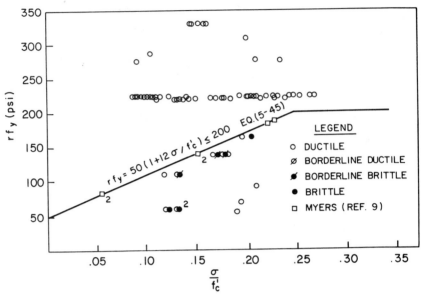

Fig. 5.13. Comparison of Eq. (5–45) with test data.

of Eq. (5–45), six beams were tested by Myers[9] according to this equation. The test points for these beams are plotted as hollow squares in Fig. 5.13. All six beams exhibited ductile failure.

It should be pointed out that the minimum web reinforcement requirement of Eq. (5–45) has been developed for rectangular members. For hollow box sections it appears reasonable to modify this equation by the reduction factor, $4h/x$, resulting in:

$$2A_t + A_v = \frac{50 b_w s}{f_y} \left(1 + 12 \frac{\sigma}{f'_c}\right) \left(\frac{4h}{x}\right) \qquad (5\text{--}49)$$

where $x/10 \le h \le x/4$. For prestressed I-sections Eq. (5–45) may be modified by a factor $\sqrt{d/b_w}$ giving:

$$2A_t + A_v = \frac{50 b_w s}{f_y} \left(1 + 12 \frac{\sigma}{f'_c}\right) \sqrt{\frac{d}{b_w}} \qquad (5\text{--}50)$$

Equations (5–49) and (5–50) appear to be logical but have not yet been conclusively substantiated owing to the lack of test information.

It should be mentioned that the detailing of web reinforcement is quite important to ensure the ductility of prestressed I-sections. The following three design details are recommended:

1. For web reinforcement x_1 shall be greater than $0.75b_w$.
2. Stirrup spacing in the web shall be less than $h_w/2$, where h_w is the height of the web. For a tapered flange thickness, h_w may be the total height less the average thickness of the top plus bottom flanges.
3. Closed transverse reinforcement should be provided in the flanges.

5.2.5 Generalized ACI Torsion Design Procedures

The ACI torsion design procedures in the 1977 or 1983 code Section 11.6 for nonprestressed members can be easily generalized to include prestressed members by incorporating the four modifications given below. These modifications are included in "Tentative Recommendations for the Design of Prestressed and Nonprestressed Members to Resist Torsion" given in Appendix B. The four modifications are highlighted by the asterisk symbols (*) at the right margin of the pages.

(1) Sections 11.6.1 and 11.6.3.
In Section 11.6.1 the factored torsional moment, $\phi(0.5\sqrt{f'_c}\,\Sigma x^2 y)$, above which torsional effects must be considered, should be multiplied by the prestress factor γ. Similarly, in Section 11.6.3 a maximum factored torsional moment, $\phi(4\sqrt{f'_c}\,\Sigma x^2 y/3)$, is allowed after redistribution of internal forces in a statically indeterminate structure. This maximum factored torsional moment should also be multiplied by the prestress factor γ.

(2) Section 11.6.6.1.
For the torsional moment strength provided by concrete, Eq. (11–22) has been generalized to become:

$$T_c = \frac{T_{c0}}{\sqrt{1 + \left(\dfrac{T_{c0}V_u}{V_{c0}T_u}\right)^2}} \qquad (11\text{–}22)$$

and the conjugate Eq. (11–5) in Section 11.3.1.4 is:

$$V_c = \frac{V_{c0}}{\sqrt{1 + \left(\dfrac{V_{c0}T_u}{T_{c0}V_u}\right)^2}} \qquad (11\text{–}5)$$

where

$T_{c0} = 0.8\sqrt{f_c'}\,\Sigma x^2 y(2.5\gamma - 1.5)$.

$V_{c0} = V_{ci}$ from Eq. (11–11), or V_{cw} from Eq. (11–13), whichever is the lesser.

For nonprestressed concrete $V_{c0} = 2\sqrt{f_c'}\,b_w d$ and $T_{c0}/V_{c0} = 0.4\Sigma x^2 y/b_w d = 0.4/C_t$.

(3) Sections 11.6.9.2. and 11.6.9.3. For the minimum area of closed stirrups, Eq. (11–16) in Section 11.5.5.5 should be:

$$A_v + 2A_t = 50\frac{b_w s}{f_y}\left(1 + 12\frac{\sigma}{f_c'}\right) \le 200\frac{b_w s}{f_y} \qquad (11\text{--}16)$$

and the term $(50b_w s/f_y)(1 + 12\sigma/f_c')$ in Eq. (11–16) must be multiplied by $4h/x$ for box sections and by $\sqrt{d/b_w}$ for I-sections.

(4) Section 11.6.9.4. The maximum permissible factored torque has been generalized as:

$$\frac{4\sqrt{f_c'}\,\gamma_2 \Sigma x^2 y}{\sqrt{1 + \left(\dfrac{0.4\gamma_2 V_u}{C_t T_u}\right)^2}}$$

where $\gamma_2 = (1 - 0.833\sigma/f_c')\sqrt{1 + 10\sigma/f_c'}$. For nonprestressed concrete $\gamma_2 = 1.0$.

Finally, it should be emphasized that these generalized ACI torsion design criteria have been proposed by Zia and the author. They are being studied by the ASCE-ACI Committee 445, Shear and Torsion, but should not be regarded as a committee-endorsed document.

5.2.6 Design Example 5.1

Design Example 4.4 in Chapter 4 illustrates the design procedures for a nonprestressed reinforced concrete beam. This same beam (shown in Fig. 4.24) will now be designed as a post-tensioned beam. The use of the same structure and loadings should facilitate the comparison of the proposed generalized design procedures for prestressed concrete (See Appendix B) with the current ACI design procedures for nonprestressed concrete.

Figure 4.24 shows an 18 in. by 28 in. beam with 30 ft clear span supporting a 12 in. by 28 in. cantilever at midspan. A concentrated load P is applied at the end of the cantilever at a distance 2 ft from the center line of the beam. This superimposed load is made up of a service dead load of 42 kips and a service live load of 36 kips. Design the reinforcement for the beam according to the 1977 or 1983 ACI Building Code, except that the shear and torsional reinforcement should be designed according to the generalized procedures for prestressed concrete in Appendix B. The material strengths are f'_c = 4,000 psi, f_y = 60,000 psi for mild steel, and 270K seven-wire strands for prestressed tendons.

Solution

Flexural Moments, Shears, and Torques:
Dead load:

Beam dead weight:

$$w_d = 150\frac{18(28)}{144} = 525\text{ plf}$$

Superimposed concentrated D.L. = 42 k

Bending moment at face of column (assumed fixed ends):

$$M_d = \frac{42(30)(12)}{8} + \frac{0.525(30)^2(12)}{12} = 2,363\text{ in.-k}$$

Bending moment at midspan:

$$M_d = \frac{42(30)(12)}{8} + \frac{0.525(30)(12)}{24} = 2,126\text{ in.-k}$$

Shear at face of column:

$$V_d = \frac{42}{2} + \frac{0.525(30)}{2} = 28.9\text{ k}$$

Shear at a distance $h/2$ from face of column:

$$V_d = \frac{42}{2} + 0.525\left(15 - \frac{14}{12}\right) = 28.3\text{ k}$$

Shear at midspan:

$$V_d = \frac{42}{2} = 21.0 \text{ k}$$

Torsional moment throughout:

$$T_d = \frac{42(2)(12)}{2} = 504 \text{ in.-k}$$

Live load:

Superimposed concentrated L.L. = 36 k

Bending moment at support and at midspan (assumed fixed ends):

$$M_L = \frac{36(30)(12)}{8} = 1,620 \text{ in.-k}$$

Shear throughout:

$$V_L = \frac{36}{2} = 18.0 \text{ k}$$

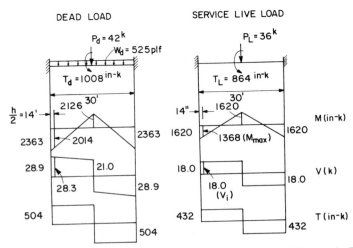

Fig. 5.14. Flexural moment, shear, and torque diagrams for dead and live loads (Example 5.1).

Torsional moment throughout:

$$T_L = \frac{36(2)(12)}{2} = 432 \text{ in.-k}$$

The flexural moment, shear, and torque diagrams for dead and live loads are given in Fig. 5.14. The flexural moment diagram for dead load should be slightly curved, but has been taken as two straight lines for simplicity.

Design of Prestress Tendons:
The prestress tendons are provided to balance the dead load moment given in Fig. 5.14. Using the load balancing concept there will be no deflection under the dead load. The tendon profile consists of two straight lines as shown in Fig. 5.15 (a). The positive eccentricity at midspan and the negative

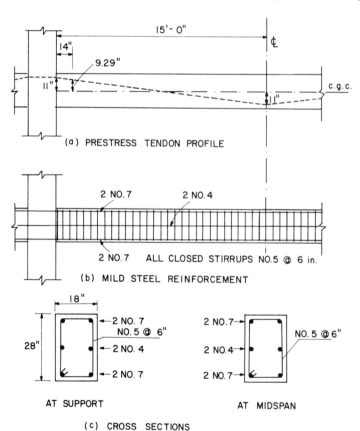

Fig. 5.15. Reinforcement details (Example 5.1).

eccentricity at support are both 11 in., which is the maximum possible in construction. The sag h and the effective prestress F_{se} are:

$$h = 2(11) = 22 \text{ in.}$$

$$F_{se} = \frac{2,363 + 2,126}{22} = 204 \text{ k}$$

Using ½ in., 270K strands and assuming a permissible stress of $0.70f_{pu}$ and a 22% loss of prestress, the number of strands, N, is:

$$N = \frac{204}{0.153(0.7)(270)(0.78)} = 9.04 \qquad \text{Use 9 strands}$$

$$\text{Actual } F_{se} = 9(0.153)(0.7)(270)(0.78) = 203 \text{ k}$$

Check the positive eccentricity at midspan to prevent excessive tensile stress at top fiber:

Allowable tensile stress $\sigma_{ti} = 3\sqrt{f'_{ci}} = 3\sqrt{3,000} = 164$ psi

Section modulus $Z_t = \dfrac{18(28)^2}{6} = 2,352$ in.^{3}

Kern distance $k_b = \dfrac{28}{6} = 4.67$ in.

Initial prestress $F_{si} = \dfrac{203}{0.78} = 260$ k

Midspan dead load $M_d = 2,126$ in.-k

$$k_b + \frac{\sigma_{ti}Z_t}{F_{si}} + \frac{M_d}{F_{si}} = 4.67 + \frac{0.164(2,352)}{260} + \frac{2,126}{260} = 14.3 \text{ in.} > 11 \text{ in.} \quad \text{O.K.}$$

Check the negative eccentricity at support to prevent excessive tensile stress at bottom fiber:

$\sigma_t = 6\sqrt{f'_c} = 3\sqrt{4,000} = 380$ psi
$Z_b = 2,352$ in.^{3}
$k_t = 4.67$ in.
$F_{si} = 260$ k
$M_d = 2,363$ in.-k at support

$$k_t + \frac{\sigma_t Z_b}{F_{si}} + \frac{M_d}{F_{si}} = 4.67 + \frac{0.38(2,352)}{260} + \frac{2,363}{260} = 17.2 \text{ in.} > 11 \text{ in.} \qquad \text{O.K.}$$

Check ultimate strength, assuming that the post-tensioning tendons are grouted (ACI Eq. 18–3):

$$\rho_p = \frac{A_{ps}}{bd} = \frac{9(0.153)}{18(25)} = 0.00306$$

$$f_{ps} = f_{pu} \left(1 - 0.5 \, \rho_p \, \frac{f_{pu}}{f'_c} \right) = 270 \left[1 - 0.5(0.00306) \frac{270}{4} \right] = 242 \text{ ksi}$$

$$\omega_p = \frac{\rho_p f_{ps}}{f'_c} = \frac{0.00306(242)}{4} = 0.185 < 0.3 \qquad \text{O.K.}$$

$$\phi M_n = \phi A_{ps} f_{ps} d(1 - 0.59 \, \omega_p)$$
$$= 0.9(9)(0.153)(242)(25)[1 - 0.59 \, (0.185)]$$
$$= 6,682 \text{ in.-k}$$
$$M_u = 1.4 M_d + 1.7 M_L = 1.4(2,363) + 1.7(1,620) = 6,062 \text{ in.-k}$$
$$< 6,682 \text{ in.-k} \qquad \text{O.K.}$$

Shear and Torsional Web Reinforcement:

$$\Sigma x^2 y = 18^2(28) = 9,072 \text{ in.}^3$$
$$b_w d = 18(25) = 450 \text{ in.}^2$$

Eq. (4–39):
$$C_t = \frac{b_w d}{\Sigma x^2 y} = \frac{450}{9,072} = 0.0496 \, \frac{1}{\text{in.}}$$

$$\sigma = \frac{F_{se}}{A_c} = \frac{203,000}{18(28)} = 403 \text{ psi}$$

$$\gamma = \sqrt{1 + 10 \, \frac{\sigma}{f'_c}} = \sqrt{1 + 10 \, \frac{403}{4,000}} = 1.417$$

At a distance d from face of column $V_u = 1.4(28.3) + 1.7(18) = 70.2$ k and $T_u = 1.4(504) + 1.7(432) = 1440$ in.-k.

Section 11.6.1:

$$\phi \gamma (0.5\sqrt{f'_c} \, \Sigma x^2 y) = 0.85(1.417)(0.5\sqrt{4,000})(9,072)$$
$$= 345,534 \text{ in.-lb.} = 346 \text{ in.-k} < 1440 \text{ in.-k}$$

Torsion cannot be neglected in design.

Section 11.6.9.4:

$$\gamma_2 = (1 - 0.833\sigma/f_c')\sqrt{1 + 10\sigma/f_c'}$$
$$= \left(1 - 0.833\frac{403}{4,000}\right) 1.417 = 1.30$$

$$T_{n,\,max} = \frac{4\sqrt{f_c'}\,\gamma_2\Sigma x^2 y}{\sqrt{1 + \left(\dfrac{0.4\gamma_2 V_u}{C_t T_u}\right)^2}} = \frac{4\sqrt{4,000}(1.30)(9,072)}{\sqrt{1 + \left[\dfrac{0.4(1.30)(70.2)}{0.0496(1,440)}\right]^2}}$$
$$= 2,656,699 \text{ in.-lb} = 2,657 \text{ in.-k} > 1,440 \text{ in.-k} \qquad \text{O.K.}$$

The size of the beam is sufficient.

$$T_{co} = 0.8\sqrt{f_c'}\,\Sigma x^2 y\,(2.5\gamma - 1.5) = 0.8\sqrt{4,000}\,(9,072)(2.5 \times 1.417 - 1.5)$$
$$= 937,529 \text{ in.-lb} = 938 \text{ in.-k}$$

$$V_{cw} = (3.5\sqrt{f_c'} + 0.3f_{pc})b_w d + V_p$$
$$= \frac{[3.5\sqrt{4,000} + 0.3(403)](18)(25)}{1,000} + \frac{22}{180}(203) = 178.8 \text{ k}$$

At a distance $h/2$ from column face $e = 11 - (22/180)14 = 9.29$ in. (see Fig. 5.15); $M_d = 2,363 - (14/180)(2,363 + 2,126) = 2,014$ in.-k; $M_{max} = 1,620 - (14/90)(1,620) = 1,368$ in.-k, $V_d = 28.3$k, and $V_i = 18.0$k (see Fig. 5.14).

$$f_{pe} = \frac{F_{se}}{A_c} + \frac{F_{se}e}{Z_t} = 403 + \frac{203,000(9.29)}{2,352} = 1,205 \text{ psi}$$

$$f_d = \frac{M_d}{Z_t} = \frac{2,014,000}{2,352} = 856 \text{ psi}$$

$$M_{cr} = Z_t(6\sqrt{f_c'} + f_{pe} + f_d)$$
$$= 2,352(6\sqrt{4,000} + 1,205 - 856) = 1,713,369 \text{ in.-lb} = 1,713 \text{ in.-k}$$

$$V_{ci} = 0.6\sqrt{f_c'}\,b_w d + V_d + \frac{V_i M_{cr}}{M_{max}}$$
$$= \frac{0.6\sqrt{4,000}(18)(25)}{1,000} + 28.3 + \frac{(18)(1,713)}{1,368}$$
$$= 17.1 + 28.3 + 22.5 = 67.9 \text{ k} > 1.7\sqrt{f_c'}\,b_w d = 48.4 \text{ k}$$

Since $V_{ci} < V_{cw}$, $V_{co} = V_{ci} = 67.9$ k.

Eq. (11–22): $T_c = \dfrac{T_{co}}{\sqrt{1 + \left(\dfrac{T_{co}}{V_{co}}\dfrac{V_u}{T_u}\right)^2}} = \dfrac{938}{\sqrt{1 + \left[\dfrac{938(70.2)}{67.9(1,440)}\right]^2}} = 778$ in.-k

Eq. (11–5): $V_c = \dfrac{V_{co}}{\sqrt{1 + \left(\dfrac{V_{co}}{T_{co}}\dfrac{T_u}{V_u}\right)^2}} = \dfrac{67.9}{\sqrt{1 + \left[\dfrac{67.9(1,440)}{938(70.2)}\right]^2}} = 37.9$ k

$T_s = \dfrac{T_u}{\phi} - T_c = \dfrac{1,440}{0.85} - 778 = 916$ in.-k

Use #5 bars (concrete cover 1.5 in.):

$x_1 = 18 - 2(1.5 + 0.3125) = 14.38$ in.

$y_1 = 28 - 2(1.5 + 0.3125) = 24.38$ in.

$\alpha_t = 0.66 + 0.33\dfrac{y_1}{x_1} = 0.66 + 0.33\dfrac{24.38}{14.38} = 1.22 < 1.5$

Eq. (11–23): $\dfrac{A_t}{s} = \dfrac{T_s}{\alpha_t x_1 y_1 f_y} = \dfrac{916}{1.22(14.38)(24.38)(60)} = 0.0357$ in.^{2}/in.

$V_s = \dfrac{V_u}{\phi} - V_c = \dfrac{70.2}{0.85} - 37.9 = 44.7$ k

Eq. (11–17): $\dfrac{A_v}{s} = \dfrac{V_s}{d f_y} = \dfrac{44.7}{25(60)} = 0.0298$ in.^{2}/in.

Total web reinforcement:

$\dfrac{A_t}{s} + \dfrac{1}{2}\dfrac{A_v}{s} = 0.0357 + \dfrac{0.0298}{2} = 0.0506$ in.^{2}/in.

$s = \dfrac{0.31}{0.0506} = 6.13$ in. < 12 in.

Section 11.6.8.1:

$s_{max} = \dfrac{x_1 + y_1}{4} = \dfrac{14.38 + 24.38}{4} = 9.69$ in. > 6.13 in.   O.K.

$\dfrac{3h}{4} = \dfrac{3(28)}{4} = 21$ in. > 6.13 in.   O.K.

Eq. (11–16): $\dfrac{50b_w}{f_y}\left(1 + 12\dfrac{\sigma}{f'_c}\right) = \dfrac{50(18)}{60,000}\left(1 + 12\dfrac{403}{4,000}\right)$

$$= 0.0331 \text{ in.}^2/\text{in.} < 2(0.0506) \qquad \text{O.K.}$$

Use No. 5 closed stirrups at 6-in. spacing throughout the beam.

Torsional Longitudinal Reinforcement:

Eq. (11–24): $\hat{A}_l = \dfrac{A_t}{s}\,2(x_1 + y_1) = 0.0357(2)(14.38 + 24.38) = 2.77 \text{ in.}^2$

Eq. (11–25):

$$\hat{A}_l = \left[\frac{400xs}{f_y}\frac{T_u}{\left(T_u + \dfrac{V_u}{3C_t}\right)} - 2A_t\right]\frac{x_1 + y_1}{s}$$

$$= \left[\frac{400(18)(6.13)}{60,000}\frac{1,440}{1,440 + \dfrac{70.2}{3(0.0496)}} - 2(0.31)\right]\frac{(14.38 + 24.38)}{6.13}$$

$$= -0.417 \text{ in.}^2 < 2.77 \text{ in.}^2 \qquad \text{O.K.}$$

Use four No. 7 corner bars and two No. 4 bars at mid-length to satisfy the 12-in. spacing.

$$\hat{A}_l = 4(0.60) + 2(0.31) = 2.80 \text{ in.}^2 > 2.77 \text{ in.}^2 \qquad \text{O.K.}$$

Check Stirrups at Midspan Joint (see Eq. 9–53):

$$V_n = \frac{b_f d_s}{s}A_v f_y = \frac{12 + 25}{6}(2)(0.31)(60) = 229 \text{ k}$$

$$V_u/\phi = 120/0.85 = 141 \text{ k} < 229 \text{ k} \qquad \text{O.K.}$$

Arrangement of Reinforcement:
Detailed arrangement of longitudinal bars and stirrups is shown in Fig. 5.15. Comparison of this design with the design of nonprestressed members in Fig. 4.25 shows clearly that (1) the 4.94 in.² of 60 ksi flexural reinforcement at support and midspan for nonprestressed members has been replaced by

nine 270K ½-in. strands of 1.377 in.,² and (2) the prestress has reduced the closed stirrups from No. 6 at 6 in. to No. 5 at 6 in. and the torsional longitudinal steel area from 3.85 in.² to 2.77 in.²

REFERENCES

1. Cowan, H. J., "Strength of Reinforced Concrete Under the Action of Combined Stresses at the Representation of the Criterion of Failure by a Space Model," *Nature* (London), Vol. 169, 1952, p. 663.

2. Cowan, H. J. and S. Armstrong, "Experiments on the Strength of Reinforced and Prestressed Concrete Beams and of Concrete-Encased Steel Joists in Combined Bending and Torsion," *Magazine of Concrete Research,* Vol. 7, No. 19, March 1955, pp. 3–20.

3. Humphreys, R., "Torsional Properties of Prestressed Concrete," *The Structural Engineer* (London), Vol. 35, No. 6, June 1957, pp. 213–224.

4. Nylander, H., "Torsion and Torsional Restraint of Concrete Structures" (Vridning Vridningsinspanning vid Betongkongstruktioner), Statens Kommitté för Byggnadsforskning, Stockholm, Bulletin No. 3, 1945.

5. Hsu, T. T. C., "Torsion of Structural Concrete—Uniformly Prestressed Rectangular Members Without Web Reinforcement," *Journal of the Prestressed Concrete Institute,* Vol. 13, No. 2, April 1968, pp. 34–44. Also PCA Research and Development Laboratories, Bulletin D140.

6. Private communication to ACI Committee 438, Torsion, from Thomas T. C. Hsu, October 17, 1967, reporting a series of test results at the Structural Laboratory of the Portland Cement Association, Skokie, Illinois.

7. Zia, P. and W. D. McGee, "Torsion Design of Prestressed Concrete," *Journal of the Prestressed Concrete Institute,* Vol. 19, No. 2, March–April 1974, pp. 46–65.

8. Zia, P. and T. T. C. Hsu, "Design for Torsion and Shear in Prestressed Concrete," Proceedings, Symposium on Shear and Torsion, Fall Convention of the American Society of Civil Engineers, Chicago, October 16–20, 1978.

9. Myers, G., "Minimum Torsional Web Reinforcement for Prestressed Concrete," M. S. thesis under the supervision of Thomas T. C. Hsu, Dept. of Civil Engineering, University of Miami, Coral Gables, Florida, 1981.

10. Gangarao, H. V. S. and P. Zia, "Rectangular Prestressed Concrete Beams Under Combined Bending and Torsion," Civil Engineering Dept. Report, North Carolina State University, Raleigh, April 1970.

11. Henry R. L. and P. Zia, "Behavior of Rectangular Prestressed Concrete Beams Under Combined Torsion, Bending and Shear," Civil Engineering Dept. Report, North Carolina State University, Raleigh, 1971.

12. McGee, W. D. and P. Zia, "Prestressed Concrete Members Under Torsion, Shear and Bending," Civil Engineering Dept. Report, North Carolina State University, Raleigh, 1973.

13. Mukherjee, P. R. and E. L. Kemp, "Ultimate Torsional Strength of Plain, Prestressed and Reinforced Concrete Members of Rectangular Cross Section," Civil Engineering Studies No. 2003, University of West Virginia, Morgantown, 1970.

14. Mukherjee, P. R. and J. Warwarak, "Prestressed Concrete Beams with Web Reinforcement Under Combined Loading," Structural Engineering Report No. 24, University of Alberta, Edmonton, 1970.

15. Woodhead, H. R. and A. E. McMullen, "A Study of Prestressed Concrete Under Combined Loading," Research Report No. CE72–43, University of Calgary, Calgary, 1972.

Plate 6: Collapse of a six story reinforced concrete building due to the shear and torsion failure of the reinforced concrete ribbed raft foundation, 1979. One corner of the building has sunk 3.5 m.

6
Skew-Bending Theories for Combined Loadings

6.1 INTRODUCTION

This chapter deals with the strength of reinforced concrete members subjected to torsion, bending, and shear. The methods of analysis and design included in this chapter belong to a general category normally referred to as skew-bending theories. There is another prominent category of theories known as truss analogy. This second group of theories will be discussed in Chapter 7.

The basic characteristic of skew-bending theories is the assumption of a skew failure surface. This surface is initiated by a helical crack on three faces of a rectangular beam, while the ends of this helical crack are connected by a compression zone near the fourth face. The failure surface intersects both the longitudinal reinforcement bars and the closed stirrups. The forces in this reinforcement provide the internal forces and moments to resist the external forces and moments. At the failure of a beam, the two parts of the beam separated by the failure surface rotate against each other about a neutral axis on the inside edge of the compression zone. It is often assumed that both the longitudinal steel and stirrups will yield at the collapse of the beam. Based on this assumption it is possible to write equilibrium conditions and to use these equations for the purpose of design or analysis.

In 1958 Lessig[1,2] first proposed the skew-bending theory in connection with two modes of failure. Mode 1 failure has the compression zone near the top face of the beam, whereas in mode 2 failure the compression zone is along a side face. For each mode of failure, she took two equilibrium conditions, one equilibrium of moments about the neutral axis and another equilibrium of forces along the normal to the compression zone. To determine the angle of inclination of the helical crack that initiates the failure surface, she minimized the strength of the member and obtained a third equation. In this way she produced a set of three basic equations for the analysis of each mode of failure. These equations can be solved by a trial-and-error procedure.

Lessig's theory was simplified and incorporated into the Russian Code in 1962.[3] The first simplification consisted of neglecting some minor contribution of the stirrups to the torsional resistance. The second simplification resulted in the elimination of the trial-and-error procedures. Even with these two simplifications, the method of analysis in the Russian Code was quite tedious. This code also recognized the shear failure mode (when shear and torsion predominated) and proposed an empirical equation to guard against its occurrence. Empirical limits were also given to prevent the crushing of concrete before the yielding of steel and to avoid partially overreinforced beams.

By making an assumption that uncoupled Lessig's set of three equations, Collins et al.[4,5] were able to combine these three equations into one equation for each failure mode. This one equation could also be expressed in a nondimensional form, and the first nondimensional interaction curve was achieved. Mode 1 failure produced a torsion–bending interaction curve, whereas mode 2 failure gave a torsion–shear interaction curve. Collins et al. also discovered a third mode of failure. This mode 3 failure has a compression zone near the bottom face of the beam. It occurred in an unsymmetrically reinforced member subjected to a small bending moment and a large torsional moment. The three interaction curves for the three modes of failures formed an interaction surface between torsion, bending, and shear. This interaction surface was also modified by an empirical equation for shear failure, which may be considered as the fourth mode of failure. When compared to tests, this theory was found to be unconservative in the torsion–bending interaction[6] and to be somewhat too conservative in the torsion–shear interaction.[4]

Collins et al.'s theory served as a basis for the Australian Code of 1973.[7-9] This code utilized the concept of an equivalent bending moment for torsion. That is, torsional moment was converted into an equivalent bending moment that was added to the actual bending moment for the design of longitudinal steel according to the conventional flexural mechanics. The simple expression for equivalent bending moment was derived from mode 1 and mode 3 failures by specifying an optimum ratio of longitudinal steel to stirrups. This optimum ratio was obtained by minimizing the total weight of longitudinal steel and stirrups to resist a certain torque. Similarly, the Australian Code also converted a torsional moment into an equivalent shear force that could be added to the flexural shear force for the design of web reinforcement according to the conventional code method. The expression for the equivalent shear force, however, had been derived empirically based on tests in shear failure. Although the concepts of equivalent bending moment and equivalent shear force were quite simple and straightforward in design, the code also called for two checks. One check was aimed at preventing mode 2 failure, and the other check was to make sure that the web reinforcement was sufficient for mode 1 failure based on the optimum ratio of longitudinal steel to stirrups. In 1962, shortly after the appearance of Lessig's theory, Yudin pointed out[10] that Lessig employed only two equilibrium conditions. One force equilibrium condition was used to determine the depth of the compression zone. The other single moment equilibrium condition would be insufficient, in the case of analysis, to determine the stresses in both the longitudinal reinforcement and the stirrups. In the case of design, this single moment equilibrium condition is insufficient to determine the volume ratio of the longitudinal steel to stirrup steel. In essence, Lessig allowed the weakness in one type of reinforcement to be compensated by the other type. By introducing two moment equilibrium conditions instead of one, Yudin was able to derive two design equations, one for the longitudinal steel and another for the stirrups. Assuming a crack inclination of 45°, these two equations would require an equal volume of both types of steel in the case of pure torsion. It is interesting to note that Yudin's equations and requirements were identical to those of Rausch's theory (Section 3.2.1) which was based on truss analogy with 45° concrete compression struts. The fact that the skew-bending theory and the truss analogy method could produce the same results was most satisfying.

The third equilibrium condition was also used by Goode and Helmy[11] in the analysis of a given beam under combined loading. By using three equilibrium conditions, Lessig's or Yudin's mode 1 failure could be separated into two types of partial-yielding failure. One type of failure is due to the yielding of longitudinal steel only, and the other is due to the yielding of stirrups alone. It is also interesting to note here that the use of three equilibrium conditions would actually satisfy the six general equilibrium conditions shown by Silva et al.[12] Silva et al. also showed that Yudin's theory resulted in an interaction curve composed of straight lines.

Despite the theories of Yudin and Silva et al., tests had consistently shown that yielding of both the longitudinal reinforcement and the stirrups could occur at failure even if the ratio of the two types of steel varied widely. This phenomenon could be explained by Elfgren et al.'s theory[13,14] This theory assumed a variable angle of inclination for the cracks that could be adjusted such that both the longitudinal reinforcement and stirrups could yield at failure. The angle of inclination of the cracks could also be different at the two side faces in mode 1 and mode 3 failures, thus creating a net internal vertical force to balance the external shear. In this way, an additional equilibrium equation of the vertical forces was introduced. This fourth equilibrium equation gave rise to explicit interaction between the bending moment and the shear force. As a result, the most general and complete interaction surface for torsion, bending, and shear was produced. Elfgren et al.'s interaction surface has been compared to tests.[6,14] In general, it can satisfactorily predict the trend of the test data points. This theory, however, is somewhat unconservative when torsion predominates.

It is interesting to examine the above skew-bending theories in terms of the theory of plasticity. For structural analysis, the plasticity theory provides two general approaches, namely, the statics approach and the kinematic approach. The former searches for a stress distribution that is everywhere in equilibrium internally and balances the external load without violating the yield criteria. As a result, it produces a lower-bound solution. The latter approach searches for a deformation mechanism that satisfies the geometric boundary conditions and for which the internal dissipation of energy equals the expenditure of energy due to external load. This approach produces an upper-bound solution.

The kinematic approach can also be divided into two methods. The first method is known as the "equilibrium method." In this method, the conditions of equilibrium are used to find the deformation mechanism that gives the lowest load-carrying capacity. The second method is called the "work method" because the principle of virtual work is used to find the deformation mechanism that gives the lowest load-carrying capacity.

In general, all the skew-bending theories that assume the yield of both longitudinal reinforcement and stirrups belong to the kinematic approach. Elfgren et al.'s theory is the equilibrium method, whereas Lessig's theory and Collins et al.'s theory can be shown to be similar to the work method.[14] Hence, all these three methods should give upper-bound solutions. In contrast, the truss model theories, which will be presented in Chapter 7, belong to the statics approach and theoretically should give a lower-bound solution.

It has been pointed out by Müller[15] that all the above skew-bending theories, which assume a continuous failure surface, actually violate the kinematic condition and the flow rule required by the theory of plasticity. To satisfy these conditions, he has proposed a "discontinuous collapse mechanism." This subject, however, is beyond the scope of this book.

6.2 LESSIG'S THEORY[1]

In testing reinforced concrete rectangular members subjected to torsion in combination with bending and shear, Lessig observed that diagonal cracks occur on all four faces of the beam. These cracks will be joined from one face to the other in a helical manner. When failure is approached due to yielding of steel, a helical crack on three faces will open up and separate the beam into two parts. These two parts of the beam rotate about a neutral axis that joins the two ends of the helical crack near the fourth face.

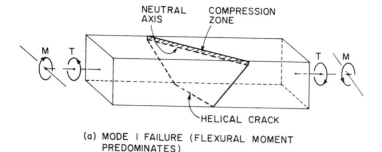

(a) MODE I FAILURE (FLEXURAL MOMENT PREDOMINATES)

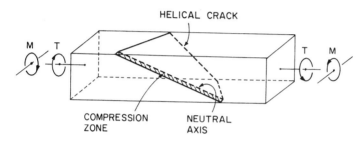

(b) MODE 2 FAILURE (TORSIONAL MOMENT PREDOMINATES)

Fig. 6.1. Lessig's skew-bending failure modes.

When a positive flexural moment predominates, cracks appear first on the bottom face due to flexural tension and then develop into the side faces. A helical crack on these three faces creates a skew failure surface as shown in Fig. 6.1 (a). The neutral axis, which joins the end of the helical cracks near the top of the vertical face, is approximately parallel to the top face. Above the neutral axis is a compression zone inclining at an angle to the longitudinal axis of the beam. Such a failure mode is referred to as a mode 1 failure.

When torsional moment and shear force predominate, cracks appear first at an angle of about 45° to the longitudinal axis of the beam on the vertical face in which the principal stresses due to torsion and shear are additive. These cracks later spread to the bottom and top faces and form a failure surface, as shown in Fig. 6.1 (b). The neutral axis and the compression zone are near the other vertical face, in which the principal stresses due to torsion and shear are subtractive. Such a failure mode is called the mode 2 failure.

Before those two modes of failure are analyzed, we shall first state the assumptions made in the analysis:

1. Both the longitudinal bars and stirrups, which intersect the failure surface, yield at failure of the beam.
2. The tensile strength of concrete is neglected.
3. The spacing of the stirrups is constant within the failure zone.
4. No external loads are present within the failure zone.
5. The shear stresses in the compression zone have no effect on the strength of a member.
6. The dowel forces of reinforcement are neglected.
7. The twisting component of torque on the failure surface has no effect on the ultimate strength.

The last three assumptions were not explicitly stated by Lessig, but they are implied in the derivation.

6.2.1 Mode 1 Failure

Figure 6.2 (a) shows the mode 1 failure surface with the helical crack $BDEA$ inclined on each face at an angle to the longitudinal axis of the beam. This skew failure surface intersects the longitudinal and transverse steel bars. The figure also shows the cross section of the beam and the dimensions of the stirrups. The following symbols have been defined:

$a, a_1, a_2 =$ depth of the equivalent rectangular compression stress block in the compression zone, the subscripts 1 and 2 indicating the smallest and largest depth at the edges, respectively

$A_{l1} =$ area of bottom tension longitudinal steel in mode 1 failure

$A_w =$ area of one leg of a closed stirrup

$b =$ width of a cross section

$c =$ length of the longitudinal projection of the compression zone in mode 1 failure

$d =$ effective depth of cross section in mode 1 failure, measured from the top extreme compression fiber to the centroid of bottom tension reinforcement

$d' =$ distance from the extreme compression fiber to the centroid of compression reinforcement

$f_c'' =$ average stress of the compression stress block ($f_c'' = 0.85\ f_c'$)

$f_{ly} =$ yield stress of longitudinal steel

$f_{sy} =$ yield stress of stirrups

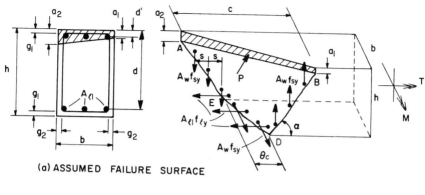

(a) ASSUMED FAILURE SURFACE

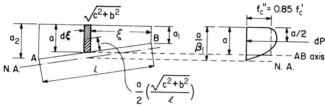

(b) COMPRESSION ZONE

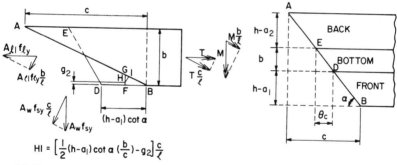

$$HI = \left[\frac{1}{2}(h-a_1)\cot a\left(\frac{b}{c}\right)-g_2\right]\frac{c}{\ell}$$

(c) TOP VIEW OF FAILURE SURFACE AND
PROJECTIONS OF FORCES AND MOMENTS

(d) UNFOLDING THREE
FACES INTO PLANE

Fig. 6.2. Mode 1 failure.

$g_1, g_2 =$ concrete covers in vertical and horizontal directions, respectively, measured from the surface to the center line of the nearest stirrup bar

$h =$ height of a cross section

$\ell =$ length of line AB in mode 1 failure

$M =$ bending moment

$r = A_w f_{sy} b / A_{11} f_{1y} s$, a measure of the ratio of stirrup force to longitudinal steel force

$s =$ spacing of stirrups

$T =$ torque

$\alpha =$ angle of the helical crack $BDEA$ with respect to the longitudinal axis of the beam (mode 1 failure)

$\theta =$ ratio of the longitudinal projection of the bottom crack DE to the longitudinal projection of the compression zone in mode 1 failure ($\theta \approx b/(2h + b)$)

$\phi = T/M$, ratio of torque to bending moment

$\Delta = d - a/2$

$\overline{\Delta} = [\theta(h - g_1 - a/2) + (1 - \theta)(b/4)(1 - \theta - 4g_2/b)]$

Using the defined symbols, the force for each stirrup is $A_w f_{sy}$, and the total force in the bottom tensile longitudinal steel according to mode 1 failure is $A_{11} f_{1y}$. Figure 6.2 (a) also shows that the compression zone has a trapezoidal shape, with a resultant compression force P acting at its centroid. In analyzing this mode 1 failure, two equilibrium conditions are considered by Lessig. They are: (a) the equilibrium of moment about line AB, where line AB is the bottom edge of the equivalent rectangular compression stress block, and (b) the equilibrium of forces along an axis perpendicular to the compression zone.

6.2.1.1 Equilibrium of Moment about Line *AB*.

The internal moment resistance is composed of moments from four sources: (1) compression force in the compression zone, (2) tension in the bottom longitudinal bars, (3) tension in the horizontal branches of the stirrups, and (4) tension in the vertical branches of the stirrups.

1. Moment about Line AB due to Forces in the Compression Zone (M_c). Figure 6.2 (b) shows the compression zone and a typical compression stress block for a typical shaded area $a(d\xi)$. The actual distribution of the compression stress is indicated by the solid curve, while the equivalent rectangular stress block is shown by the dotted line. The equivalent rectangular stress block has a stress of $f_c'' = 0.85 f_c'$ and a depth of a. The depth from the compression surface to the neutral axis becomes a/β_1, where $\beta_1 = 0.85$ for 4,000 psi concrete and should be reduced by 0.05 for every 1,000 psi beyond 4,000 psi.

The moment of the shaded area $a(d\xi)$ in the compression zone about line AB is:

$$dM_c = a(d\xi)f_c'' \left(\frac{a}{2} \frac{\sqrt{c^2 + b^2}}{l} \right)$$

$$M_c = \frac{f_c'' \sqrt{c^2 + b^2}}{2l} \int_0^{\sqrt{c^2 + b^2}} a^2 d\xi \qquad (6\text{-}1)$$

But:

$$a = a_1 + \frac{a_2 - a_1}{\sqrt{c^2 + b^2}} \xi \qquad (6\text{-}2)$$

Substituting Eq. (6–2) into Eq. (6–1) and integrating give:

$$M_c = \frac{f_c'' (c^2 + b^2)}{6l} (a_1^2 + a_1 a_2 + a_2^2) \qquad (6\text{-}3)$$

2. Moment about Line AB due to Bottom Longitudinal Bars (M_l). Referring to Fig. 6.2 (c), the projection of the force of the bottom longitudinal bars on the normal to the compression zone is $A_{l1}f_{ly}(b/l)$. Multiplying this force by the lever arm $d - (a_1 + a_2)/2$ gives the moment about line AB:

$$M_l = A_{l1}f_{ly} \left(\frac{b}{l} \right) \left(d - \frac{a_1 + a_2}{2} \right) \qquad (6\text{-}4)$$

3. Moment about Line AB due to Horizontal Branches of Stirrups (M_h). Figure 6.2 (c) also shows the force $A_w f_{sy}$ in a horizontal branch of the stirrups. The projection of this force on the normal to the compression zone is $A_w f_{sy}(c/l)$. The number of horizontal stirrups intersected by the failure surface is taken as $(\theta c/s)$. Since the lever arm for the force is $h - g_1 - (a_1 + a_2)/2$, the moment about line AB due to the horizontal branches of stirrups is:

$$M_h = \left(\frac{\theta c}{s} \right) A_w f_{sy} \left(\frac{c}{l} \right) \left(h - g_1 - \frac{a_1 + a_2}{2} \right)$$

$$= \frac{A_w f_{sy}}{s} \frac{c^2}{l} \theta \left(h - g_1 - \frac{a_1 + a_2}{2} \right) \qquad (6\text{-}5)$$

4. Moment about line AB due to Vertical Branches of Stirrups (M_v). Referring to Figs. 6.2 (a) and (c), the horizontal projection of the crack BD is $(h - a_1)$ cot α. The centroid of the vertical branch of the stirrups along crack BD is designated as point H in Fig. 6.2 (c). Point H is assumed to be located at a distance g_2 from the midpoint F of the crack BD. The distance HI from the centroid of the stirrups to the line AB is:

$$HI = HG\left(\frac{c}{l}\right) = (FG - g_2)\left(\frac{c}{l}\right) = \left(\frac{1}{2}BD\frac{b}{c} - g_2\right)\left(\frac{c}{l}\right)$$

$$= \left[\frac{1}{2}(h - a_1)\cot\alpha\left(\frac{b}{c}\right) - g_2\right]\left(\frac{c}{l}\right) \tag{6-6}$$

The moment about line AB due to both layers of vertical stirrups near the BD and AE cracks is:

$$M_v = \frac{(h - a_1)\cot\alpha}{s}A_{w}f_{sy}\left[\frac{1}{2}(h - a_1)\cot\alpha\left(\frac{b}{c}\right) - g_2\right]\frac{c}{l}$$

$$+ \frac{(h - a_2)\cot\alpha}{s}A_{w}f_{sy}\left[\frac{1}{2}(h - a_2)\cot\alpha\left(\frac{b}{c}\right) - g_2\right]\frac{c}{l} \tag{6-7}$$

Cot α in Eq. (6–7) can be related to the longitudinal length c if we look at the geometric relationship in Fig. 6.2 (d). In this figure the two vertical faces are unfolded onto the horizontal plane, and the helical crack $BDEA$ becomes a straight line. Then, cot α can be expressed approximately as:

$$\cot\alpha = \frac{(1 - \theta)c}{(2h - a_1 - a_2)} \tag{6-8}$$

Inserting cot α into Eq. (6–7) and simplifying result in:

$$M_v = \frac{A_{w}f_{sy}}{s}\frac{c^2}{l}(1 - \theta)\left[\frac{b}{2}(1 - \theta)\frac{(h - a_1)^2 + (h - a_2)^2}{(2h - a_1 - a_2)^2} - g_2\right] \tag{6-9}$$

Now, let us turn to the external moments M and T in Fig. 6.2 (a) and (c). The projections of these two moment vectors along the line AB are $M(b/l)$ and $T(c/l)$. Taking the equilibrium of external and internal moments about the line AB and utilizing Eqs. (6–3), (6–4), (6–5), and (6–9), we have:

$$M\frac{b}{l} + T\frac{c}{l} = M_c + M_l + M_h + M_v$$

$$Mb + Tc = \frac{f_c''}{6}(c^2 + b^2)(a_1^2 + a_1a_2 + a_2^2) + A_{11}f_{ly}b\left(d - \frac{a_1 + a_2}{2}\right)$$

$$+ \frac{A_wf_{sy}}{s}c^2\theta\left(h - g_1 - \frac{a_1 + a_2}{2}\right) \qquad (6\text{-}10)$$

$$+ \frac{A_wf_{sy}}{s}c^2(1 - \theta)\left[\frac{b}{2}(1 - \theta)\frac{(h - a_1)^2 + (h - a_2)^2}{(2h - a_1 - a_2)^2} - g_2\right]$$

Notice in Eq. (6–10) that the moment resistance is a function of c, θ, a_1, and a_2.

Let us consider the depth of the compression zone a_1 and a_2. They should be chosen such that the moment resistance becomes a minimum. Differentiating Eq. (6–10) with respect to a_1 and a_2 separately and equating them to zero, we obtain two equations. Since a_1 and a_2 are symmetrical in Eq. (6–10), the two equations after differentiation are identical, except that a_1 and a_2 have changed places. For these two equations to be equal, it is necessary that $a_1 = a_2$. In other words, $a_1 = a_2$ is a necessary condition to make the moment resistance of Eq. (6–10) a minimum.

Substituting $a_1 = a_2 = a$ into Eq. (6–10) gives:

$$Mb + Tc = \underbrace{\frac{f_c''}{2}(c^2 + b^2)a^2}_{\text{Concrete}} + \underbrace{A_{11}f_{ly}b(d - a)}_{\text{Longit. steel}}$$

$$+ \frac{A_wf_{sy}}{s}c^2\left[\underbrace{\theta(h - g_1 - a)}_{\substack{\text{Horiz.}\\\text{stirrups}}} + \underbrace{(1 - \theta)\frac{b}{4}\left(1 - \theta - 4\frac{g_2}{b}\right)}_{\substack{\text{Vert.}\\\text{stirrups}}}\right] \qquad (6\text{-}11)$$

Equation (6–11) shows clearly that the moment resistance in mode 1 failure stems from four sources, namely, the concrete in the compression zone, the bottom tension longitudinal steel, the bottom horizontal branches of stirrups, and the vertical branches of stirrups near both vertical faces. The first term contributed by concrete can be eliminated from Eq. (6–11) by considering the equilibrium of forces along the normal to the compression zone.

6.2.1.2 Equilibrium of Forces along the Normal to the Compression Zone.

As shown in Fig. 6.2 (a), there are three forces that have projections along the normal to the compression zone, namely, the concrete compression force P, the bottom longitudinal steel force $A_{11}f_{1y}$, and the horizontal branches of stirrups $A_{w}f_{sy}(\theta c/s)$. Summing the projections of these three forces and equating them to zero, we have:

$$f_c''(c^2 + b^2)a - A_{11}f_{1y}b - \frac{A_{w}f_{sy}}{s}\theta c^2 = 0 \qquad (6\text{--}12)$$

Solving for a gives:

$$a = \frac{A_{11}f_{1y}b\left(1 + r\theta\dfrac{c^2}{b^2}\right)}{f_c''(c^2 + b^2)} \qquad (6\text{--}13)$$

where:

$$r = \frac{A_{w}f_{sy}b}{A_{11}f_{1y}s} \qquad (6\text{--}14)$$

The quantity r is a measure of the ratio of stirrup force to longitudinal steel force in mode 1 failure.*

Multiplying Eq. (6–12) by $a/2$ and subtracting it from Eq. (6–11):

$$Mb + Tc = \underbrace{A_{11}f_{1y}b\left(d - \frac{a}{2}\right)}_{\substack{\text{Longit.}\\\text{steel}}}$$

$$+ \frac{A_{w}f_{sy}}{s}c^2\left[\underbrace{\theta\left(h - g_1 - \frac{a}{2}\right)}_{\substack{\text{Horiz.}\\\text{stirrups}}} + \underbrace{(1 - \theta)\frac{b}{4}\left(1 - \theta - 4\frac{g_2}{b}\right)}_{\substack{\text{Vert.}\\\text{stirrups}}}\right] \qquad (6\text{--}15)$$

---

* In Lessig's paper[1] a parameter $p = A_{w}f_{sy}h/A_{11}f_{1y}s$ was used, and the resulting expressions of her equations were slightly different.

Let:

$$\phi = \frac{T}{M} \tag{6-16}$$

ϕ is the ratio of torsional to bending moments. Equation (6–15) can be written as:

$$
\begin{aligned}
M\left(1 + \frac{c}{b}\phi\right) = T&\left(\frac{1}{\phi} + \frac{c}{b}\right) \\
= A_{11}f_{1y}&\left\{\left(d - \frac{a}{2}\right) + r\left(\frac{c}{b}\right)^2\left[\theta\left(h - g_1 - \frac{a}{2}\right)\right.\right. \\
&\left.\left. + (1 - \theta)\frac{b}{4}\left(1 - \theta - 4\frac{g_2}{b}\right)\right]\right\}
\end{aligned}
\tag{6-17}
$$

Let us now examine θ in Eq. (6–17). It is assumed that θ is a constant for a given cross section. If the depth of the compression zone, a, is neglected in Fig. 6.2 (d), geometry shows that θ can be expressed as:

$$\theta = \frac{\theta c}{c} = \frac{b}{2h + b} \tag{6-18}$$

In Eq. (6–17) we can define:

$$\Delta_1 = d - \frac{a}{2} \tag{6-19}$$

$$\bar{\Delta}_1 = \left[\theta\left(h - g_1 - \frac{a}{2}\right) + (1 - \theta)\frac{b}{4}\left(1 - \theta - 4\frac{g_2}{b}\right)\right] \tag{6-20}$$

Δ_1 is the lever arm of the bottom longitudinal bars, while $\bar{\Delta}_1$ is a cross-sectional parameter involving the lever arms of the horizontal and vertical branches of stirrups. Neither Δ_1 nor $\bar{\Delta}_1$ is a function of the length c. Therefore, Eq. (6–17) can be written as:

$$T = A_{11}f_{1y}\frac{\Delta_1 + r\bar{\Delta}_1\left(\frac{c}{b}\right)^2}{\frac{1}{\phi} + \frac{c}{b}} \tag{6-21}$$

6.2.1.3 Minimum Torsional Resistance. To find the value of c that gives the torque T a minimum, we differentiate Eq. (6–21) and equate it to zero:

$$\left(\frac{1}{\phi}+\frac{c}{b}\right)\frac{r\overline{\Delta}_1}{b^2}2c - \left(\Delta_1 + \frac{r\overline{\Delta}_1}{b^2}c^2\right)\frac{1}{b} = 0 \qquad (6\text{--}22)$$

Solving for c in Eq. (6–22) gives:

$$\frac{c}{b} = -\frac{1}{\phi} + \sqrt{\frac{1}{\phi^2} + \frac{1}{r}\frac{\Delta_1}{\overline{\Delta}_1}} \qquad (6\text{--}23)$$

Equation (6–23) shows that the c value, which corresponds to a minimum torque, is a function of the moment ratio ϕ, the reinforcement ratio r, and the cross-sectional properties, $\Delta_1/\overline{\Delta}_1$. The longitudinal length, c, however, should be limited to a maximum value of:

$$c_{\max} = 2h + b \qquad (6\text{--}24)$$

This is so because the inclinations of cracks have never been found by tests to be less than 45° even for the case of pure torsion. Equation (6–24) can be understood from Fig. 6.2 (d) if $\alpha = 45°$ and the depth of the compression zone, a, is neglected.

The trial-and-error procedures to find the minimum torsional resistance due to mode 1 failure can be summarized as follows:

1. The cross-sectional properties (b, h, A_{l1}, A_w, s, g_1, g_2, r) and moment ratio $\phi = T/M$ are given.
2. Calculate θ from Eq. (6–18).
3. Assume the depth of the compression zone a and calculate the cross-sectional parameters Δ_1 and $\overline{\Delta}_1$ from Eqs. (6–19) and (6–20).
4. Calculate the longitudinal length c from Eq. (6–23) or Eq. (6–24), whichever is smaller.
5. Calculate a from Eq. (6–13). If a is not the same as that assumed in step (3), repeat steps (3), (4), and (5) until a satisfactory accuracy is reached.
6. Calculate the minimum torsional resistance T from Eq. (6–21).

In step (3) an approximate value of a should be assumed. Lessig recommended the following values:

$$\text{For } \phi \leq 0.1, \qquad a = \frac{A_{11}f_{1y}}{f_c'' b} \qquad (6\text{-}25)$$

$$\text{For } \phi > 0.1, \qquad a = 0 \qquad (6\text{-}26)$$

The expression of a in Eq. (6-25) is identical to that used in pure flexural analysis. This would be the upper limit of a when $\alpha = 90°$. Consequently, Eq. (6-25) should be a good approximation when torsion is small. In contrast, when ϕ approaches infinity, the value of a should become very small. Hence, $a = 0$ should be a good approximation when torsion is large. Since the effect of the value a on the ultimate strength of the beam is small, the approximate value of a given by Eqs. (6-25) and (6-26) will usually produce sufficiently accurate results without a second cycle of trial and error.

6.2.2 Mode 2 Failure

Figure 6.3 (a) shows the skew failure surface formed by the helical crack in mode 2 failure. The following additional symbols have been defined:

A_{12} = area of the tensile longitudinal steel near a vertical face in mode 2 failure

c_2 = length of the longitudinal projection of the compression zone in mode 2 failure

d_2 = effective depth in mode 2 failure, measured from a side face to the centroid of tension reinforcement on the opposite side face

l_2 = the length of the line AB in mode 2 failure

$r_2 = A_{w}f_{sy}h/A_{12}f_{1y}s$, a measure of the ratio of stirrup force to longitudinal steel force in mode 2 failure

V = shear force

α_2 = angle of the helical crack with respect to the longitudinal axis of the beam (mode 2 failure)

$\delta = Vb/2T$

θ_2 = ratio of the longitudinal projection of the vertical crack DE to the longitudinal projection of the compression zone in mode 2 failure $[\theta_2 \approx h/(2b + h)]$

$\Delta_2 = d_2 - a/2$

$\overline{\Delta}_2 = [\theta_2(b - g_2 - a/2) + (1 - \theta_2)(h/4)(1 - \theta_2 - 4g_1/h)]$

Similarly to mode 1 failure, two equilibrium conditions are considered.

6.2.2.1 Equilibrium of Moments about Line AB. The external moment about line AB is contributed by the external torque T and the shear

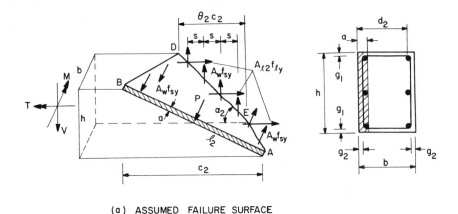

(a) ASSUMED FAILURE SURFACE

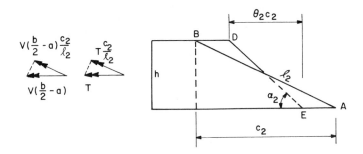

(b) SIDE VIEW OF FAILURE SURFACE AND PROJECTIONS
OF EXTERNAL MOMENTS ALONG AB

Fig. 6.3. Mode 2 failure.

force V. The flexural moment M does not enter into consideration because the moment vector of M is perpendicular to line AB. The projection of T and $V[(b/2) - a]$ along AB is shown in Fig. 6.3 (b), giving the external moment:

$$T\frac{c_2}{\ell_2} + V\left(\frac{b}{2} - a\right)\frac{c_2}{\ell_2}$$

On the other hand, the internal moment resistance can be obtained from Eq. (6–11) for mode 1 failure by interchanging b and h, g_1 and g_2, and by replacing A_{l1}, d, c, and θ by A_{l2}, d_2, c_2, and θ_2, respectively. Equating the external and internal moments results in:

$$Tc_2 + V\left(\frac{b}{2} - a\right)c_2 = \frac{f_c''}{2}(c_2^2 + h^2)a^2 + A_{12}f_{1y}h(d_2 - a)$$

$$+ \frac{A_w f_{sy}}{s}c_2^2\left[\theta_2(b - g_2 - a)\right. \tag{6-27}$$

$$\left. + (1 - \theta_2)\frac{h}{4}\left(1 - \theta_2 - 4\frac{g_1}{h}\right)\right]$$

6.2.2.2 Equilibrium of Forces along the Normal to the Compression Zone.

In addition to the three internal forces, P, $A_{12}f_{1y}$, and $A_w f_{sy}(\theta c_2/s)$, the external shear force V also has a projection along the normal to the compression zone. Summing the projections of these four forces and equating them to zero, we have:

$$f_c''(c_2^2 + h^2)a - A_{12}f_{1y}h - \frac{A_w f_{sy}}{s}\theta_2 c_2^2 + Vc_2 = 0 \tag{6-28}$$

Solving for a gives:

$$a = \frac{A_{12}f_{1y}h\left(1 + r_2\theta_2\frac{c_2^2}{h^2}\right) - Vc_2}{f_c''(c_2^2 + h^2)} \tag{6-29}$$

where $r_2 = \dfrac{A_w f_{sy}h}{A_{12}f_{1y}s}$. $\tag{6-30}$

Multiplying Eq. (6–28) by $a/2$ and subtracting it from Eq. (6–27) give:

$$Tc_2 + V\left(\frac{b}{2} - \frac{a}{2}\right)c_2 = A_{12}f_{1y}h\left(d_2 - \frac{a}{2}\right)$$

$$+ \frac{A_w f_{sy}}{s}c_2^2\left[\theta_2\left(b - g_2 - \frac{a}{2}\right) + (1 - \theta_2)\frac{h}{4}\left(1 - \theta_2 - 4\frac{g_1}{h}\right)\right] \tag{6-31}$$

Since a is much smaller than b, the term $(Va/2)c_2$ on the left-hand side can be neglected. This simplification should be conservative. Let us also denote:

$$\delta = \frac{Vb}{2T} \tag{6-32}$$

Equation (6–31) becomes:

$$T(1 + \delta)\frac{c_2}{h} = V\left(1 + \frac{1}{\delta}\right)\frac{bc_2}{2h}$$

$$= A_{12}f_{1y}\left\{\left(d_2 - \frac{a}{2}\right) + r_2\left(\frac{c_2}{h}\right)^2\left[\theta_2\left(b - g_2 - \frac{a}{2}\right)\right.\right.$$

$$\left.\left. + (1 - \theta_2)\frac{h}{4}\left(1 - \theta_2 - \frac{4g_1}{h}\right)\right]\right\} \tag{6–33}$$

Let us also assume:

$$\theta_2 = \frac{h}{2b + h} \tag{6–34}$$

and denote:

$$\Delta_2 = d_2 - \frac{a}{2} \tag{6–35}$$

$$\overline{\Delta}_2 = \left[\theta_2\left(b - g_2 - \frac{a}{2}\right) + (1 - \theta_2)\frac{h}{4}\left(1 - \theta_2 - \frac{4g_1}{h}\right)\right] \tag{6–36}$$

Δ_2 and $\overline{\Delta}_2$ are cross-sectional parameters and are not functions of c_2. Using Δ_2 and $\overline{\Delta}_2$, Eq. (6–33) can be expressed as:

$$T = A_{12}f_{1y}\frac{\Delta_2 + r_2\overline{\Delta}_2\left(\dfrac{c_2}{h}\right)^2}{(1 + \delta)\left(\dfrac{c_2}{h}\right)} \tag{6–37}$$

6.2.2.3 Minimum Torsional Resistance. The value of c_2 that corresponds to a minimum torsional resistance can be obtained by differentiating Eq. (6–37) with respect to c_2 and equating it to zero:

$$c_2\left(r_2\overline{\Delta}_2\frac{2c_2}{h^2}\right) - \left(\Delta_2 + r_2\overline{\Delta}_2\frac{c_2^2}{h^2}\right) = 0 \tag{6–38}$$

Solving c_2 from Eq. (6–38) we arrive at:

$$\frac{c_2}{h} = \sqrt{\frac{1}{r_2}\left(\frac{\Delta_2}{\overline{\Delta}_2}\right)} \tag{6–39}$$

Equation (6–39) shows that the c_2 value, which corresponds to a minimum torsional resistance, is a function of the reinforcement ratio r_2 and the cross-sectional properties $\Delta_2/\overline{\Delta}_2$. It is not a function of the shear-to-torque ratio, δ. The value of c_2 should be limited to:

$$c_{2,\max} = 2b + h \qquad (6\text{–}40)$$

The trial-and-error procedures to find a minimum torsional resistance in mode 2 failure can be summarized:

1. The cross-sectional properties (b, h, A_{12}, A_w, s, g_1, g_2, and r_2) and shear-to-torque ratio $\delta = Vb/2T$ are given.
2. Calculate θ_2 from Eq. (6–34).
3. Assume the depth of the compression zone a and calculate the cross-sectional parameters Δ_2 and $\overline{\Delta}_2$ from Eqs. (6–35) and (6–36).
4. Calculate the longitudinal length c_2 from Eq. (6–39) or (6–40), whichever is less.
5. Calculate a from Eq. (6–29). If a is not the same as that assumed in step (3), repeat steps (3), (4), and (5) until a satisfactory accuracy is reached.
6. Calculate the minimum torsional resistance T from Eq. (6–37).

Since the angle α is always close to 45° in mode 2 failure, it would be simple to assume $c_2 = c_{2,\max}$ in step (4) and to use Eq. (6–40) directly. Substituting c_2 and θ_2 into Eq. (6–29), a can be calculated in a straightforward manner without the trial-and-error process.

When the moment, shear, and torque are all large, it would be difficult to determine the failure modes of the beam. In such cases, both the mode 1 failure and the mode 2 failure should be considered, and the lesser strength will govern.

6.2.3   Russian Code of 1962[2,3]

The equations derived in the above two sections for mode 1 and mode 2 failures were simplified and adopted in the Russian Code of 1962.* Two approximations were made:

1. Neglect the vertical branches of stirrups in mode 1 failure and the horizontal branches in mode 2 failure.

---

* The basic equations for mode 1 and mode 2 failures continue in the current 1976 Russian Code, but the new code has included the mode 3 failure and the simplifications for the depth of the compression zone used by Collins et al. in Section 6.3.

2. The center lines of the stirrups coincide with the center line of the longitudinal bars; i.e., $h - g_1 = d$, and $b - g_2 = d_2$.

As a result of these two assumptions the internal forces in the mode 1 and mode 2 failures are as shown in Fig. 6.4.

6.2.3.1 Mode 1 Failure. Using these two approximations, Eqs. (6–19) and (6–20) can be simplified as follows:

$$\Delta_1 = d - \frac{a}{2} \tag{6–41}$$

$$\overline{\Delta}_1 = \theta \left(d - \frac{a}{2} \right) \tag{6–42}$$

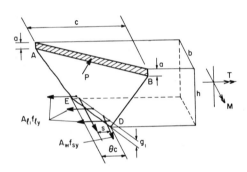

(a) MODE I FAILURE

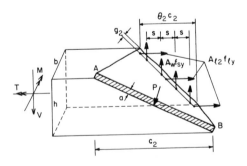

(b) MODE 2 FAILURE

Fig. 6.4. Simplified internal forces in failure surface of Russian Code.

Substituting these values of Δ_1 and $\overline{\Delta}_1$, plus θ from Eq. (6–18), into Eqs. (6–23), (6–13), and (6–21) gives:

$$\frac{c}{b} = -\frac{1}{\phi} + \sqrt{\frac{1}{\phi^2} + \left(\frac{1 + 2h/b}{r}\right)} \leq 1 + 2\frac{h}{b} \tag{6–43}$$

$$a = \frac{A_{11}f_{1y}\left[1 + \left(\frac{r}{1 + 2h/b}\right)\left(\frac{c}{b}\right)^2\right]}{f_c'' b\left[1 + \left(\frac{c}{b}\right)^2\right]} \tag{6–44}$$

$$T = A_{11}f_{1y}\left(d - \frac{a}{2}\right)\frac{\left[1 + \left(\frac{r}{1 + 2h/b}\right)\left(\frac{c}{b}\right)^2\right]}{\frac{1}{\phi} + \frac{c}{b}} \tag{6–45}$$

It should be emphasized that c in Eq. (6–43) is no longer a function of a. Consequently, a can be found from Eq. (6–44) without the tedious trial-and-error processes. The procedures for analysis are simply:

1. Given the cross section properties (h, b, and r) and torque-to-bending moment ratio, ϕ, calculate the ratio c/b from Eq. (6–43).
2. Substitute c/b into Eq. (6–44) to obtain the depth of compression zone a.
3. Substitute c/b and a into Eq. (6–45) to obtain the torque T.

6.2.3.2 Mode 2 Failure. Using the two stated approximations, Eqs. (6–35) and (6–36) become:

$$\Delta_2 = d_2 - \frac{a}{2} \tag{6–46}$$

$$\overline{\Delta}_2 = \theta_2\left(d_2 - \frac{a}{2}\right) \tag{6–47}$$

Substituting these values of Δ_2 and $\overline{\Delta}_2$ plus θ_2 from Eq. (6–34) into Eqs. (6–39), (6–29), and (6–37) gives:

$$\frac{c_2}{h} = \sqrt{\frac{1 + 2b/h}{r_2}} \leq 1 + 2\frac{b}{h} \tag{6–48}$$

$$a = \frac{A_{12}f_{1y}\left[1+\left(\dfrac{r_2}{1+2b/h}\right)\left(\dfrac{c_2}{h}\right)^2\right]-V\dfrac{c_2}{h}}{f_c''h\left[1+\left(\dfrac{c_2}{h}\right)^2\right]} \tag{6–49}$$

$$T = A_{12}f_{1y}\left(d_2-\frac{a}{2}\right)\frac{\left[1+\left(\dfrac{r_2}{1+2b/h}\right)\left(\dfrac{c_2}{h}\right)^2\right]}{(1+\delta)\left(\dfrac{c_2}{h}\right)} \tag{6–50}$$

Again, c_2 in Eq. (6–48) is not a function of a, so that the trial-and-error process is eliminated.

The six Eqs. (6–43) through (6–45) and (6–48) through (6–50) appear in the 1962 Russian Code in somewhat different expressions. This old version of the Russian Code did not utilize the parameters $r = A_w f_{sy} b/A_{11} f_{1y} s$ and $r_2 = A_w f_{sy} h/A_{12} f_{1y} s$. The parameters r and r_2 were employed later in the 1976 Russian Code. The use of these parameters will also facilitate further simplification in Section 6.3.

6.2.3.3 Empirical Limitations. Four empirical limits were adopted by the Russian Code and were explained in Ref. 2.

1. Limits to Prevent Crushing of the Concrete Compression Zone before the Yielding of Reinforcement. Two provisions were given for this purpose, one for mode 1 failure under a small torque-to-bending moment ratio, $\phi \leq 0.2$. Another is for mode 2 failure.

Mode 1 Failure ($\phi \leq 0.2$). It can be seen from Eq. (6–43) that the longitudinal projection of the compression zone c increases gently with the increase of the torque-to-bending moment ratio ϕ. This relationship can be replaced by a simpler equation suggested by Lessig:

$$\frac{c}{b} = \sqrt{5\phi} \tag{6–51}$$

In the boundary case of pure bending when $\phi = 0$, Eq. (6–51) gives $c = 0$, which is correct. Substituting c from Eq. (6–51) into Eq. (6–44) gives:

$$a = \frac{A_{11}f_{1y}}{f_c''b(1+5\phi)}\left[1+5\phi\left(\frac{r}{1+2h/b}\right)\right] \tag{6–52}$$

When $\phi \leq 0.2$, the term $5\phi r/(1 + 2h/b)$ is much less than unity and therefore can be neglected:

$$a = \frac{A_{11}f_{1y}}{f_c'' b(1 + 5\phi)} \tag{6-53}$$

If $a > 2d'$, compression steel also needs to be considered. Then:

$$a = \frac{A_{11}f_{1y} - A_{13}f_{1y}}{f_c'' b(1 + 5\phi)} \geq 2d' \tag{6-53a}$$

where A_{13} is the area of longitudinal steel near the top face. The depth of the compression zone a calculated by Eq. (6–53) or (6–53a) should be limited to a maximum of:

$$a_{max} = (0.55 - 0.7\sqrt{\phi})\, d \tag{6-54}$$

Notice that when ϕ approaches zero, a_{max} approaches 0.55, which is the value given by the Russian Code for the case of pure bending. Equation (6–54) has been substantiated by tests in Fig. 6.5 for ϕ values up to 0.2.

Mode 2 Failure. To prevent concrete failure in the compression zone of mode 2 failure, Lessig suggested that the torque should be limited to a maximum value of:

$$T_{max} = 0.07 f_c'' x^2 y \tag{6-55}$$

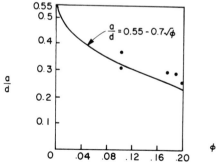

Fig. 6.5. Determination of depth of compression zone by tests.

where x and y are the smaller and larger dimensions, respectively, of a rectangular cross section. The parameters x^2y and f_c'' are similar to those in Eq. (4–43) of the ACI Code, except that f_c'' does not have a square root. This limit also does not take into account the interaction of shear and torsion.

Equations (6–54) and (6–55) are used to check the size of a cross section subjected to torsion in combination with bending moment and shear forces.

2. *Ratios of Longitudinal Force to Stirrup Force to Ensure the Yielding of Both Types of Steel.* For the reinforcement along the width of a cross section (flexural tension side) in mode 1 failure:

$$0.5 \leq r \left(1 + \frac{2}{\phi} \sqrt{\frac{b}{2h + b}}\right) \leq 1.5 \qquad (6\text{–}56)$$

For the reinforcement along the height of a cross section in mode 2 failure:

$$0.5 \leq r_2 \leq 1.5 \qquad (6\text{–}57)$$

Equation (6–57) is similar to the author's Eq. (3–41a) in Chapter 3.

3. *Shear Failure.* In the Russian Code the shear strength without torsion, V_0, is given by the empirical expression:

$$V_0 = \sqrt{0.6 f_c'' b d^2 \left(\frac{A_w f_{sy}}{s}\right)} - A_w f_{sy} \qquad (6\text{–}58)$$

Shear failure (instead of skew bending in mode 2) often governs when shear is large and torsion is small. To prevent shear failure in combined shear and torsion a simple empirical rule was given:

$$V = \frac{V_0}{1 + \dfrac{1.5}{\delta}} \qquad (6\text{–}59)$$

where

$V =$ Shear strength in combined shear and torsion
$V_0 =$ shear strength without torsion, Eq. (6–58)
$\delta = Vb/2T$

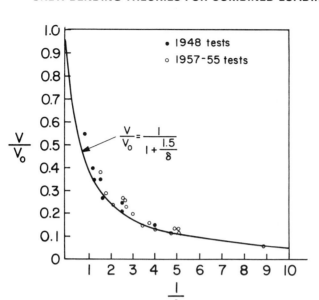

Fig. 6.6. Relationship between V/V_0 and $1/\delta$.

Equation (6–59) is plotted in Fig. 6.6 with the supporting test data points. When torque vanishes ($\delta = \infty$), Eq. (6–59) becomes identical to Eq. (6–58).

4. Reinforcing Details. Finally, it is interesting to point out two details regarding stirrups in the Russian Code. First, when $\phi \leq 0.2$, it is permissible to omit the transverse reinforcement near the side of width b (perpendicular to the bending plane) compressed by bending, if the following condition is satisfied:

$$T \leq \left(\frac{1}{2} - \frac{1}{6}\frac{b}{h}\right) b^2 h f'_t \qquad (6\text{–}60)$$

where f'_t is the tensile strength of concrete. Equation (6–60) is the cracking torque based on plastic theory (see Eq. 2–7). In essence, the Russian Code permits the use of open stirrups, if the torque does not produce cracks. Second, the maximum stirrup spacing depends on the shear force in the Russian Code. If shear and torsion are small, this code allows stirrup spacing up to x_1 on the shorter sides and up to y_1 on the larger sides, where x_1

and y_1 are the shorter and larger dimensions of stirrups, measured from the center lines.

6.2.4 Example 6.1

The structure and loadings used in Design Example 4.4 will now be used to illustrate the application of the 1962 Russian Code. The design process for this structure according to the ACI Code has been given in Section 4.5 and in Figs. 4.24 and 4.25. There are two advantages in employing this example. First, since this member has been designed by the ACI Code, a re-analysis of this member by the Russian Code will furnish a comparison between the two codes. Second, the Russian Code provides a method of analysis to check a given reinforced concrete member for safety, not a method to design the size and the reinforcement of a member. Therefore, the well-designed member in Example 4.4 (Figs. 4.24 and 4.25) provides a good starting point.

Some difficulties arise in applying the Russian Code. To begin with, the concrete strength grade is defined quite differently in the Russian Code. Instead of using the 6 in. by 12 in. cylinders in compression tests, the Russian Code specifies 20-cm cubes. The conversion factor from cube strength to cylinder strength is unclear. Next, from a statistical point of view, the definitions of standard compression strength of concrete, f'_c, the standard yield strength of reinforcement, f_y, and the acceptance criteria for these materials are all quite different in the two codes. Consequently, it is a formidable task to evaluate the material reduction factor, ϕ, for the Russian Code in terms of the ACI standard material properties. Furthermore, the minimum design load for a structure is also quite different in the two countries, so that a comparison of the load factors in the two codes becomes quite meaningless as far as the total safety of a structure is concerned. To avoid all these difficulties, it is decided simply to use the ACI load factors and the ACI material reduction factors, ϕ_{ACI}, in connection with the Russian Code. In this way, the total factor of safety will be the same in the two examples, and any difference of calculation will be caused by the natures of the two methods alone.

Let us check the critical section at the support. Figure 4.24 shows that at this section $M_u = 6,062$ in.-kips, $V_u = 71.0$ kips, and $T_u = 1,440$ in.-kips.

$$\phi = \frac{T_u}{M_u} = \frac{1,440}{6,062} = 0.238$$

$$\delta = \frac{V_u b}{2T_u} = \frac{71.0(18)}{2(1,440)} = 0.444$$

$A_w = 0.44$ in. (for No. 6 stirrups)

$A_{11} = 5(1.27) = 6.35$ in.^{2} (for five No. 10 bars)

$A_{13} = 2(1.27) = 2.54$ in.^{2} (for two No. 10 bars)

$$A_{12} = \frac{A_{11} + A_{13}}{2} + 0.60 = \frac{6.35 + 2.54}{2} + 0.60 = 5.05 \text{ in.}^2$$

$$r = \frac{A_w f_{sy} b}{A_{11} f_{1y} s} = \frac{0.44(60)(18)}{6.35(60)(6)} = 0.208$$

$$r_2 = \frac{A_w f_{sy} h}{A_{12} f_{1y} s} = \frac{0.44(60)(28)}{5.05(60)(6)} = 0.407$$

$f_c'' = 0.85 f_c' = 0.85(4,000) = 3,400$ psi

$$d = 28 - 1.5 - 0.75 - \frac{1.27}{2} = 25.12 \text{ in.}$$

$$d_2 = 18 - 1.5 - 0.75 - \frac{1.27}{2} = 15.12 \text{ in.}$$

Mode 1 Failure:

Eq. (6–43):
$$\frac{c}{b} = -\frac{1}{\phi} + \sqrt{\frac{1}{\phi^2} + \left(\frac{1 + 2h/b}{r}\right)}$$

$$= -\frac{1}{0.238} + \sqrt{\frac{1}{0.238^2} + \left(\frac{1 + 2(28/18)}{0.208}\right)} = 1.92$$

$$1 + 2\frac{h}{b} = 1 + 2\frac{28}{18} = 4.11 > 1.92 \qquad \text{O.K.}$$

Eq. (6–44):
$$a = \frac{A_{11} f_{1y}\left[1 + \left(\dfrac{r}{1 + 2h/b}\right)\left(\dfrac{c}{b}\right)^2\right]}{f_c'' b\left[1 + \left(\dfrac{c}{b}\right)^2\right]}$$

$$= \frac{6.35(60)\left[1 + \dfrac{0.208}{1 + 2(28/18)}1.92^2\right]}{3.4(18)[1 + 1.92^2]} = 1.58 \text{ in.}$$

Eq. (6–45): $T_n = A_{11} f_{1y} \left(d - \dfrac{a}{2} \right) \dfrac{\left[1 + \left(\dfrac{r}{1 + 2h/b} \right) \left(\dfrac{c}{b} \right)^2 \right]}{\dfrac{1}{\phi} + \dfrac{c}{b}}$

$$= 6.35(60) \left(25.12 - \frac{1.58}{2} \right) \frac{\left[1 + \dfrac{0.208}{1 + 2(28/18)} \, 1.92^2 \right]}{\dfrac{1}{0.238} + 1.92}$$

$$= 1{,}796 \text{ in.-k}$$

$$\frac{T_u}{\phi_{\text{ACI}}} = \frac{1{,}440}{0.85} = 1{,}694 \text{ in.-k} < 1{,}796 \text{ in.-k} \, (T_n) \qquad \text{O.K.}$$

Since T_u/ϕ_{ACI} is quite close to T_n, the degree of safety in the Russian method is apparently quite close to the ACI method.

Mode 2 Failure:

Eq. (6–48): $\dfrac{c_2}{h} = \sqrt{\dfrac{1 + 2b/h}{r_2}} = \sqrt{\dfrac{1 + 2(18/28)}{0.407}} = 2.37$

$$1 + 2\frac{b}{h} = 1 + 2\frac{18}{28} = 2.29 < 2.37 \qquad \text{Use 2.29}$$

Eq. (6–49): $a = \dfrac{A_{12} f_{12} \left[1 + \left(\dfrac{r_2}{1 + 2b/h} \right) \left(\dfrac{c_2}{h} \right)^2 \right] - V_u \left(\dfrac{c_2}{h} \right)}{f_c'' h \left[1 + \left(\dfrac{c_2}{h} \right)^2 \right]}$

$$= \frac{5.05(60) \left[1 + \dfrac{0.407}{1 + 2(18/28)} \, 2.29^2 \right] - 71.0(2.29)}{3.4(28)(1 + 2.29^2)} = 0.711 \text{ in.}$$

Eq. (6–50): $T_n = A_{12} f_{1y} \left(d_2 - \dfrac{a}{2} \right) \dfrac{\left[1 + \left(\dfrac{r_2}{1 + 2b/h} \right) \left(\dfrac{c_2}{h} \right)^2 \right]}{(1 + \delta) \left(\dfrac{c_2}{h} \right)}$

$$= 5.05(60) \left(15.12 - \frac{0.711}{2} \right) \frac{\left[1 + \dfrac{0.407}{1 + 2(18/28)} \, 2.29^2 \right]}{(1 + 0.444)(2.29)}$$

$$= 2{,}613 \text{ in.-kips.} > 1{,}694 \text{ in.-kips} \left(\frac{T_u}{\phi_{\text{ACI}}} \right) \qquad \text{O.K.}$$

T_n for mode 2 failure is considerably greater than T_u/ϕ_{ACI}. This is so because A_{12} has been calculated from $A_{11}/2 + A_{13}/2 + 0.60$, including one-half of all the flexural tensile steel, $A_{11}/2$. An assumption has been made that all the flexural tensile steel, A_{11}, is equally distributed at the two corners, a condition that helps to prevent mode 2 failure. If only the one No. 10 corner bar has been included in the area A_{12}, T_n will be smaller and closer to the T_u/ϕ_{ACI} value. In the Russian method, there is always a doubt as to the correct way of calculating A_{12}.

Check Compression Failure:
Compression failure in mode 1:

Since $\phi = 0.238 > 0.2$, the 1962 Russian Code does not require the checking of compression failure in mode 1. Nevertheless, we shall do it, as an illustration of the method. To be conservative, the flexural compression steel will be neglected in the calculation of the depth of the compression zone, a.

Eq. (6–53): $a = \dfrac{A_{11}f_{1y}}{f_c'' b(1 + 5\phi)} = \dfrac{6.35(60)}{3.4(18)[1 + 5(0.238)]} = 2.84$ in.

Eq. (6–54): $a_{max} = (0.55 - 0.7\sqrt{\phi})d = (0.55 - 0.7\sqrt{0.238})(25.12)$
$$= 5.24 \text{ in.} > 2.84 \text{ in.} \qquad \text{O.K.}$$

Compression failure in mode 2:

Eq. (6–55) $\quad T_{max} = 0.07f_c'' \, x^2y = 0.07(3.4)(18)^2(28) = 2{,}159$ in.-k

$$\dfrac{T_u}{\phi_{ACI}} = \dfrac{1{,}440}{0.85} = 1{,}694 \text{ in.-k} < 2{,}159 \text{ in.-k} \qquad \text{O.K.}$$

Check Yielding of Both Longitudinal Steel and Stirrups:
On the width of a cross section (flexural tension side):

Eq. (6–56) $\quad r\left(1 + \dfrac{2}{\phi}\sqrt{\dfrac{b}{2h + b}}\right) = 0.208\left(1 + \dfrac{2}{0.238}\sqrt{\dfrac{18}{2(28) + 18}}\right) = 1.07$
$$0.5 < 1.07 < 1.5 \qquad \text{O.K.}$$

On the height of a cross section:

Eq. (6–57): $\qquad\qquad r_2 = 0.407 < 0.5$

Although r_2 is less than the limit of 0.5, this violation is acceptable because r_2 has been calculated from $A_{12} = 5.05$ in.^{2} which includes one-half of the flexural tensile steel, $A_{11}/2$. If only one No. 10 corner bar of the flexural tensile steel is included, then $A_{12} = 2(1.27) + 0.60 = 3.14$ in.^{2} and:

$$r_2 = \frac{A_w f_{sy} h}{A_{12} f_{1y} s} = \frac{0.44(60)(28)}{3.14(60)(6)} = 0.654 > 0.5 \qquad \text{O.K.}$$

Check Shear Failure:

Eq. (6–58): $\quad V_0 = \sqrt{0.6 f_c'' b d^2 \left(\frac{A_w f_{sy}}{s}\right)} - A_w f_{sy}$

$$= \sqrt{0.6(3.4)(18)(25.12)^2 \left(\frac{2(0.44)(60)}{6}\right)} - 2(0.44)(60)$$

$$= 399 \text{ kips}$$

Eq. (6–59): $\qquad V = \dfrac{V_0}{1 + \dfrac{1.5}{\delta}} = \dfrac{399}{1 + \dfrac{1.5}{0.444}} = 91.1 \text{ k}$

$$\frac{V_u}{\phi_{\text{ACI}}} = \frac{71.0}{0.85} = 83.5 \text{ k} < 91.1 \text{ k} \qquad \text{O.K.}$$

Even though the member should not fail in shear, it will be shown later, in Section 6.3.4 and Fig. 6.9 (c), that Eq. (6–59) is excessively conservative. In fact, Eq. (6–59) has disappeared from the 1976 Russian Code.

Check Reinforcement Details:
Assume $f_t' = 5\sqrt{f_c'} = 5\sqrt{4,000} = 316$ psi.

Eq. (6–60): $\quad T = \left(\frac{1}{2} - \frac{1}{6}\frac{b}{h}\right) b^2 h f_t' = \left(\frac{1}{2} - \frac{1}{6}\frac{18}{28}\right)(18)^2(28)(0.316)$

$$= 1,126 \text{ in.-k} < 1,694 \text{ in.-k} \ (T_u/\phi_{\text{ACI}})$$

Closed stirrups must be provided.

6.3 COLLINS ET AL.'S THEORY[4,5]

6.3.1 Mode 1 Failure

In examining the three Eqs. (6–43) through (6–45) in the Russian Code for mode 1 failure, Collins et al. observed that they can be reduced to one simple interaction equation, if the following conservative assumption is made: The depth of the compression zone is assumed to be equal to that under pure bending. That is, Eq. (6–44) is replaced by:

$$a = \frac{A_{11}f_{1y}}{f_c'' b} \qquad (6\text{–}61)$$

so that the expression $A_{11}f_{1y}\left(d - \frac{a}{2}\right)$ in Eq. (6–45) becomes the pure bending strength, M_{01}, in mode 1 failure. We are now left with only the two Eqs. (6–43) and (6–45).

Substituting Eq. (6–43) into Eq. (6–45) and denoting:

$$\chi = \frac{r}{1 + 2h/b} \qquad (6\text{–}62)$$

we have:

$$T = M_{01} \frac{1 + \chi\left[-\dfrac{1}{\phi} + \sqrt{\dfrac{1}{\phi^2} + \dfrac{1}{\chi}}\right]^2}{\sqrt{\dfrac{1}{\phi^2} + \dfrac{1}{\chi}}} \qquad (6\text{–}63)$$

where $M_{01} = A_{11}f_{1y}\left(d - \frac{a}{2}\right)$, in which a is calculated by Eq. (6–61). Simplifying gives:

$$T = M_{01}\, 2\chi\left[\sqrt{\frac{1}{\phi^2} + \frac{1}{\chi}} - \frac{1}{\phi}\right] \qquad (6\text{–}64)$$

Substituting ϕ by T/M, rearranging, and squaring to eliminate the radical sign result in:

$$\left(\frac{T}{M_{01}}\right)^2 + 4\chi\frac{M}{M_{01}} = 4\chi \qquad (6\text{–}65)$$

Let's denote T_{01} as the pure torsional strength of a member in mode 1 failure. It can be obtained by setting $M = 0$ in Eq. (6–65):

$$T_{01} = 2\sqrt{\chi}M_{01} \qquad (6\text{–}66)$$

Replacing M_{01} by $T_{01}/2\sqrt{\chi}$ in the first term of Eq. (6–65), we arrive at:

$$\left(\frac{T}{T_{01}}\right)^2 + \frac{M}{M_{01}} = 1 \qquad (6\text{–}67)$$

This nondimensional interaction curve, Eq. (6–67), is plotted in Fig. 6.7 (a). The curve approaches a circular configuration near the moment axis and approaches a linear one near the torsion axis.

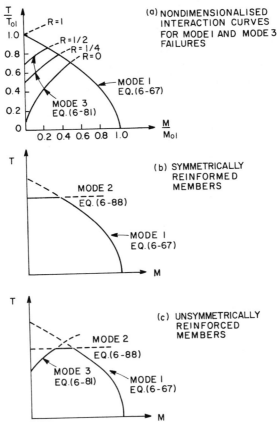

Fig. 6.7. Collins et al.'s interaction curves between torsional and bending moments.

6.3.2   Mode 3 Failure

A third mode of failure, shown in Fig. 6.8, was discovered by Collins et al., who referred to it as mode 3 failure. This failure mode will occur in an unsymmetrically reinforced beam (where the top longitudinal steel is significantly less than the bottom longitudinal steel) subjected to a large torque and a small bending moment. Under such circumstances, the yielding of top longitudinal steel will precede the yielding of bottom longitudinal steel. In mode 3 failure, the skew failure surface is initiated by a helical crack on the top and the two side faces. The compression zone is near the bottom face and is inclined at an angle to the longitudinal axis of the beam. The following additional symbols are defined:

A_{l3} = area of the tensile longitudinal steel near the top face in mode 3 failure

c_3 = length of the longitudinal projection of the compression zone in mode 3 failure

d_3 = effective depth in mode 3 failure, measured from the bottom extreme compression fiber to the centroid of top tension reinforcement

$M_{01} = A_{11}f_{1y}(d - a/2)$, pure bending strength in mode 1 failure

$M_{03} = A_{l3}f_{1y}(d_3 - a/2) \approx RM_{01}$, pure bending strength in mode 3 failure

$R = A_{l3}/A_{l1}$, ratio of area of top longitudinal steel to area of bottom longitudinal steel

$r = A_w f_{sy}b/A_{l1}f_{1y}s$, used in mode 1 failure

$r_3 = A_w f_{sy}b/A_{l3}f_{1y}s = r/R$ for mode 3 failure

$T_{01} = 2\sqrt{\chi}\ M_{01}$, pure torsional strength in mode 1 failure

$T_{03} = \sqrt{R}\ T_{01}$, pure torsional strength in mode 3 failure

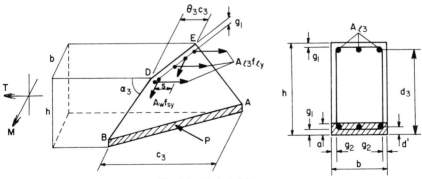

Fig. 6.8. Mode 3 failure.

α_3 = angle of the helical crack with respect to the longitudinal axis of the beam (mode 3 failure)

θ_3 = ratio of the longitudinal projection of the top crack DE to the longitudinal projection of the compression zone in mode 3 failure $(\theta_3 \approx b/(2h + b) = \theta)$

$\chi = r/(1 + 2h/b)$, used in mode 1 failure

$\chi_3 = r_3/(1 + 2h/b) = \chi/R$ for mode 3 failure

An equation for mode 3 failure can be obtained from the corresponding equation for mode 1 failure by the following operation: Replace d, c, θ, A_{11}, r, χ, M, and ϕ in equations for mode 1 failure by d_3, c_3, θ_3, A_{13}, r_3, χ_3, $-M$, and $-\phi$, respectively. Hence, the three basic Eqs. (6–43), (6–44), and (6–45) for mode 1 failure can be converted into three basic equations for mode 3 failure as follows:

$$\frac{c_3}{b} = \frac{1}{\phi} + \sqrt{\frac{1}{\phi^2} + \left(\frac{1 + 2h/b}{r_3}\right)} \leq 1 + 2\frac{h}{b} \qquad (6\text{–}68)$$

$$a = \frac{A_{13}f_{1y}\left[1 + \left(\dfrac{r_3}{1 + 2h/b}\right)\left(\dfrac{c_3}{b}\right)^2\right]}{f_c'' b\left[1 + \left(\dfrac{c_3}{b}\right)^2\right]} \qquad (6\text{–}69)$$

$$T = A_{13}f_{1y}\left(d_3 - \frac{a}{2}\right)\frac{\left[1 + \left(\dfrac{r_3}{1 + 2h/b}\right)\left(\dfrac{c_3}{b}\right)^2\right]}{-\dfrac{1}{\phi} + \dfrac{c_3}{b}} \qquad (6\text{–}70)$$

Again, Collins et al. assumed that the value of a in Eq. (6–69) can be replaced by the value of a for pure bending analysis:

$$a = \frac{A_{13}f_{1y}}{f_c'' b} \qquad (6\text{–}71)$$

Therefore, the expression $A_{13}f_{1y}(d_3 - a/2)$ in Eq. (6–70) becomes the pure bending strength M_{03} in mode 3 failure.

Before we simplify Eq. (6–68) and (6–70), let us express the symbols r_3, χ_3, and M_{03} for mode 3 failure in terms of r, χ, and M_{01} for mode 1 failure. Let:

$$R = \frac{A_{13}}{A_{11}} \tag{6-72}$$

Then:

$$r_3 = \frac{A_w f_{sy} b}{A_{13} f_{1y} s} = \frac{A_w f_{sy} b}{A_{11} f_{1y} s} \frac{A_{11}}{A_{13}} = \frac{r}{R} \tag{6-73}$$

and:

$$\chi_3 = \frac{r_3}{1 + 2h/b} = \frac{r}{1 + 2h/b} \left(\frac{r_3}{r}\right) = \frac{\chi}{R} \tag{6-74}$$

$$M_{03} = A_{13} f_{1y} \left(d_3 - \frac{a}{2}\right) = R A_{11} f_{1y} \left(d_3 - \frac{a}{2}\right)$$

$$\approx R A_{11} f_{1y} \left(d - \frac{a}{2}\right) = R M_{01} \tag{6-75}$$

Substituting Eqs. (6–73) through (6–75) into Eqs. (6–68) and (6–70) we have:

$$\frac{c_3}{b} = \frac{1}{\phi} + \sqrt{\frac{1}{\phi^2} + \frac{R}{\chi}} \tag{6-76}$$

$$T = M_{01} \frac{R + \chi \left(\dfrac{c_3}{b}\right)^2}{-\dfrac{1}{\phi} + \dfrac{c_3}{b}} \tag{6-77}$$

Substituting c_3/b from Eq. (6–76) into Eq. (6–77) gives:

$$T = M_{01} \, 2\chi \left[\sqrt{\frac{1}{\phi^2} + \frac{R}{\chi}} + \frac{1}{\phi}\right] \tag{6-78}$$

Replacing ϕ by T/M in Eq. (6–78) and simplifying results in:

$$\left(\frac{T}{M_{01}}\right)^2 - 4\chi \frac{M}{M_{01}} = 4\chi R \tag{6-79}$$

We can find the pure torsional strength in mode 3 failure, T_{03}, by setting $M = 0$ in Eq. (6–79) and remembering Eq. (6–66):

$$T_{03} = \sqrt{R}\,(2\sqrt{\chi}\,M_{01}) = \sqrt{R}\,T_{01} \qquad (6\text{--}80)$$

Inserting $T_{01}/2\sqrt{\chi}$ for M_{01} in the first term of Eq. (6–79), we arrive at:

$$\left(\frac{T}{T_{01}}\right)^2 - \frac{M}{M_{01}} = R \qquad (6\text{--}81)$$

It is interesting to point out that Eqs. (6–78), (6–79), and (6–81) for mode 3 failure can actually be obtained, respectively, from Eqs. (6–64), (6–65), and (6–67) for mode 1 failure by a very simple operation. That operation is to replace χ, M_{01}, T_{01}, ϕ, and M in the latter three equations by χ/R, RM_{01}, $\sqrt{R}\,T_{01}$, $-\phi$, and $-M$, respectively.

Equation (6–81) is also plotted in Fig. 6.7 (a) for R values of 0, ¼, ½, and unity. It can be seen that mode 3 failure could indeed precede mode 1 failure if the values of R and M/M_{01} were small. The transition point where the mode 3 strength is equal to the mode 1 strength can be derived by subtracting Eq. (6–81) from Eq. (6–67):

$$\frac{M}{M_{01}} = \frac{1-R}{2} \qquad (6\text{--}82)$$

According to Eq. (6–82), the transition points for $R = 0$, ¼, ½, and 1 are located at the M/M_{01} values of 0.5, 0.375, 0.25, and 0, respectively.

6.3.3 Mode 2 Failure

The three basic Eqs. (6–48), (6–49), and (6–50) for mode 2 failure in the Russian Code can also be reduced into one interaction curve, if we replace Eq. (6–49) by the pure flexural equation:

$$a = \frac{A_{12}f_{1y}}{f_c'' h} \qquad (6\text{--}83)$$

Then the expression $A_{12}f_{1y}\left(d_2 - \dfrac{a}{2}\right)$ in Eq. (6–50) becomes M_{02}, the pure flexural strength in mode 2 failure. Substituting Eq. (6–48) into Eq. (6–50) gives:

$$T = \frac{M_{02}}{1+\delta}\,2\,\sqrt{\frac{r_2}{(1+2b/h)}} \qquad (6\text{--}84)$$

To convert M_{02} and r_2 for mode 2 failure to M_{01} and r for mode 1 failure, we define:

$$R_2 = \frac{M_{02}}{M_{01}} = \frac{A_{12}f_{1y}\left(d_2 - \frac{a}{2}\right)}{A_{11}f_{1y}\left(d - \frac{a}{2}\right)}$$

$$= \left(\frac{A_{12}}{A_{11}}\right)\left(\frac{d_2 - \frac{a}{2}}{d - \frac{a}{2}}\right) \approx \left(\frac{A_{12}}{A_{11}}\right)\left(\frac{b}{h}\right) \tag{6-85}$$

Then:

$$r_2 = \frac{A_{w}f_{sy}h}{A_{12}f_{1y}s} = \frac{A_{w}f_{sy}b}{A_{11}f_{1y}s}\frac{h}{b}\frac{A_{11}}{A_{12}} = \frac{r}{R_2} \tag{6-86}$$

Inserting $M_{02} = R_2 M_{01}$ and $r_2 = r/R_2$ into Eq. (6-84), we arrive at:

$$T = \frac{M_{01}}{1 + \delta}\sqrt{\frac{4rR_2}{1 + 2b/h}} \tag{6-87}$$

The pure torsional strength T_{02} for mode 2 failure can be obtained by setting $V = 0$; i.e., $\delta = Vb/2T = 0$.

$$T_{02} = M_{01}\sqrt{\frac{4rR_2}{1 + 2b/h}} \tag{6-88}$$

T_{02} is a constant for a given cross section and is plotted as horizontal straight lines in Fig. 6.7 (b) and (c). Figure 6.7 (b) shows the T–M intersection curve for a symmetrically reinforced member. The curve consists of two solid segments, one for mode 1 failure, Eq. (6-67), and the other for mode 2 failure, Eq. (6-88). Figure 6.7 (c) gives the interaction curve for a typical unsymmetrically reinforced member. The solid curve includes three segments, plotted from Eqs. (6-67), (6-88), and (6-81) for modes 1, 2, and 3, respectively.

Inserting Eq. (6-88) into Eq. (6-87) and remembering $\delta = Vb/2T$, we have:

$$T + V\frac{b}{2} = T_{02} \tag{6-89}$$

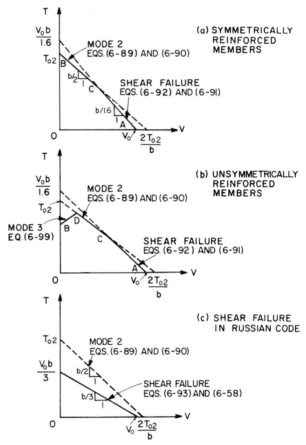

Fig. 6.9. Collins et al.'s interaction curves for torsion and shear.

The torsion–shear interaction curves for this equation are plotted in Fig. 6.9 (a) for symmetrical reinforced members and in Fig. 6.9 (b) for unsymmetrical reinforced members. Note that the shear strength without torsion, V_0, is obtained by setting $T = 0$ in Eq. (6–89):

$$V_0 = \frac{2T_{02}}{b} = \frac{2M_{01}}{b} \sqrt{\frac{4rR_2}{1 + 2b/h}} \qquad (6\text{–}90)$$

6.3.4 Shear Failure

In members subjected to shear and bending without torsion, the member may fail in shear. The shear strength without torsion is given in many codes (including the ACI Code and the Australian) as:

$$V_0 = V_c + \frac{A_v f_y d}{s} \qquad (6\text{-}91)$$

where V_c is the shear resistance contributed by concrete, and $A_v f_y d/s$ is the shear resistance contributed by web reinforcement. V_c may be specified differently in different codes.

The value of V_0 calculated from Eq. (6-91) is often much less than that calculated from Eq. (6-90), meaning that the latter is often unconservative. In other words, when shear is large and torsion is small, failure is likely to be caused by shear instead of by skew bending in the second mode. To prevent shear failure under combined shear and torsion an empirical equation was suggested in the Australian Code of 1973:

$$V + \frac{1.6T}{b} = V_0 \qquad (6\text{-}92)$$

The straight lines representing Eq. (6-92) are also included in Fig. 6.9 (a) and (b). Figure 6.9 (a) illustrated clearly that shear failure will govern when shear is large and torsion is small. In contrast, skew-bending failure in mode 2 will govern when torsion is large and shear is small.

It would be interesting to compare Eq. (6-92) to Eq. (6-59) of the 1962 Russian Code. To do this we rewrite Eq. (6-59) as:

$$V + \frac{3T}{b} = V_0 \qquad (6\text{-}93)$$

It can be seen that Eq. (6-93) is very conservative because the constant of 3 is considerably greater than the constant of 1.6 used in Eq. (6-92).

Equation (6-93) for shear failure is also compared to Eq. (6-89) for mode 2 failure in Fig. 6.9 (c). This figure shows that shear failure, Eq. (6-93), will govern in all cases, including the range where shear is small and torsion is large because the slope of $b/3$ for the shear failure curve is less than the slope of $b/2$ for the mode 2 failure curve. This is obviously incorrect, as mode 2 failure, Eq. (6-89), should theoretically govern in the region where torsion predominates. Hence, the applicability of this very conservative Eq. (6-93) should be restricted only to the narrow region where shear predominates.

6.3.5 Interaction Surface

The interaction between torsion and bending moments was illustrated in Fig. 6.7. The absence of shear in this interaction is physically possible. However, the interaction between torsion and shear, as shown in Fig. 6.9, is

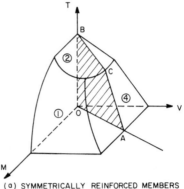

(a) SYMMETRICALLY REINFORCED MEMBERS

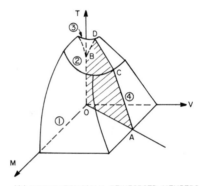

(b) UNSYMMETRICALLY REINFORCED MEMBERS

Fig. 6.10. Collins et al.'s interaction surface for combined torsion, bending, and shear (Ref. 4).

not as simple because shear forces are always associated with bending moments. Even if failure occurs at the inflection point, the bending moment will still be involved because the failure surface spreads over a significant length along the beam. Theoretically, the case of combined torsion and shear without bending does not exist. So what is the physical meaning of the interaction curves between torsion and shear in Fig. 6.9? To understand this, we must have a broader concept of the interaction surface.

The interaction of torque, shear, and bending moment on the strength of a member can be represented by a surface in a three-dimensional space as shown in Fig. 6.10. The three axes of the space represent the torque, T, the shear force, V, and the bending moment, M. This concept of a three-dimensional interaction surface was first employed by the author for members without web reinforcement.[16] Here, an interaction surface for members with web reinforcement will be constructed according to Collins et al.'s theory.[4,5] Take the case of a symmetrically reinforced member. The T–M interaction curve for strength in Fig. 6.7 (b) has been transferred to the T–M plane of the three-dimensional space in Fig. 6.10 (a). In the M–V plane the bending strength and the shear strength of a member are assumed to be independent of each other, resulting in the two straight lines indicated. A diagonal line from O to A on the M–V plane would mean that the bending moment and shear force increase proportionally during the loading history. When a torque is added to the proportionally increasing bending moment and shear, the interaction can be represented by a point on the shaded plane $OACB$ in Fig. 6.10 (a). If this shaded plane is projected onto the T–V plane, we obtain the T–V interaction curve shown in Fig. 6.9 (a).

The whole interaction surface for a symmetrically reinforced member is also sketched in Fig. 6.10 (a). It consists of three regions, labeled ①, ②, and ④, representing mode 1 failure, mode 2 failure, and shear failure, respectively.

Now, let us examine the interaction surface for unsymmetrically reinforced members shown in Fig. 6.10 (b). The T–M interaction curve on the T–M plane was borrowed from Fig. 6.7 (c). In the M–V plane we again assume two straight lines, implying the independence of bending moment and shear force from each other. The shaded plane $OACDB$ represents a state of combined torsion, shear, and bending moment in which bending moment always increases proportionally with the shear force. When this shaded plane is projected onto the T–V plane, we obtain the T–V interaction curve in Fig. 6.9 (b).

Notice that the T–V interaction curve for unsymmetrically reinforced members in Fig. 6.9 (b) includes a segment BD for mode 3 failure. Such failure can indeed occur when torque is large and shear and bending moment are small. To derive this segment of the interaction curve we have to first assume a certain ratio of moment to shear, M/V, and then convert the T–M interaction relationship for mode 3 failure into a T–V interaction relationship.

Take the case of a simply supported beam, where failure occurs as close to the end as possible. The center of the failure surface should be located at a distance $c_3/2$ from the face of the support. Consequently, the ratio of bending moment to shear force is:

$$\frac{M}{V} = \frac{V(c_3/2)}{V} = \frac{c_3}{2} \tag{6–94}$$

Using this assumed M/V ratio:

$$\phi = \frac{T}{M} = \frac{2T}{Vc_3} = \left(\frac{2T}{Vb}\right)\left(\frac{b}{c_3}\right) = \frac{1}{\delta\left(\dfrac{c_3}{b}\right)} \tag{6-95}$$

Substituting this value of ϕ into Eq. (6–76) for mode 3 failure gives:

$$\frac{c_3}{b} = \sqrt{\frac{R}{\chi}\frac{1}{(1-2\delta)}} \tag{6-96}$$

Notice that $\delta = Vb/2T$ will always be small when mode 3 failure governs. Neglecting δ in Eq. (6–96), c_3/b can be assumed as a constant for a given cross section:

$$\frac{c_3}{b} \approx \sqrt{\frac{R}{\chi}} \tag{6-97}$$

Substituting Eq. (6–97) into Eq. (6–77) and remembering $\chi = r/(1 + 2h/b)$, we arrive at:

$$T = \frac{M_{01}}{(1-\delta)} \sqrt{\frac{4rR}{1+2h/b}} \tag{6-98}$$

or:

$$T - V\left(\frac{b}{2}\right) = M_{01} \sqrt{\frac{4rR}{1+2h/b}} \tag{6-99}$$

It can be seen from Eq. (6–99) that the relationship between T and V is linear for a given cross section. In actuality, of course, this relationship should be somewhat nonlinear, if $(1-2\delta)$ was not neglected in Eq. (6–96). Equation (6–99) has been plotted in Fig. 6.9 (b).

The complete interaction surface for an unsymmetrically reinforced member is sketched in Fig. 6.10 (b). The surface includes four regions, designated (1), (2), (3), and (4), representing mode 1, mode 2, mode 3, and shear-type failures, respectively.

The torsion–bending interaction curve based on skew-bending theory was first verified experimentally by McMullen and Warwaruk.[6] Two series of their tests are presented in Fig. 6.11. Figure 6.11 (a) gives data points for

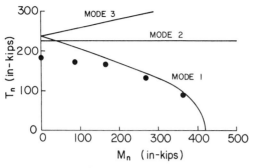

(a) SYMMETRICALLY REINFORCED MEMBERS
(GROUP 2)

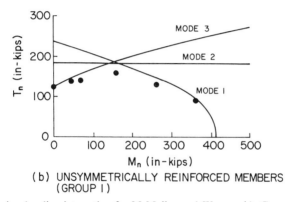

(b) UNSYMMETRICALLY REINFORCED MEMBERS
(GROUP I)

Fig. 6.11. Torsion–bending interaction for McMullen and Warwaruk's Groups 1 and 2 tests.

a series of beams with symmetrical reinforcement, whereas Fig. 6.11 (b) is for a series of unsymmetrically reinforced beams. It can be seen that these tests substantiated the general trend of the predicted interaction curves. However, the theory is unconservative when pure torsion is approached (in the case of symmetrically reinforced beams) or at the vicinity of the transition point (in the case of unsymmetrically reinforced beams). This confirms the author's previous observation (see Fig. 3.13) that Lessig's theory overestimates considerably the pure torsional strength of symmetrically reinforced concrete members.

Experimental verifications of the torsion–shear interaction curve has been given by Collins et al.[4] in Fig. 6.12. For this typical series of beams, the general trend of the data points appears to be quite conservative.

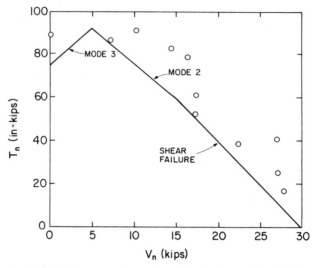

Fig. 6.12. Torsion–shear interaction for Collins et al.'s Series V tests.

6.3.6 Australian Code of 1973[7]

The Australian Code of 1973 was based on the four basic equations derived in Collins et al.'s theory for the four modes of failure. Let us first summarize these equations:

Mode 1 failure, Eq. (6–64):

$$T = M_{01} 2\chi \left[\sqrt{\frac{1}{\phi^2} + \frac{1}{\chi}} - \frac{1}{\phi} \right] \qquad (6\text{--}100)$$

Mode 2 failure, Eq. (6–87):

$$T = M_{01} \frac{1}{1 + \delta} \sqrt{\frac{4rR_2}{(1 + 2b/h)}} \qquad (6\text{--}101)$$

Mode 3 failure, Eq. (6–78):

$$T = M_{01} 2\chi \left[\sqrt{\frac{1}{\phi^2} + \frac{R}{\chi}} + \frac{1}{\phi} \right] \qquad (6\text{--}102)$$

Shear failure, Eq. (6–92):

$$V + \frac{1.6T}{b} = V_0 \qquad (6\text{--}103)$$

In these equations:

$$\chi = \frac{r}{(1 + 2h/b)} \qquad (6\text{--}104a)$$

$$r = \frac{A_w f_{sy} b}{A_{l1} f_{ly} s} \qquad (6\text{--}104b)$$

$$R = \frac{A_{l3}}{A_{l1}} \approx \frac{M_{03}}{M_{01}} \qquad (6\text{--}104c)$$

$$R_2 = \frac{M_{02}}{M_{01}} \approx \frac{1 + R}{2} \frac{b'}{d'} \qquad (6\text{--}104d)$$

b' = width of a closed stirrup, measured between center lines

d' = depth of a closed stirrup, measured between center lines

$$\phi = \frac{T}{M} \qquad (6\text{--}104e)$$

$$\delta = \frac{Vb}{2T} \qquad (6\text{--}104f)$$

6.3.6.1   Equivalent and Effective Shear Forces.

In Eq. (6–103) the term $1.6T/b$ for torsion can be considered as an equivalent shear force; and the two terms on the left-hand side, $V + 1.6T/b$, have been called the effective shear force V'. That is:

$$V' = V + \frac{1.6T}{b} \qquad (6\text{--}105)$$

When the effective shear force V' is less than the shear resistance V_0 specified by the Australian Code, shear failure of a member will be prevented.

6.3.6.2   Equivalent and Effective Bending Moments for Mode 1 Failure.

The concept of equivalent and effective shear forces has been extended to the design for bending moments. This is possible if the design is based on an "optimum r ratio." As discussed in Section 3.2.3, the longitudinal steel and stirrups can both yield in an underreinforced member if the volume ratio of longitudinal steel to stirrups does not exceed a certain range, i.e., if r varies within a certain limit. It is, therefore, economical to specify an r ratio that produces a minimum weight for the total amount of steel. This r ratio is referred to as the "optimum r ratio."

To find the optimum r ratio we first express the total volume of reinforcement per unit length of the beam, W, as:

$$W = (A_{l1} + RA_{l1}) + \frac{A_w}{s} 2(b' + d') \qquad (6\text{--}106a)$$

or:

$$W = C_1 A_{l1}(1 + kr) \qquad (6\text{--}106b)$$

where:

$$k = \frac{f_{ly}}{f_{sy}} \frac{2}{(1 + R)b}(b' + d') \qquad (6\text{--}106c)$$

$$C_1 = 1 + R \qquad (6\text{--}106d)$$

For a given size of beam and a given ratio of top to bottom steel, C_1 and k are constant. The area of steel to prevent the most common type of failure, mode 1, is from Eq. (6–100):

$$A_{l1} = \frac{T}{f_{ly}\left(d - \dfrac{a}{2}\right)\dfrac{2r}{1 + 2h/b}\left[\sqrt{\dfrac{1}{\phi^2} + \dfrac{(1 + 2h/b)}{r}} - \dfrac{1}{\phi}\right]} \qquad (6\text{--}107)$$

Substituting A_{l1} from Eq. (6–107) into Eq. (6–106b):

$$W = C_2 \frac{1 + kr}{r\left[\sqrt{\left(\dfrac{1}{\phi}\right)^2 + \dfrac{1 + 2h/b}{r}} - \dfrac{1}{\phi}\right]} \qquad (6\text{--}108)$$

where $C_2 = C_1 T (1 + 2h/b)/2 f_{ly}\left(d - \dfrac{a}{2}\right)$ is a constant for a given cross section and a given T. Differentiating Eq. (6–108) with respect to r and equating it to zero, we obtain an expression for r that corresponds to a minimum of W:

$$r = \frac{1}{k + \dfrac{2\sqrt{k}}{\phi\sqrt{1 + 2h/b}}} \qquad (6\text{--}109a)$$

Equation (6–106c) shows that k is a function of R, h/b, and the concrete cover. The value of k normally varies from 2 to 7 and for design purpose can be taken conveniently as 4. Equation (6–109a) becomes:

$$r = \cfrac{1}{4 + \cfrac{4}{\phi\sqrt{(1 + 2h/b)}}} \qquad (6\text{–}109b)$$

Substituting r from Eq. (6–109b) into Eq. (6–100) and observing that:

$$\frac{1}{\chi} = \frac{1 + 2h/b}{r} = 4\left(1 + 2\frac{h}{b}\right) + \frac{4}{\phi}\sqrt{1 + 2\frac{h}{b}}$$

$$\sqrt{\frac{1}{\phi^2} + \frac{1}{\chi} - \frac{1}{\phi}} = 2\sqrt{1 + 2\frac{h}{b}}$$

we obtain:

$$T = M_{01}\frac{1}{\left(\sqrt{1 + 2\dfrac{h}{b}} + \dfrac{1}{\phi}\right)} \qquad (6\text{–}110)$$

Noting that $\phi = T/M$, Eq. (6–110) is rearranged in the form of:

$$T\sqrt{1 + 2\frac{h}{b}} + M = M_{01}$$

Let:

$$T' = T\sqrt{1 + 2\frac{h}{b}} \qquad (6\text{–}111)$$

Then:

$$T' + M = M_{01} \qquad (6\text{–}112)$$

T' is called the equivalent bending moment, and $T' + M$ would be the effective bending moment. $T' + M$ can be used to design the longitudinal tension reinforcement in mode 1 failure. The simplicity of this approach is appealing. This design utilizes only simple flexural mechanics and avoids the complex analysis of Eq. (6–100).

Using the definition of equivalent bending moment T' in Eq. (6–111) and the definition of effective bending moment $T' + M$ in Eq. (6–112), we can express the optimum value of r in Eq. (6–109b) in a much simpler form:

$$r = \frac{T'}{4M_{01}} = \frac{T'}{4A_{11}f_{1y}\left(d - \dfrac{a}{2}\right)} \qquad (6\text{–}113)$$

Equating this optimum value of r to the definition of r in Eq. (6–104b), taking $b(d - a/2) = x_1 y_1$, and introducing the materials reduction factor ϕ, we arrive at:

$$A_w = \frac{T's}{\phi 4 x_1 y_1 f_{sy}} \qquad (6\text{–}114)$$

Equation (6–114) provides the required area of the closed stirrups to prevent mode 1 failure. A_w is the area of one leg of a closed stirrup. x_1 and y_1 are the smaller and larger center-line dimensions, respectively, of a closed stirrup.

6.3.6.3 Mode 3 Failure. The tensile longitudinal steel area required to prevent mode 3 failure is designed such that the torsional strength in mode 3 failure is equal to the torsional strength in mode 1 failure. This condition has been expressed by Eq. (6–82), where $R \approx M_{03}/M_{01}$:

$$\frac{M}{M_{01}} = \frac{1}{2}\left(1 - \frac{M_{03}}{M_{01}}\right) \qquad (6\text{–}115)$$

Inserting $M_{01} = T' + M$ from Eq. (6–112) into Eq. (6–115) gives:

$$T' - M = M_{03} \qquad (6\text{–}116)$$

Equation (6–116) shows that tension longitudinal steel for mode 3 failure is required when $T' > M$. This steel is required to resist an effective bending moment of $T' - M$. This design is based on conventional flexural mechanics and avoids the complex analysis of Eq. (6–102). In the case when $T' < M$, mode 3 failure is impossible, and no tensile steel is required for this mode of failure.

6.3.6.4 Mode 2 Failure. After all the steel reinforcements are designed, the member should be checked to prevent mode 2 failure. The formula for this check can be derived by substituting the optimum value of r from Eq. (6–113) and the definition of $R_2 = M_{02}/M_{01}$ into Eq. (6–101):

$$T' \leq M_{02} \frac{1}{(1 + \delta)^2} \frac{(1 + 2h/b)}{(1 + 2b/h)} \qquad (6\text{--}117)$$

It is interesting to point out that Eq. (6–117) has been used to check the mode 2 failure in the 1978 SAA Prestressed Concrete Code, but not in the 1973 SAA Concrete Structures Code. Instead of using Eq. (6–117), the latter code has the following provisions:

1. In any region where M_u is less than $0.5\,T_u$, the cross section of the beam and the area of longitudinal reinforcement shall be such that the beam can withstand an equivalent transverse bending moment, M_2 (not acting simultaneously with either M_1 or M_3), given by:

$$M_2 = T_u \sqrt{1 + 2\frac{b}{h}} \qquad (6\text{--}117a)$$

2. The area A_w given by Eq. (6–114) shall not be taken to be less than:

$$A_w = \frac{M_2 s}{\phi 4 x_1 y_1 f_{sy}} \qquad (6\text{--}117b)$$

The logic of these two equations is unclear.

6.3.6.5 Design Procedures.
According to the 1973 Australian Code for reinforced concrete, the scope of design for members subjected to combined loadings can be summarized as:

1. Design the web reinforcement for shear failure by the effective shear force $V + 1.6\ T/b$, Eq. (6–105).
2. Design the closed stirrups for mode 1 failure according to Eq. (6–114). These closed stirrups can be considered as part of the web reinforcement required in step (1). The remainder of the web reinforcement, if any, may be in the form of open stirrups or closed stirrups.
3. Design tension longitudinal steel for mode 1 failure using the effective bending moment $T' + M$, where $T' = T\sqrt{1 + 2h/b}$, Eq. (6–111).
4. Design tension longitudinal steel for mode 3 failure using $T' - M$. If $T' < M$, no reinforcement is required.
5. Check Eqs. (6–117a) and (6–117b) for mode 2 failure, when $M_u < 0.5\,T_u$.

The above design procedures for nonprestressed concrete have been extended to prestressed concrete.[8,9] The main difference between the two design criteria is the design of web reinforcement to prevent shear failure. In contrast to the effective shear force used in nonprestressed members, the provisions for prestressed concrete follow an approach similar to the generalized ACI method discussed in Chapter 5.

6.3.7 Example 6.2

The application of the 1973 Australian Code will also be illustrated by the structures and the loadings of Example 4.4 (see Section 4.5 and Figs. 4.24 and 4.25). As in the treatment of the Russian Code in Example 6.1, we will use the ACI load factors and the ACI materials reduction factor (ϕ factor) in the calculation.

Shear and Torsion Web Reinforcement:
The critical section for shear and torsion was specified in the Australian Code to be at a distance d from the support. At this section, $V_u = 69.5$ kips, and $T_u = 1,440$ in.-k. The effective shear force, V'_u, and the equivalent bending moment, T'_u, are:

Eq. (6–105): $\quad V'_u = V_u + \dfrac{1.6 T_u}{b} = 69.5 + \dfrac{1.6(1,440)}{18} = 197.5$ kips

Eq. (6–111): $\quad T'_u = T_u \sqrt{1 + 2\dfrac{h}{b}} = 1,440 \sqrt{1 + 2\dfrac{28}{18}} = 2,920$ in.-k

The contribution of concrete in shear, v_c, is given in a table of the 1973 Australian Code as a function of concrete strength, f'_c, and the percentage of flexural tensile steel, ρ. For $f'_c = 4,000$ psi and $\rho = 1.40\%$, v_c is given as 127 psi.

$$V_c = v_c b_w d = 0.127(18)(25.12) = 57.4 \text{ kips}$$
$$V_s = \frac{V'_u}{\phi} - V_c = \frac{197.5}{0.85} - 57.4 = 175 \text{ kips}$$
$$\frac{A_v}{s} = \frac{V_s}{d f_y} = \frac{175}{15.12(60)} = 0.116 \text{ in.}^2/\text{in.}$$

Assume No. 6 stirrups:

$$x_1 = 18 - 2(1.5 + 0.375) = 14.25 \text{ in.}$$
$$y_1 = 28 - 2(1.5 + 0.375) = 24.25 \text{ in.}$$

Eq. (6–114): $\dfrac{A_w}{s} = \dfrac{T'_u}{\phi 4x_1 y_1 f_{sy}} = \dfrac{2,920}{0.85(4)(14.25)(24.25)(60)}$

$= 0.0414$ in.^{2}/in.

This amount of web reinforcement must be supplied by closed stirrups. The remainder of stirrups is:

$$\frac{1}{2}\frac{A_v}{s} - \frac{A_w}{s} = \frac{0.116}{2} - 0.0414 = 0.0166 \text{ in.}^2/\text{in.}$$

This remainder of web reinforcement could theoretically be provided by vertical bars or open stirrups. However, in practice, it would be more convenient to use all closed stirrups. Thus:

$$A_v = 2(0.44) = 0.88 \text{ in.}^2$$

$$s = \frac{0.88}{0.116} = 7.6 \text{ in.} < 12 \text{ in.}$$

$$s_{max} = \frac{x_1 + y_1}{4} = \frac{14.25 + 24.25}{4} = 9.625 \text{ in.} > 7.6 \text{ in.} \qquad \text{O.K.}$$

$$\frac{d}{2} = \frac{25.12}{2} = 12.56 \text{ in.} > 7.6 \text{ in.} \qquad \text{O.K.}$$

Use No. 6 closed stirrups at 7½-in. spacing throughout the member. Comparing this 7½-in. spacing with the 6-in. spacing designed by ACI Code, it can be seen that the 1973 Australian Code is somewhat more liberal.

Flexural and Torsional Longitudinal Steel:

The longitudinal tensile steel for mode 1 failure, A_{l1}, should be designed at the column face. At this section, $M_u = 6,062$ in.-k and $T_u = 1,440$ in.-k.* The equivalent bending moment, T', and the effective bending moment, M'_u, are:

Eq. (6–111): $\quad T'_u = T_u \sqrt{1 + 2\frac{h}{b}} = 1,440 \sqrt{1 + 2\frac{28}{18}} = 2,920$ in.-k

$$M'_u = M_u + T'_u = 6,062 + 2,920 = 8,982 \text{ in.-k}$$

---

* It is not clear from the 1973 Australian Code whether this torque should be taken at the column face or should be reduced to the value at a distance d from the column face (similar to shear). This question is irrelevant here, since the applied torques are the same at these two sections.

Assume $a = 7.7$ in.:

$$A_{l1} = \frac{M'_u}{\phi f_y \left(d - \dfrac{a}{2}\right)} = \frac{8,982}{0.9(60)\left(25.12 - \dfrac{7.7}{2}\right)} = 7.82 \text{ in.}^2$$

$$a = \frac{A_{l1} f_{ly}}{0.85 f'_c b} = \frac{7.82(60)}{0.85(4)(18)} = 7.67 \text{ in.} \approx 7.7 \text{ in.} \qquad \text{O.K.}$$

$$a_{\max} = 0.75\,\beta_1 d \,\frac{87,000}{87,000 + f_y} = 0.75(0.85)(25.12)\,\frac{87,000}{87,000 + 60,000}$$

$$= 9.48 \text{ in.} > 7.67 \text{ in.} \qquad \text{O.K.}$$

Use 5 No. 10 + 2 No. 8 = 5(1.27) + 2(0.79) = 7.93 in.^{2} > 7.82 in.^{2} O.K.

This longitudinal tensile steel for mode 1 failure at the column face is slightly more conservative than that required by the ACI Code. According to the ACI Code in Example 4.4, we have designed five No. 10 bars in the flexural tension side plus two No. 7 bars at mid-height. It can be seen that by using the equivalent bending moment, the Australian Code does not allow us to calculate torsional longitudinal bars at mid-height. To control cracking, additional bars will have to be added.

The longitudinal tensile steel for mode 3 failure, A_{l3}, can be calculated at the inflection point as follows:

$$T'_u = 2,920 \text{ in.-k}$$
$$M_u = 0$$

Eq. (6–116) $M'_u = T'_u - M_u = 2,920 - 0 = 2,920 \text{ in.-k}$

Assume $a = 2.2$:

$$A_{l3} = \frac{M'_u}{\phi f_y \left(d - \dfrac{a}{2}\right)} = \frac{2,920}{0.9(60)\left(25.12 - \dfrac{2.2}{2}\right)} = 2.25 \text{ in.}^2$$

$$a = \frac{A_{l3} f_{ly}}{0.85 f'_c b} = \frac{2.25(60)}{0.85(4)(18)} = 2.21 \text{ in.} \approx 2.2 \text{ in.} \qquad \text{O.K.}$$

Use two No. 10 bars 2(1.27) = 2.54 in.^{2} > 2.25 in.^{2} O.K.

The two No. 10 corner bars throughout the beam are just adequate.

Check Mode 2 Failure:

Let us check mode 2 failure near inflection point where $M_u < 0.5 \, T_u$.

Eq. (6–117a): $M_2 = T_u \sqrt{1 + 2\dfrac{b}{h}} = 1{,}440 \sqrt{1 + 2\dfrac{18}{28}} = 2{,}177$ in.-k

$$A_{12} = \frac{A_{11} + A_{13}}{2} = \frac{7.93 + 2.54}{2} = 5.24 \text{ in.}^2$$

$$a = \frac{A_{12}f_{1y}}{0.85f'_c h} = \frac{5.24(60)}{0.85(4)(28)} = 3.30 \text{ in.}$$

$$d_2 = 18 - 1.5 - 0.75 - \frac{1.27}{2} = 15.12 \text{ in.}$$

$$M_{02} = A_{12}f_{1y}\left(d_2 - \frac{a}{2}\right) = 5.24(60)\left(15.12 - \frac{3.30}{2}\right) = 4{,}235 \text{ in.-k}$$
$$> 2{,}177 \text{ in.-k } (M_2) \qquad \text{O.K.}$$

Eq. (6–117b): $\dfrac{A_w}{s} = \dfrac{M_2}{\phi 4 x_1 y_1 f_{sy}} = \dfrac{2{,}177}{0.85(4)(14.25)(24.25)(60)}$
$$= 0.0309 \text{ in.}^2/\text{in.} < 0.0414 \text{ in.}^2/\text{in.} \qquad \text{O.K.}$$

Let us also check mode 2 failure at column face according to Eq. (6–117), which is not required by the 1973 Australian Code:

$$\delta = \frac{V_u b}{2T_u} = \frac{71.0(18)}{2(1{,}440)} = 0.444$$

$$M_{02}\frac{1}{(1+\delta)^2}\frac{1 + 2(h/b)}{1 + 2(b/h)} = 4{,}235\,\frac{1}{(1+0.444)^2}\frac{1 + 2(28/18)}{1 + 2(18/28)}$$
$$= 3{,}653 \text{ in.-k}$$

$$\frac{T'_u}{\phi} = \frac{2{,}920}{0.85} = 3{,}435 \text{ in.-k} < 3{,}653 \text{ in.-k} \qquad \text{O.K.}$$

6.4 YUDIN-TYPE THEORY[10]

In 1962 Yudin pointed out that in Lessig's analysis (Section 6.2) only two equilibrium equations were used, namely, the equilibrium of moments about the neutral axis *AB* and the equilibrium of forces along the normal to the compression zone. These two equations are insufficient in design to determine

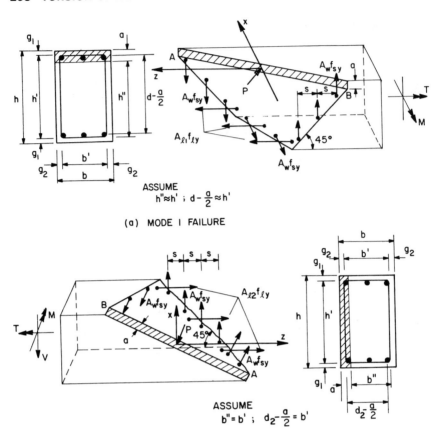

(a) MODE I FAILURE

(b) MODE 2 FAILURE

Fig. 6.13. Yudin-type failure surfaces.

the three unknowns, which are the longitudinal steel area, A_l, the web steel area, A_w, and the depth of the compression zone, a. In Lessig's theory, the weakness in stirrups could be compensated by an excess amount of longitudinal steel, and vice versa. In other words, the volume ratio of longitudinal steel to stirrups, m, could be arbitrarily changed.

In order to determine the ratio m such that both longitudinal steel and stirrups will yield simultaneously in design, Yudin employed three equilibrium equations:

1. Equilibrium of moments about an axis through the centroid of the compression zone and parallel to the longitudinal axis of the beam [z-axis in Fig. 6.13 (a) and (b)].

2. Equilibrium of moments about an axis through the centroid of the compression zone and perpendicular to the z-axis. This x-axis is horizontal in mode 1 failure, Fig. 6.13 (a), and is vertical in mode 2 failure, Fig. 6.13 (b).
3. Equilibrium of forces along the normal to the compression zone.

Yudin's 1962 analysis preceded the discovery of Collins et al.'s mode 3 failure in 1965.[4] Hence, Yudin's analysis was limited to the special case of symmetrically reinforced members. However, it is not difficult to include the mode 3 failure and to extend Yudin's analysis to the more general case of unsymmetrically reinforced members.[12] This general case will be presented here and will be called the Yudin-type theory.

In addition to Lessig's assumptions 1 through 6 (Section 6.2), Yudin made the two following assumptions:

1. All the cracks are inclined at an angle 45° to the longitudinal axis.
2. $h'' = h - g_1 - a = h'$; $d - a/2 = h'$ (see Fig. 6.13a).
   $b'' = b - g_2 - a = b'$; $d_2 - a/2 = b'$ (see Fig. 6.13b).

6.4.1 Mode 1 and Mode 3 Failures

Referring to mode 1 failure in Fig. 6.13 (a), the equilibrium of internal and external moments about the z-axis can be written as:

$$T = \underbrace{A_w f_s \frac{b'}{s} \left(d - \frac{a}{2} \right)}_{\text{Horiz. stirrups}} + \underbrace{2 A_w f_s \frac{h''}{s} \frac{b'}{2}}_{\text{Vert. stirrups}}$$

$$T = 2b'h' \frac{A_w f_s}{s} \tag{6–118}$$

If yielding of stirrups occurs:

$$T_{wy} = 2b'h' \frac{A_w f_{sy}}{s} \tag{6–119}$$

Assuming the yielding of bottom longitudinal steel, the equilibrium of internal and external moments about the x-axis is:

$$M = A_{11}f_{1y}\underbrace{\left(d - \frac{a}{2}\right)}_{\text{Longit. steel}} - \underbrace{A_w f_s \frac{h''}{s}(b' + h')}_{\text{Vert. stirrups}}$$

$$M = A_{11}f_{1y}h' - A_w f_s \frac{h'}{s}(b' + h') \qquad (6\text{-}120)$$

Equation (6–118) can be inserted into Eq. (6–120) to obtain a failure interaction relationship between T and M:

$$T\frac{b' + h'}{2b'} + M = A_{11}f_{1y}h' \qquad (6\text{-}121)$$

Let:

$$T_{01} = \frac{2b'h'A_{11}f_{1y}}{b' + h'} \qquad (6\text{-}122a)$$

$$M_{01} = A_{11}f_{1y}h' \qquad (6\text{-}122b)$$

Equation (6–121) can be put in the form of a nondimensional interaction curve for yielding of the bottom longitudinal steel:

$$\frac{T}{T_{01}} + \frac{M}{M_{01}} = 1 \qquad (6\text{-}123)$$

Equations (6–123) and (6–119) are plotted in Fig. 6.14 (a) and (b).

In the case when both the longitudinal steel and stirrups yield simultaneously, we can substitute T from Eq. (6–119) and T_{01} from Eq. (6–122a) into Eq. (6–123), resulting in:

$$\underbrace{\frac{A_w 2(b' + h')}{2A_{11}s}\frac{f_{sy}}{f_{1y}}}_{\dfrac{1}{m}} + \frac{M}{M_{01}} = 1 \qquad (6\text{-}124)$$

For symmetrically reinforced members the total area of longitudinal steel is $2A_{11}$, and $2A_{11}s/A_w2(b' + h') = m$, the volume ratio of longitudinal steel to stirrups. If f_{1y}/f_{sy} is taken as unity, Eq. (6–124) can be written as:

$$M = M_{01} \left(1 - \frac{1}{m} \right) \qquad (6\text{--}125)$$

Equation (6–125) states that the moment that corresponds to the simultaneous yielding of longitudinal steel and stirrups depends on the volume ratio m (see Fig. 6.14a). In the case of pure torsion ($M = 0$), Eq. (6–125) requires $m = 1$, which represents equal volumes of longitudinal steel and stirrups. It is interesting to point out that this "equal volume" requirement for simultaneous yielding of longitudinal steel and stirrups in pure torsion is identical to Rausch's requirement of Eq. (3–17). Furthermore, Yudin's Eq. (6–119) is the same as Rausch's Eq. (3–18).

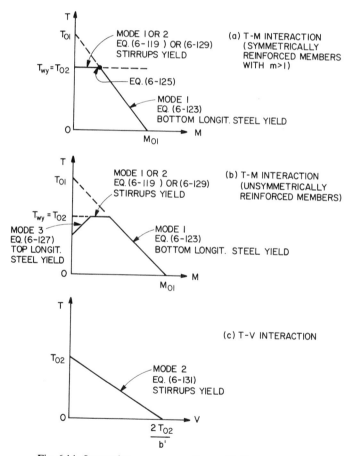

Fig. 6.14. Interaction curves according to Yudin-type theory.

It is also interesting to note that the third equilibrium condition of forces along the normal to the compression zone has not been used here. This equilibrium condition can be used to determine the depth of the compression zone. In view of the second assumption, which gives approximate values of h'', b'', $d - a/2$, and $d_2 - a/2$, this force equilibrium condition becomes irrelevant.

For unsymmetrically reinforced members failed in mode 3, Eq. (6–119) remains valid, and Eq. (6–121) requires only two minor modifications. The moment M and the steel area A_{l1} should be replaced by $-M$ and RA_{l1}, respectively, where $R = A_{l3}/A_{l1}$. Then we have:

$$T\frac{b' + h'}{2b'} - M = RA_{l1}f_{ly}h' \qquad (6\text{–}126)$$

and:

$$\frac{T}{T_{01}} - \frac{M}{M_{01}} = R \qquad (6\text{–}127)$$

Equation (6–127) is plotted in Fig. 6.14 (b) for yielding of top longitudinal steel.

6.4.2 Mode 2 Failure

For mode 2 failure the equilibrium of internal and external moments about the z-axis in Fig. 6.13 (b) is:

$$T + V\frac{b'}{2} = 2b'h'\frac{A_w f_{sy}}{s} \qquad (6\text{–}128)$$

Let:

$$T_{02} = 2b'h'\frac{A_w f_{sy}}{s} \qquad (6\text{–}129)$$

$$V_{02} = 4h'\frac{A_w f_{sy}}{s} = \frac{2T_{02}}{b'} \qquad (6\text{–}130)$$

Then Eq. (6–128) becomes:

$$\frac{T}{T_{02}} + \frac{V}{V_{02}} = 1 \tag{6-131}$$

Equation (6–131) is plotted in Fig. 6.14 (c). It will govern when failure is due to stirrup yielding. Equation (6–129) for pure torsion in mode 2 failure is also given in Fig. 6.14 (a) and (b). Equation (6–129) for mode 2 failure is actually identical to Eq. (6–119) for mode 1 failure.

The equilibrium of moments about the x-axis in Fig. 6.13 (b) is:

$$A_{12}f_{1y}b' - A_{w}f_{sy}\frac{b'}{s}(b' + h') = 0 \tag{6-132}$$

Substituting $A_{w}f_{sy}$ from Eq. (6–128) into Eq. (6–132) gives:

$$T + V\frac{b'}{2} = \frac{2b'h'A_{12}f_{1y}}{b' + h'} \tag{6-133}$$

Equation (6–133) can also be expressed in a linear, nondimensional form like Eq. (6–131). It will govern when failure is due to yielding of longitudinal steel.

The simplicity of Yudin-type equations is very conducive to practical design. The design procedures may be as follows:

1. Design bottom longitudinal steel by Eq. (6–121).
2. Design stirrups by Eq. (6–128).
3. Design top longitudinal steel by Eq. (6–126).
4. Check total longitudinal steel by Eq. (6–133).

It should be cautioned that this theory does not recognize shear failure and thus will be unsafe when shear predominates.

6.5 ELFGREN ET AL.'S INTERACTION SURFACE[13,14]

In Yudin's theory the yielding of both the longitudinal steel and the web reinforcement requires a specific m ratio. However, tests have shown consistently that both types of steel can yield at failure within a wide range of the m ratio. This phenomenon can be explained by Elfgren et al.'s refined failure surface, incorporating the variable angle of the concrete compressive struts at each face.

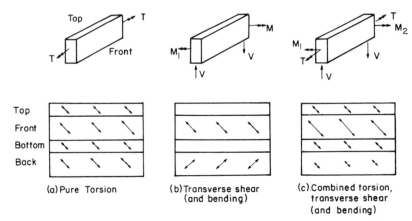

Fig. 6.15. Diagonal tensile stresses due to torsion and transverse shear for a rectangular beam (views of developed beam surfaces).

Failure Surfaces. For a member subjected to pure torque, T, the diagonal tensile stresses will spiral around the four beam faces as shown in Fig. 6.15 (a). For a beam loaded with a vertical shear force, V, diagonal tensile stresses will occur only on the two vertical faces as shown in Fig. 6.15 (b). If a beam is subjected to both torque T and shear V, the diagonal tensile stresses will be additive on one of the vertical faces and subtractive on the other (see Fig. 6.15c). Only diagonal tensile stresses due to torsion exist on the horizontal faces.

Figure 6.16 shows the mode 1 skew failure surface of a beam element subjected to torsion, bending, and shear. Owing to the different diagonal tensile stresses in the different faces of the beam, the inclination of the cracks and that of the concrete compression struts between the cracks will vary from face to face. The inclination may also change as the load is increased such that yielding will occur both in the longitudinal steel and in the stirrups at failure. However, the angle of the cracks is not sensitive to the change of the loading condition because an old crack, once formed, cannot change its direction. Only new cracks can adjust their inclination to satisfy the new equilibrium condition, and, therefore, to modify somewhat the general pattern of the cracks. In theory, we can imagine a concrete compression strut that can vary its angle at will to satisfy the equilibrium condition. This idealization is very convenient and will be assumed in this analysis.

For simplicity the rectangular beam in Fig. 6.16 is reinforced with four longitudinal corner bars and closed stirrups. The four faces of the beam are denoted r (right-hand side), b (bottom), ℓ (left-hand side), and t (top). For mode 1 failure the inclinations of the concrete compression struts to

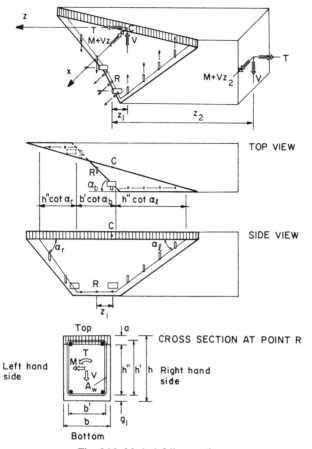

Fig. 6.16. Mode 1 failure surface.

the horizontal beam axis are designated α_r, α_b, and α_l for the right, bottom, and left faces, respectively. For mode 2 failure the angles are α_b, α_r, and α_l in Fig. 6.17.

Assumptions. The assumptions used in Elfgren et al.'s theory include the first six assumptions in Lessig's theory (Section 6.2) plus Yudin's second simplification as follows:

$$h'' = h - g_1 - a = h'; \quad d - a/2 = h' \text{ (see Fig. 6.16 for definitions of symbols).}$$

$$b'' = b - g_2 - a = b'; \quad d_2 - a/2 = b' \text{ (see Fig. 6.17).}$$

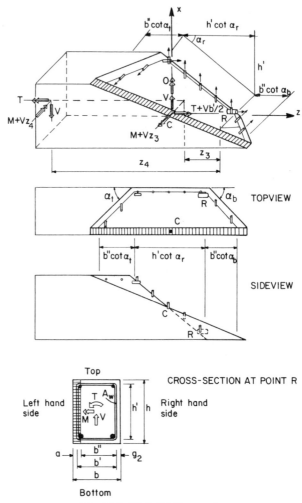

Fig. 6.17. Mode 2 failure surface.

6.5.1 Mode 1 and Mode 3 Failures

6.5.1.1 Equilibrium Conditions. Four equilibrium conditions are required to balance the forces and moments in Elfgren et al.'s beam elements shown in Fig. 6.16 for mode 1 failure:

1. Equilibrium of moments about the z-axis
2. Equilibrium of moments about the x-axis

3. Equilibrium of forces along the normal to the compression zone.
4. Equilibrium of forces in the vertical direction

The first three equilibrium equations are identical to those used by Yudin (Section 6.4). The third equilibrium condition, as explained previously, is not needed because the depth of the compression zone has been rendered unnecessary by Yudin's second simplifying assumption. The fourth equilibrium condition, however, should be taken into account in Elfgren et al.'s theory because their beam element has a failure surface with a different angle on each face. The net vertical internal forces along the two vertical faces can be made to balance the external shear force. This last equilibrium condition was not utilized in Yudin's theory because the crack angles were assumed to be 45° on both vertical faces, and the internal vertical forces are self-balancing in the vertical direction.

It is interesting to note here that elementary mechanics requires a spacial force system to satisfy six equilibrium equations. The two additional equations are the equilibrium of moments about a vertical axis through the centroid of the compression zone C and the equilibrium of forces along a horizontal axis in the plane of the compression zone through the point C. The former equation is automatically satisfied because the longitudinal bars are arranged symmetrically about the vertical axis and the latter can be satisfied by taking into account a horizontal shear force in the compression zone. This shear force is assumed to have no effect on the strength of the member.

Referring to Fig. 6.16, the equilibrium of internal and external moments about the z-axis can be readily written as:

$$T = A_w f_{sy} \left[\frac{b' \cot \alpha_b}{s} h' + \frac{h''}{s} (\cot \alpha_l + \cot \alpha_r) \frac{b'}{2} \right] \qquad (6\text{--}134)$$

Then we shall examine the second equilibrium condition of moments about the x-axis. As a vertical shear force, V, is present, the external bending moment varies along the length of the beam. It has a value, M, at the reference point, R, where the longitudinal center line of the bottom longitudinal bars intersects the failure surface.

The longitudinal distance between the point R and the centroid of the compression zone C is z_1. Consequently, the external moment about the x-axis will be $M + Vz_1$. The second equation gives:

$$M + Vz_1 = A_{11} f_{ly} h' - A_w f_{sy} \frac{h'' \cot \alpha_l}{s} \left(\frac{b'}{2} \cot \alpha_b + \frac{h''}{2} \cot \alpha_r \right)$$
$$- A_w f_{sy} \frac{h'' \cot \alpha_r}{s} \left(\frac{b'}{2} \cot \alpha_b + \frac{h''}{2} \cot \alpha_l \right) \qquad (6\text{--}135)$$

The third equilibrium equation of forces in the vertical direction is:

$$V = A_w f_{sy} \frac{h''}{s} (\cot \alpha_l - \cot \alpha_r) \qquad (6\text{-}136)$$

It should be noted that the vertical resistance force of the compression zone does not appear in this equation. It has been neglected for simplicity.

6.5.1.2 Inclination of Concrete Compression Struts. To simplify the above three equilibrium equations, we shall now examine the angle of inclination, α, of the concrete compression struts. Let us make a horizontal cut in one wall of a hollow box section that has a wall thickness of t (see Fig. 6.18). The length of the cut is s, which is the spacing of the stirrups. Let:

F_c = the compressive force in the concrete strut over the length s
F_w = the force in one stirrup leg (= $A_w f_{sy}$ at yielding)
τ = the shear stress along the horizontal cut (the shear force along the length s is $\tau t s$)

The equilibrium of the three forces, $\tau t s$, F_c, and F_w, is given by the force polygon in Fig. 6.18. It can be seen that:

$$\cot \alpha = \tau \frac{ts}{F_w} \qquad (6\text{-}137)$$

Since t, s, and F_w are constants, Eq. (6-137) states that $\cot \alpha$ is directly proportional to the shear stress, τ. Define:

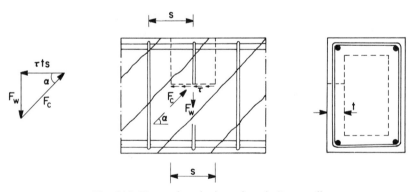

Fig. 6.18. Forces along horizontal cut in beam wall.

τ_T = shear stress caused by a torque, T, alone
τ_V = shear stress caused by a flexural shear, V, alone
α_T = angle of inclination for a beam subjected to pure torsion
α_V = angle of inclination for a beam subjected to flexural shear without torsion

Then Eq. (6–137) becomes:

$$\cot \alpha_T = \tau_T \frac{ts}{F_w} \qquad (6\text{–}138)$$

$$\cot \alpha_V = \tau_V \frac{ts}{F_w} \qquad (6\text{–}139)$$

For the three tension faces in mode 1 failure (Fig. 6.16), the left face is subjected to flexural shear and torsional shear acting in the same direction. In the right face, these two types of shear stresses are acting in the opposite direction. In the bottom face, only torsional shear stress exists. Consequently we can write:

$$\cot \alpha_l = \cot \alpha_T + \cot \alpha_V \qquad (6\text{–}140)$$
$$\cot \alpha_b = \cot \alpha_T \qquad (6\text{–}141)$$
$$\cot \alpha_r = \cot \alpha_T - \cot \alpha_V \qquad (6\text{–}142)$$

The horizontal distance z_1 along the beam axis between points R and C (Fig. 6–16) can be written as:

$$z_1 = \frac{1}{2}(h'' \cot \alpha_r + b' \cot \alpha_b + h'' \cot \alpha_l) - (h'' \cot \alpha_r + \frac{b'}{2} \cot \alpha_b)$$

$$= \frac{1}{2} h''(\cot \alpha_l - \cot \alpha_r) \qquad (6\text{–}143)$$

Inserting Eqs. (6–140) and (6–142) into Eq. (6–143), z_1 can be expressed in a very simple form:

$$z_1 = h'' \cot \alpha_V \qquad (6\text{–}144)$$

6.5.1.3 Interaction Equation. For $h'' = h'$ the equilibrium Eqs. (6–134) and (6–136) can be written using Eqs. (6–140) through (6–142):

$$T = 2b'h' \frac{A_w f_{sy}}{s} \cot \alpha_T \qquad (6\text{–}145)$$

$$V = 2h' \frac{A_w f_{sy}}{s} \cot \alpha_V \qquad (6\text{-}146)$$

In a similar way Eq. (6–135) becomes:

$$M + V z_1 = A_{11} f_{1y} h' - \frac{A_w f_{sy} h'}{2s} [(\cot \alpha_T + \cot \alpha_V)(b' \cot \alpha_T$$
$$+ h' \cot \alpha_T - h' \cot \alpha_V) + (\cot \alpha_T - \cot \alpha_V)(b' \cot \alpha_T$$
$$+ h' \cot \alpha_T + h' \cot \alpha_V)]$$

$$M + V z_1 = A_{11} f_{1y} h' - \frac{A_w f_{sy} h'}{s} [(b' + h') \cot^2 \alpha_T - h' \cot^2 \alpha_V] \qquad (6\text{-}147)$$

Substituting Eqs. (6–144), (6–145), and (6–146) into Eq. (6–147) to eliminate z_1, $\cot \alpha_T$, and $\cot \alpha_V$ results in:

$$\frac{M}{A_{11} f_{1y} h'} + \frac{T^2}{(2b'h')^2} \frac{s}{A_w f_{sy}} \frac{(b' + h')}{A_{11} f_{1y}} + \frac{V^2}{(2h')^2} \frac{s}{A_w f_{sy}} \frac{h'}{A_{11} f_{1y}} = 1 \qquad (6\text{-}148)$$

Let:

$$M_{01} = A_{11} f_{1y} h' \qquad (6\text{-}149)$$

$$T_{01} = 2b'h' \frac{A_w f_{sy}}{s} \sqrt{\frac{A_{11} f_{1y}}{b' + h'} \frac{s}{A_w f_{sy}}} \qquad (6\text{-}150)$$

$$V_{01} = 2h' \frac{A_w f_{sy}}{s} \sqrt{\frac{A_{11} f_{1y}}{h'} \frac{s}{A_w f_{sy}}} \qquad (6\text{-}151)$$

Then Eq. (6–148) becomes:

$$\frac{M}{M_{01}} + \left(\frac{T}{T_{01}}\right)^2 + \left(\frac{V}{V_{01}}\right)^2 = 1 \qquad (6\text{-}152)$$

This nondimensionalized interaction surface is plotted in Fig. 6.19 (a). It can be seen that the interaction curves on the T–M and M–V planes are second-degree parabolas, while the curve on the T–V plane is an ellipse.

6.5.1.4 Comparison with Yudin-Type Theory.
Elfgren et al.'s torsion–bending interaction curve in the T–M plane is compared to that of the Yudin-type theory, Eqs. (6–119) and (6–123), in Fig. 6.19 (a). It can be seen that the former always predicts higher failure strengths than the

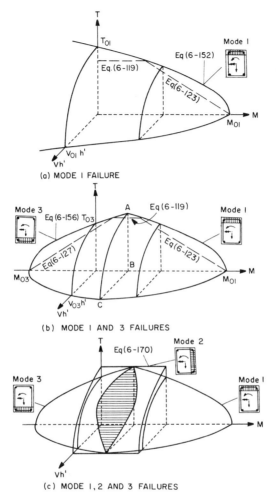

(a) MODE I FAILURE

(b) MODE I AND 3 FAILURES

(c) MODE 1,2 AND 3 FAILURES

Fig. 6.19. Elfgren et al.'s interaction surface.

latter. This is so because Elfgren et al.'s theory allows the inclination of the concrete compression struts to vary such that both the longitudinal steel and the stirrups will yield when the beam fails. At the intersection point of Eqs. (6–119) and (6–123) both types of steel will yield according to Yudin's theory. Hence, both theories provide the same prediction.

A comparison of Eq. (6–150) with the corresponding Eq. (6–119) in the Yudin-type theory is illuminating. The expression of the former differs from that of the latter by the factor under the square root sign. This factor corrects the torsional strength of members for different ratios of longitudinal steel to stirrups. Comparing Eq. (6–150) to Eq. (6–145), it can be seen that this factor is:

$$\cot \alpha_T = \sqrt{\frac{A_{11}f_{1y}}{b'+h'}\frac{s}{A_w f_{sy}}} \tag{6-153}$$

Physically, therefore, this correction is obtained by varying the angle of inclination, α_T, of the compression struts.

Equation (6–151) for shear (without torsion) can be examined in a similar manner. Comparing it with Eq. (6–146) shows that:

$$\cot \alpha_V = \sqrt{\frac{A_{11}f_{1y}}{h'}\frac{s}{A_w f_{sy}}} \tag{6-154}$$

Also, taking $\alpha_V = 45°$ in Eq. (6–146) gives:

$$V = 2h' \frac{A_w f_{sy}}{s} \tag{6-155}$$

This is the well-known ACI Code formula for design of web reinforcement, which is derived from the 45° truss analogy (Note $d \approx h'$). This expression can also be obtained from Eq. (6–136) by assuming $\alpha_1 = -\alpha_r = 45°$.

Comparing Eq. (6–155) with Eq. (6–151), it can be seen that the square root factor in Eq. (6–151) corrects the shear strength of members with different ratios of longitudinal steel to stirrups. Equation (6–154) shows that this correction is accomplished physically by varying the angle α_V.

6.5.1.5 Mode 3 Failure.

Interaction equations for mode 3 failure can be obtained from Eqs. (6–149) through (6–152) by replacing M and A_{11} by $-M$ and A_{13}, respectively. This operation gives the interaction equation as follows:

$$-\frac{M}{M_{03}} + \left(\frac{T}{T_{03}}\right)^2 + \left(\frac{V}{V_{03}}\right)^2 = 1 \tag{6-156}$$

where:

$$M_{03} = A_{13}f_{1y}h' \tag{6-157}$$

$$T_{03} = 2b'h' \frac{A_w f_{sy}}{s} \sqrt{\frac{A_{13}f_{1y}}{b'+h'}\frac{s}{A_w f_{sy}}} \tag{6-158}$$

$$V_{03} = 2h' \frac{A_w f_{sy}}{s} \sqrt{\frac{A_{13}f_{1y}}{h'}\frac{s}{A_w f_{sy}}} \tag{6-159}$$

The interaction surface described by Eq. (6–156) for mode 3 failure is plotted in Fig. 6.19 (b) to the left of the ABC plane. To the right is the failure surface for mode 1. These two surfaces are symmetrical about the plane ABC. In the T–M plane Elfgren et al.'s interaction curves, taking $V = 0$ in Eqs. (6–152) and (6–156), are compared with the Yudin-type straight lines, represented by Eqs. (6–119), (6–123), and (6–127).

6.5.2 Mode 2 Failure

6.5.2.1 Equilibrium Equations.

Two moment equilibrium equations are required in this derivation, one about the z-axis and another about the x-axis (see Fig. 6.17). Both axes pass through the centroid of the compression zone. The compression zone is located on the left side, where flexural shear and torsional shear are subtractive. Yielding of steel will commence on the right side, where the flexural shear and torsional shear are additive.

The first equilibrium of moments about the z-axis can be observed easily:

$$T + V\frac{b'}{2} = \frac{A_w f_{sy}}{s}\left[b'' \cot \alpha_t \frac{h'}{2} + b'' \cot \alpha_b \frac{h'}{2} + h' \cot \alpha_r \, b' \right] \quad (6\text{–}160)$$

Then we shall examine the second equilibrium condition about the x-axis. Since the shear force is present, the bending moment must vary along the beam. This means that the forces in the longitudinal reinforcement will also be varying. A reference point R is chosen to be located at the intersection of the lower right-hand longitudinal bar with the failure surface. The bending moment at this point is designated M, and the forces in the top and bottom longitudinal bars in the vertical section through this point are denoted $A^t_{12}f_{1y}$ and $A^b_{12}f_{1y}$, respectively. This implies that the top right-hand longitudinal bar at its intersection with the failure surface will carry a reduced force as follows:

$$A^t_{12}f_{1y} - \frac{1}{2}\frac{\Delta M}{h'} = A^t_{12}f_{1y} - \frac{1}{2}\frac{Vh' \cot \alpha_r}{h'} = A^t_{12}f_{1y} - \frac{V}{2}\cot \alpha_r$$

The total longitudinal force along the right edge of the failure surface is:

$$A^b_{12}f_{1y} + A^t_{12}f_{1y} - \frac{V}{2}\cot \alpha_r = A_{12}f_{1y} - \frac{V}{2}\cot \alpha_r$$

The moment equation about the x-axis is therefore:

$$\left(A_{12}f_{1y} - \frac{V}{2}\cot\alpha_r\right)b' = \frac{A_w f_{sy}}{s}\, b''\cot\alpha_t \left(\frac{b''}{2}\cot\alpha_b + \frac{h'}{2}\cot\alpha_r\right)$$
$$+ \frac{A_w f_{sy}}{s}\, b''\cot\alpha_b \left(\frac{b''}{2}\cot\alpha_t + \frac{h'}{2}\cot\alpha_r\right)$$

$$(6\text{–}161)$$

6.5.2.2 Inclination of Compression Struts.

The angles of inclination α_t, α_b, and α_r of the concrete compression struts on the three faces can be expressed by the inclination caused by torsion α_T and by vertical shear α_V as follows:

$$\cot\alpha_t = \cot\alpha_T \qquad\qquad (6\text{–}162)$$

$$\cot\alpha_b = \cot\alpha_T \qquad\qquad (6\text{–}163)$$

$$\cot\alpha_r = \cot\alpha_T + \cot\alpha_V \qquad\qquad (6\text{–}164)$$

6.5.2.3 Interaction Equation.

For $b'' = b'$, Eq. (6–160) can be rewritten using Eqs. (6–162) through (6–164):

$$T + V\frac{b'}{2} = \frac{A_w f_{sy}}{s}\,[2b'h'\cot\alpha_T + h'b'\cot\alpha_V] \qquad (6\text{–}165)$$

Considering the equilibrium of forces in the vertical x-direction (Fig. 6.17) Eq. (6–165) can be split up into two equations that are identical to Eqs. (6–145) and (6–146).

Equation (6–161) can also be written as follows, noting that $b'' = b'$:

$$A_{12}f_{1y} - \frac{V}{2}(\cot\alpha_T + \cot\alpha_V)$$
$$= \frac{A_w f_{sy}}{s}\cot\alpha_T(b'\cot\alpha_T + h'\cot\alpha_T + h'\cot\alpha_V) \quad (6\text{–}166)$$

Inserting $\cot\alpha_T$ and $\cot\alpha_V$ from Eqs. (6–145) and (6–146) into Eq. (6–166), we have:

$$\left(\frac{1}{A_{12}f_{1y}}\right)\left(\frac{s}{A_w f_{sy}}\right)\left[\frac{T^2}{(2b'h')^2}(b'+h') + \frac{V^2}{(2h')^2}h' + \frac{TV}{2b'h'}\right] = 1 \quad (6\text{–}167)$$

Let:

$$T_{02} = 2b'h'\frac{A_w f_{sy}}{s}\sqrt{\frac{A_{12}f_{1y}}{(b'+h')}\frac{s}{A_w f_{sy}}} \qquad (6\text{–}168)$$

$$V_{02} = 2h'\frac{A_w f_{sy}}{s}\sqrt{\frac{A_{12}f_{1y}}{h'}\frac{s}{A_w f_{sy}}} \qquad (6\text{–}169)$$

Equation (6–167) becomes:

$$\left(\frac{T}{T_{02}}\right)^2 + \left(\frac{V}{V_{02}}\right)^2 + \frac{TV}{T_{02}V_{02}}\frac{2\sqrt{h'}}{\sqrt{b'+h'}} = 1 \qquad (6\text{–}170)$$

Equation (6–170) is plotted in Fig. 6.19 (c) as a cylindrical surface with an elliptic base in the torsion–shear plane. This surface is not a function of the bending moment, M. When the interaction surfaces for mode 1 and mode 3 failures are included, the elliptic cylinder of mode 2 cuts a slice from the interaction surfaces of modes 1 and 3. The region within the intersection has been shaded.

6.5.3 Comparison with Tests

Elfgren et al.'s theory has been compared to their own tests[14] and to the tests of McMullen and Warwaruk.[6] The comparisons are shown in Figs. 6.20 and 6.21. In these nondimensionalized diagrams the positive bending moments, M, the positive torsional moments, T, and the positive shear force, V, have been normalized by M_0, T_0, and V_0, which are the *lowest positive values* from the three possible modes of failure. For unsymmetrically reinforced members where the top reinforcement is less than the bottom reinforcement ($A_{l3} < A_{l1}$), a comparison of Eqs. (6–149) through (6–151), (6–157) through (6–159), and (6–168) and (6–169) shows that the lowest positive values are:

$$\begin{aligned} M_0 &= M_{01} \quad &(M_{03} \text{ is considered as negative}) \\ T_0 &= T_{03} \quad &(\leq T_{01} \text{ and } T_{02}) \\ V_0 &= V_{03} \quad &(\leq V_{01} \text{ and } V_{02}) \end{aligned}$$

Recalling the definition:

$$R = \frac{A_{l3}}{A_{l1}} \qquad 0 \leq R \leq 1$$

and assuming beams with four corner bars, then:

$$\frac{A_{l2}}{A_{l3}} = \frac{A_{l1} + A_{l3}}{2A_{l3}} = \frac{1 + R}{2R}$$

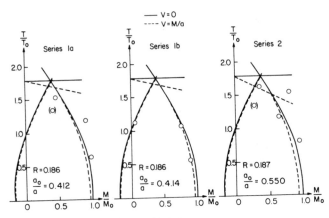

(a) Comparison of theory with tests

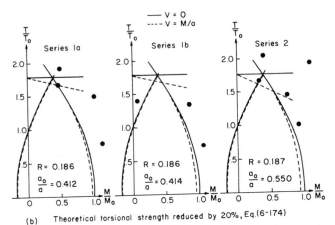

(b) Theoretical torsional strength reduced by 20%, Eq.(6-174)

Fig. 6.20. Comparison of Elfgren et al.'s theory with their own tests.

The expressions for M_{01}, T_{01}, V_{01}, M_{03}, T_{03}, V_{03}, T_{02}, and V_{02} in Eqs. (6–149) through (6–151), (6–157) through (6–159) and (6–168) and (6–169) can be converted into M_0, T_0, and V_0 as follows:

Mode 1	*Mode 3*	*Mode 2*
$M_{01} = M_0$	$M_{03} = -RM_0$	—
$T_{01} = \dfrac{T_0}{\sqrt{R}}$	$T_{03} = T_0$	$T_{02} = T_0 \sqrt{\dfrac{1+R}{2R}}$
$V_{01} = \dfrac{V_0}{\sqrt{R}}$	$V_{03} = V_0$	$V_{02} = V_0 \sqrt{\dfrac{1+R}{2R}}$

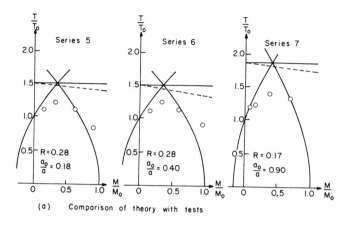

(a)   Comparison of theory with tests

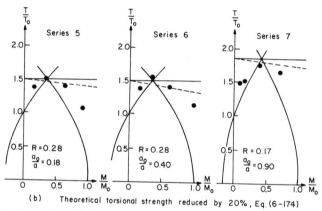

(b)   Theoretical torsional strength reduced by 20%, Eq.(6-174)

Fig. 6.21. Comparison of Elfgren et al.'s theory with McMullen and Warwaruk's tests.

The interaction Eqs. (6–152), (6–156), and (6–170) become:

Mode 1:
$$\frac{M}{M_0} + \left(\frac{T}{T_0}\right)^2 R + \left(\frac{V}{V_0}\right)^2 R = 1 \qquad (6\text{–}171)$$

Mode 3:
$$\frac{M}{M_0}\left(\frac{-1}{R}\right) + \left(\frac{T}{T_0}\right)^2 + \left(\frac{V}{V_0}\right)^2 = 1 \qquad (6\text{–}172)$$

Mode 2:
$$\left(\frac{T}{T_0}\right)^2 \frac{2R}{R+1} + \left(\frac{V}{V_0}\right)^2 \frac{2R}{R+1}$$
$$+ \left(\frac{TV}{T_0 V_0}\right)\left(\frac{2R}{R+1}\right)\frac{2}{\sqrt{1+b'/h'}} = 1 \quad (6\text{–}173)$$

In Figs. 6.20 and 6.21 the interaction curves are plotted according to the above Eqs. (6–171), (6–172), and (6–173). The solid curves in these figures represent the theoretical torsion–bending interaction curves ($V = 0$). The dotted curves represent the reduction of strength caused by the vertical shear force. No dotted curves are given in Fig. 6–21 for modes 1 and 3 failures because the effect of shear forces was shown in Fig. 6.20 to be small. Each series of beams is presented in one diagram, and the theoretical curves are calculated from the mean values of R and a_0/a for this series. The shear span symbols are $a_0 = M_0/V_0$ and $a = M/V$. The shear span ratios a_0/a given in Figs. 6.20 and 6.21 are a measure of the relative magnitude of the shear force. The smaller this ratio, the smaller the relative shear force.

Figure 6.20 (a) gives the test data points from Elfgren et al.'s series 1a, 1b, and 2. The dots in parentheses represent those beams that failed somewhat prematurely in the loading arms. It can be seen that the theory is quite reasonable, except in the vicinity of the maximum torque, where the test points fall somewhat below the theoretical prediction. This unconservatism is more clearly demonstrated by McMullen and Warwaruk's tests in Fig. 6.21 (a). All three series of their test points (series 5, 6, and 7) fall significantly below the theoretical prediction near the maximum torque.

The inconsistency between theory and tests near the maximum torque is obviously due to the overestimation of pure torsional strength of reinforced concrete beams (see Section 3.2.3 and Fig. 3.13). If we reduce the theoretical pure torsional strength T_{03} in Eq. (6–158) by 20% and take:

$$T_0 = 0.8 T_{03} \qquad (6\text{–}174)$$

the resulting comparison of theory and tests is shown in Figs. 6.20 (b) (Elfgren et al.) and 6.21 (b) (McMullen and Warwaruk). It can be seen that the agreement between theory and tests has been greatly improved.

In summary, Elfgren et al.'s theory and interaction surface are very elegant and useful. They give us a clear concept of the relationship among torsion, bending, and shear, and the effect of these quantities on the strength of reinforced concrete beams. However, the theory may overestimate the strength of a reinforced concrete beam because it follows the kinematic approach of the plasticity method. This approach will always produce an upper bound solution.

REFERENCES

1. Lessig, N. N., "Determination of Load-Carrying Capacity of Rectangular Reinforced Concrete Elements Subjected to Flexure and Torsion," Trudy No. 5, Institut Betona i Zhelezobe-

tona (Concrete and Reinforced Concrete Institute), Moscow, 1959, pp. 5–28 (in Russian). Translated by Portland Cement Association, Foreign Literature Study No. 371. Available from S.L.A. Translation Center, The John Crerar Library Translation Center, 35 W. 33rd St., Chicago, Illinois 60616.

2. Lessig, N. N., "Studies of Cases of Concrete Failure in Rectangular Reinforced Concrete Elements Subjected to Combined Flexure and Torsion," *Design of Reinforced Concrete Structures,* State Publishing Offices of Literature on Structural Engineering, Architecture and Construction Materials (Moscow), 1961, pp. 229–271 (in Russian). Translated by Portland Cement Association, Foreign Literature Study No. 398.

3. State Committee on Construction of the USSR Council of Ministers, "Structural Standards and Regulations," SNiP II-B, 1–62, State Publishing Offices for Literature on Structural Engineering, Architecture and Structural Materials, Moscow, 1962 (in Russian).

4. Collins, M. P., P. F. Walsh, F. E. Archer, and A. S. Hall, "Reinforced Concrete Beams Subjected to Combined Torsion, Bending and Shear," UNICIV Report, No. R-14, University of New South Wales, October 1965.

5. Collins, M. P., P. F. Walsh, F. E. Archer, and A. S. Hall, "Ultimate Strength of Reinforced Concrete Beams Subjected to Combined Torsion and Bending," *Torsion of Structural Concrete,* SP-18, American Concrete Institute, Detroit, 1968, pp. 279–402.

6. McMullen, A. E. and J. Warwaruk, "Concrete Beams in Bending, Torsion and Shear," *Journal of the Structural Division, ASCE,* Vol. 96, No. ST 5, May 1970, pp. 885–903.

7. Standards Association of Australia, "SAA Concrete Structures Code," A.S. CA2–1973; AS 1480–1974 (metric).

8. Standards Association of Australia, "SAA Prestressed Concrete Code," AS 1481–1978.

9. Rangan, B. V. and A. S. Hall, "Design of Prestressed Concrete Beams Subjected to Combined Bending, Shear and Torsion," *Journal of the American Concrete Institute,* Proc., Vol. 72, No. 3, March 1975, pp. 89–93.

10. Yudin, V. K., "Determination of the Load-Carrying Capacity of Rectangular Reinforced Concrete Elements Subjected to Combined Torsion and Bending," *Beton i Zhelezobeton* (Concrete and Reinforced Concrete), Moscow, No. 6, June 1962, pp. 265–269 (in Russian). Translated by Portland Cement Association, Foreign Literature Study No. 377.

11. Goode, C. D. and M. A. Helmy, "Ultimate Strength of Reinforced Concrete Beams in Bending and Torsion," *Torsion of Structural Concrete,* SP No. 18, American Concrete Institute, Detroit, 1968, pp. 357–377.

12. Silva, C. W., O. Buyukozturk, and D. N. Wormley, "Postcracking Compliance Analysis of RC Beams," *Journal of the Structural Division, ASCE,* Vol. 105, No. ST 1, January 1979, pp. 35–51.

13. Elfgren, L., I. Karlsson, and A. Losberg, "Torsion–Bending–Shear Interaction for Concrete Beams," *Journal of the Structural Division, ASCE,* Vol. 100, No. ST 8, August 1974, pp. 1657–1676.

14. Elfgren, L., "Reinforced Concrete Beams Loaded in Combined Torsion, Bending and Shear," Publication 71:3, Division of Concrete Structures, Chalmers University of Technology, Goteborg, Sweden, 1972.

15. Müller, P., "Failure Mechanism for Reinforced Concrete Beams in Torsion and Bending," *Publications (Memoire), International Association for Bridge and Structural Engineering (IABSE),* Vol. 36-II, 1976, pp. 147–163.

16. Hsu, T. T. C., "Torsion of Structural Concrete—Interaction Surface for Combined Torsion, Shear and Bending in Beams without Stirrups," *Journal of the American Concrete Institute,* Proc., Vol. 65, January 1968, pp. 51–60.

Plate 7: Prestressed concrete double-tee girders in the Dade County Mass Rapid Transit aerial guideway, Florida, under construction in 1982. Large torsional moment in the girder is caused by the wind load and by the nosing/lurching action of the vehicles.

7
Variable-Angle Truss Model

7.1 INTRODUCTION

7.1.1 Brief History of the Truss Model for Shear and Torsion

The concept of simulating the post-cracking action of a reinforced concrete member by a truss model originated with Ritter[1] and Morsch[2] at the turn of the twentieth century. For a reinforced concrete beam subjected to shear, diagonal cracks will occur and will separate the concrete into a series of concrete struts. It was assumed that this beam acts like a plane truss to carry the load. The top and bottom longitudinal bars serve as the top and bottom chords of the truss, while the transverse steel bars and the concrete struts serve as the web members. For simplicity, the inclination of the concrete struts was assumed to be 45°. Their theory, therefore, could be called the 45° truss model, or the 45° truss analogy. From equilibrium, three equations can be derived for calculating the stresses in the transverse steel, in the longitudinal steel, and in the 45° concrete struts.

The 45° truss model was extended by Rausch[3] in 1929 to simulate the post-cracking action of reinforced concrete members subjected to torsion. A concrete member reinforced with longitudinal and hoop steel is assumed to act like a tube, so that the applied torsional moment is resisted by the circulatory shear flow in the wall of the tube. Furthermore, the tube is assumed to act like a space truss in resisting the circulatory shear flow. Each straight segment of the wall is a plane truss resisting shear forces as conceived by Ritter and Morsch. A detailed discussion of Rausch's space truss analogy was given in Section 3.2.1. It should be recalled that this theory demands equal volumes of longitudinal steel and transverse steel when a member is subjected to pure torsion.

The compatibility conditions of Rausch's 45° space truss model were studied by the author. By using these compatibility conditions, in addition to Rausch's equilibrium equations, the post-cracking shear modulus and the post-cracking torsional rigidity were derived.[4] This derivation was presented in Section 3.3.1.

The 45° truss model was generalized by Lampert and Thurlimann[5] in 1969 for members subjected to torsion or to combined torsion and bending. They assumed that the angle of inclination of the concrete struts may deviate from 45°, and that the theory of plasticity is applicable to reinforced concrete members. By so doing, they were able to explain the phenomenon that both longitudinal and transverse steel may yield under pure torsion even if their volume ratios are not equal. Since the angle of the concrete struts is not 45°, they called their theory the variable-angle truss model. Three equilibrium equations were derived that incorporated this variable angle of the concrete struts. In analysis of a given member, the angle of the concrete struts is determined by the relative magnitude of the yield forces in the given transverse and longitudinal steel. In design, however, this angle may be arbitrarily chosen to achieve maximum economy in selecting transverse and longitudinal steel, provided that the serviceability conditions are met.

In studying the deformation of a member subjected to torsion, Lampert and Thurlimann also observed that a plane surface of the member before twisting becomes a hyperbolic surface after twisting. A diagonal concrete strut would, therefore, be subjected to bending in addition to compression. From geometry, they derived two compatibility equations, one relating the bending curvature of the concrete strut to the angle of twist of the whole member and another relating the curvature of the concrete strut to the extreme compressive strain at the surface of the concrete strut.

The variable-angle truss model was further applied by Elfgren[6] to members subjected to torsion, bending, and shear. He observed that this variable-angle truss model is very similar to Wagner's tensile field theory[7] for a metal girder with a thin web. After the shear buckling of the web, a metal girder will act like a truss, with the web taking only tensile stress in a diagonal direction. In a reinforced concrete member, the concrete web after cracking is assumed to take only compressive stress. For this reason, Elfgren called his theory the "compressive stress field theory." However, his formula to determine the angle of the compressive stress field was based on the plasticity theory. This is quite different from Wagner's angle for a tensile stress field, which was derived from strain compatibility. Elfgren also recognized that the angle of the compression field is different from the actual angle of the cracks.

The variable-angle truss model was developed along a somewhat different path by Collins[8] in 1973. Instead of using the theory of plasticity, he focused his attention on the strain compatibility of the truss model. As a result, he derived a compatibility equation to determine the angle of the compression stress field. This compatibility equation, which is identical to that of Wagner, enables the strain condition to be predicted by Mohr's circle. Collins called his theory the "diagonal compression field theory."

In addition to the compatibility and equilibrium equations in the variable-angle truss model, the compressive stress-strain curve of the concrete struts must be assumed. When the conventional stress-strain curve obtained from the standard concrete compression cylinder was assumed, the prediction of torsional strength was found to be very unconservative. By using a "softened" stress-strain curve, which resulted from the diagonal shear cracking, the author and Mo[17] were able to correctly predict the torsional strength as well as the deformations and strains throughout the loading history.

To clarify the names of the various theories discussed above, the author would like to suggest the following terminology: (1) Compression field theory is based on the variable-angle truss model assuming that the angle of inclination of the cracks is identical to the angle of inclination of the compression field; (2) Lampert and Thurlimann's theory and Elfgren's theory can collectively be called the plasticity compression field theory because these theories are based on the theory of plasticity; and (3) Collins's theory and the theory by the author and Mo can be called the compatibility compression field theory because it utilizes the strain compatibility of the truss model.

The plasticity compression field theory serves as a basis for the so-called accurate method in the CEB-FIP Model Code of 1978.[9-11] Tentative design recommendations have also been proposed by Collins and Mitchell[12] using the compatibility compression field theory. In this chapter we shall introduce only the official CEB-FIP Model Code.

7.1.2 Advantages of the Variable-Angle Truss Model

The advantages of the variable-angle truss model and the compression field theory can be summarized as follows:

1. The theory provides a clear concept of how a reinforced concrete beam resists shear and torsion after cracking. Consequently, the treatment of shear and torsion becomes unified. This is why we shall first study shear in this chapter before introducing torsion.
2. The interaction of shear and torsion with bending and axial load can be easily managed. The combination is quite consistent and comprehensible.
3. The effect of prestress can be included in a logical way. Hence, the whole range of prestressing from nonprestressed beams to fully prestressed beams can be unified.
4. The theory permits us to predict the deformation of a member throughout the loading history. This is a distinct advantage over the skew-bending theories, discussed in Chapter 6, which are applicable only at the ultimate load stage.
5. It provides reasonable accuracy when compared to tests and experience.
6. It can serve as a basis for the formulation of design code, as exemplified by the "accurate method" in the CEB-FIP Model Code.

As compared to ACI Code, design provisions based on the variable-angle truss model have the following advantages:

1. Such design provisions are applicable to beams that are not made up of rectangular components.
2. In a large girder, it is possible to design less web reinforcement for the side where shear and torsion are subtractive than for the side where shear and torsion are additive.
3. In the flexural compression zone of a beam, the longitudinal steel required for flexure is in compression, while the torsional steel is in tension. The simple addition of the flexural and torsional steel is obviously incorrect. The variable-angle truss model provides a more logical way to design longitudinal bars in the flexural compression zone.

7.1.3 Assumptions

The variable-angle truss model and the compression field theory are based on the following assumptions:

1. The truss model is made up of diagonal concrete struts inclined at an angle α, longitudinal bars, and transverse bars.
2. Diagonal concrete struts carry the principal compressive stress. The shear resistances of the concrete struts and the compression chord are not recognized.
3. Longitudinal and transverse bars carry only axial tension; i.e., dowel resistance is neglected.
4. The tensile strength of concrete is neglected (just as in flexural members).
6. For a solid section subjected to torsion, the concrete core does not contribute to the torsional resistance.

7.2 EQUILIBRIUM AND COMPATIBILITY OF A SHEAR ELEMENT

7.2.1 Equilibrium of a Shear Element

7.2.1.1 Equilibrium Relationship. A thin square reinforced concrete panel subjected to uniform shear flow q on all four edges is shown in Fig. 7.1 (a). The panel is reinforced with uniformly and closely spaced steel bars in both the longitudinal and the transverse directions. The steel could be imagined to smear continuously, so that the concept of steel force per unit length can be used. The percentages of steel are assumed to be different in the two directions, and the direction of the principal compressive stress is

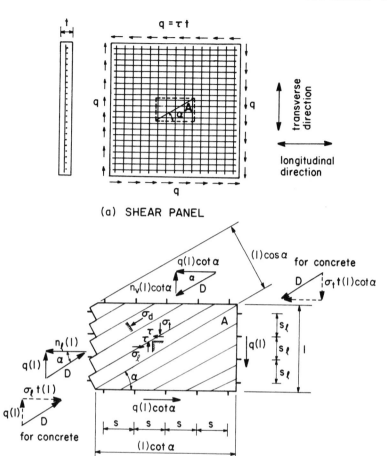

(a) SHEAR PANEL

(b) FORCES ACTING ON SHEAR ELEMENT A

Fig. 7.1. Equilibrium of a shear element.

inclined at an angle α (with respect to the longitudinal axis) other than 45°. Such a panel has actually been tested.[13]

To study the equilibrium of this reinforced concrete panel a small element A is isolated from the panel and is shown in Fig. 7.1 (b). This element A has a unit length in the transverse direction and a length of cot α in the longitudinal direction. The diagonal line connecting the lower left-hand corner to the upper right-hand corner will have an inclined angle of α and, therefore, represents the direction of the principal compressive stress (compression field). In the compression field theory this direction is assumed to be identical to that of the cracks; consequently, a series of parallel diagonal lines can be drawn that are assumed to separate the concrete into a series of diagonal concrete struts.

Element A is subjected to a shear force of $q(1)$ on the left and right edges. The shear force on the top and bottom edges is $q \cot \alpha$. Under these four external shear forces the element is in equilibrium. To find the stresses in the concrete and in the steel of element A, let us define the following notations:

q = shear flow on concrete (lb/in.)
t = thickness of the shear element (in.)
τ = shear stress on concrete = q/t (lb./in.²)
σ_t = transverse compressive normal stress on concrete
σ_l = longitudinal compressive normal stress on concrete
σ_d = diagonal compressive stress along the concrete struts
n_l = longitudinal tensile normal force of steel per unit length = $A_l f_l / s_l$
n_v = transverse (or vertical) tensile normal force of steel per unit length = $A_v f_v / s$
A_v = cross-sectional area of one transverse (or vertical) steel bar
A_l = cross-sectional area of one longitudinal steel bar
f_v = stress in the transverse (or vertical) steel bars
f_l = stress in the longitudinal steel bars
s = spacing of transverse steel bars
s_l = spacing of longitudinal steel bars

The shear force $q(1)$ acting on the left edge of the shear element in Fig. 7.1 (b) can be resolved into two components, D and $n_l(1)$, as shown by the force triangle. D is the diagonal compressive force on the concrete struts, acting on a perpendicular length of (1) cos α; and $n_l(1)$ is the longitudinal tensile normal force in the steel acting on the transverse length of unity. From the force triangle:

$$n_l = q \cot \alpha \qquad (7\text{--}1)$$

$$D = \frac{q(1)}{\sin \alpha} \qquad (7\text{--}2)$$

Similarly, the shear force $q(1) \cot \alpha$ on the top edge can be resolved into two components D and $n_v(1) \cot \alpha$, where $n_v(1) \cot \alpha$ is the transverse tensile normal force of the steel acting on the longitudinal length of $\cot \alpha$. From this force triangle:

$$n_v = q \tan \alpha \qquad (7\text{--}3)$$

The stress in the longitudinal steel, f_l, can be expressed by substituting the definition of $n_l = A_l f_l / s_l$ into Eq. (7–1):

$$f_l = \frac{qs \cot \alpha}{A_l} \tag{7–4}$$

Similarly, the stress in the transverse steel, f_v, is obtained by inserting the definition $n_v = A_v f_v / s$ into Eq. (7–3):

$$f_v = \frac{qs \tan \alpha}{A_v} \tag{7–5}$$

Now let us examine the stresses in the concrete. Equilibrium of the normal forces on the left edge requires that $\sigma_l t = n_l$. Substituting this relationship into Eq. (7–1) gives:

$$\sigma_l = \frac{q}{t} \cot \alpha = \tau \cot \alpha \tag{7–6}$$

Similarly, equilibrium of the normal forces on the top edge requires that $\sigma_t t = n_v$. As a result, Eq. (7–3) becomes:

$$\sigma_t = \frac{q}{t} \tan \alpha = \tau \tan \alpha \tag{7–7}$$

The diagonal stress in the concrete struts σ_d can be obtained from Eq. (7–2), if we observe that the diagonal force D is acting on a length of (1) cos α. Then:

$$\sigma_d = \frac{D}{t(1) \cos \alpha} = \frac{\tau}{\sin \alpha \cos \alpha} \tag{7–8}$$

7.2.1.2 Mohr's Circle for Stresses.

Equations (7–6) through (7–8) are the three basic equilibrium equations for finding stresses in the concrete of a shear panel. σ_l and σ_t can also be expressed in terms of σ_d by substituting Eq. (7–8) into Eqs. (7–6) and (7–7):

$$\sigma_l = \sigma_d \cos^2 \alpha = \frac{\sigma_d}{2} + \frac{\sigma_d}{2} \cos 2\alpha \tag{7–9}$$

$$\sigma_t = \sigma_d \sin^2 \alpha = \frac{\sigma_d}{2} - \frac{\sigma_d}{2} \cos 2\alpha \qquad (7\text{–}10)$$

Also Eq. (7–8) can be written as:

$$\tau = \sigma_d \sin \alpha \cos \alpha = \frac{\sigma_d}{2} \sin 2\alpha \qquad (7\text{–}11)$$

Equations (7–9) through (7–11) can be expressed graphically by the well-known Mohr's stress circle, shown in Fig. 7.2. In Mohr's stress circle, the horizontal and vertical axes represent the normal stress σ and the shear stress τ, respectively. A point on the circle gives the stress condition on a specifically oriented section of the shear panel. The coordinates of the point give the stresses σ and τ, while the subtending angle gives twice the angle between this section and a reference surface. The reference surface can be taken as the vertical face, on which the given stresses σ_1 and τ are acting. This reference surface is represented by point A on the circle. Point B on the opposite side of the circle (180° away) gives the stresses σ_t and τ on the horizontal face. Point C, which has a subtending angle of 2α, represents the cross section of the concrete strut. Only the stress σ_d is acting on this section. This stress is shown to be:

$$\sigma_d = \sigma_1 + \sigma_t \qquad (7\text{–}12)$$

Point O at the origin also shows that no stresses are acting on the section parallel to the direction of the cracks. When a dotted line is drawn between

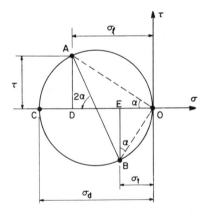

Fig. 7.2. Mohr's circle for average stresses in concrete (tensile strength of concrete neglected).

points O and A, the angle AOC will be α. Therefore, the triangle AOD embodies the relationship of Eq. (7–6). Similarly, the triangle BOE gives the relationship of Eq. (7–7).

7.2.2 Compatibility of a Shear Element

7.2.2.1 Compatibility Relationships. Let us now examine the geometric compatibility of deformations in a shear element, shown in Fig. 7.1 (b). In Section 7.2.1 we found the stress f_l in the longitudinal steel by Eq. (7–4), and the stress f_v in the transverse steel by Eq. (7–5). The tensile strains in the steel bars can be obtained from the stress–strain relationship. Assuming the stress–strain relationship to be linear and the modulus of elasticity of steel to be designated E_s, then:

$$\epsilon_l = \frac{f_l}{E_s} = \frac{\tau ts \cot \alpha}{E_s A_l} = \frac{\tau \cot \alpha}{E_s r_l} \tag{7–13}$$

where $r_l = A_l/ts$ is the longitudinal reinforcement ratio. Also:

$$\epsilon_t = \frac{f_v}{E_s} = \frac{\tau ts_l \tan \alpha}{E_s A_v} = \frac{\tau \tan \alpha}{E_s r_v} \tag{7–14}$$

where $r_v = A_v/ts_l$ is the transverse reinforcement ratio. The compressive strain in the concrete struts ϵ_d is related nonlinearly to the compressive stress σ_d, which can be calculated from Eq. (7–8). The relationship between σ_d and ϵ_d will be discussed in Section 7.2.3.

The strains in the longitudinal and transverse steel and in the concrete struts will produce shear distortions of the shear element. These distortions, due to ϵ_l, ϵ_t, and ϵ_d, are shown separately in Fig. 7.3. Figure 7.3 (a) illustrates the distortion γ_l, due to the strain of the longitudinal bars, ϵ_l. The original length of the top longitudinal edge of the element, designated AB, is equal to cot α. After elongation, the length becomes AE. The elongation BE is equal to ϵ_l cot α. To maintain compatibility of edge AE with the diagonal CB, the edge AE must rotate about point A, while the diagonal CB must rotate about point C. The rotations are completed when the end E of edge AE meets the end B of diagonal CB at point F. For small deformation, BF is assumed to be perpendicular to diagonal line CB, and EF perpendicular to edge AE. From geometry, the length $EF = \epsilon_l \cot^2 \alpha$. After deformation the shape of the shear element becomes a parallelogram as shown by the

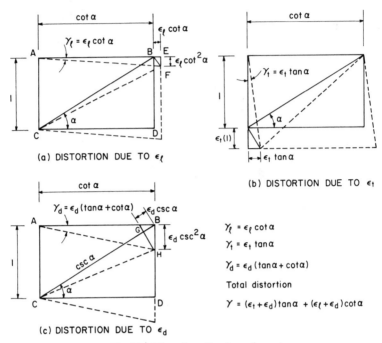

Fig. 7.3. Distortion of a shear element.

dotted lines. The shear distortion of the element γ_l is represented by the angle BAF, giving:

$$\gamma_l = \epsilon_l \cot \alpha \qquad (7\text{--}15)$$

Similarly, the shear distortion γ_t, due to ϵ_t, is illustrated in Fig. 7.3 (b), resulting in:

$$\gamma_t = \epsilon_t \tan \alpha \qquad (7\text{--}16)$$

The shear distortion γ_d, due to the strain ϵ_d, as illustrated in Fig. 7.3 (c), requires some elaboration. The original length of the diagonal CB is $\csc \alpha$, so that the shortening of the diagonal BG is $\epsilon_d \csc \alpha$. To maintain geometric compatibility of deformation, the shortened diagonal CG and the edge AB will both rotate until end G (of CG) and end B (of AB) meet at point H. The deformation BH should be $\epsilon_d \csc^2 \alpha$ from geometry. The shape of the element after deformation is given by the dotted lines, and the distortion γ_d is represented by the angle BAH. Hence:

$$\gamma_d = \frac{\epsilon_d \csc^2 \alpha}{\cot \alpha} = \frac{\epsilon_d}{\sin \alpha \cos \alpha} = \epsilon_d(\tan \alpha + \cot \alpha) \qquad (7\text{--}17)$$

The total shear distortion is obtained by adding Eqs. (7–15), (7–16), and (7–17):

$$\gamma = \gamma_l + \gamma_t + \gamma_d = (\epsilon_l + \epsilon_d) \cot \alpha + (\epsilon_t + \epsilon_d) \tan \alpha \qquad (7\text{--}18)$$

Equation (7–18) shows that the total distortion γ is a function of the angle α. The angle α that gives a minimum value of γ can be obtained by differentiating Eq. (7–18) with respect to α and equating it to zero:

$$\tan^2 \alpha = \frac{\epsilon_l + \epsilon_d}{\epsilon_t + \epsilon_d} \qquad (7\text{--}19)$$

Equation (7–19) determines the angle α that makes the total shear distortion a minimum. As will be explained below, this equation can also be derived directly as a geometric relationship among strains ϵ_l, ϵ_t, and ϵ_d.

7.2.2.2 Physical Meaning of Eq (7–19).

The geometrical meaning of Eqs. (7–18) and (7–19) will be illustrated in Fig. 7.4. Figure 7.4 (a) shows the combined shear distortion due to the tensile strains, ϵ_l and ϵ_t, in both the transverse and longitudinal steel. This distortion is obtained by adding the distortion due to ϵ_l in Fig. 7.3 (a) and the distortion due to ϵ_t in Fig. 7.3 (b). The combined distortion in Fig. 7.4 (a) now has two components, γ_l and γ_t, as given in Eqs. (7–15) and (7–16).

The distortion due to compression strain ϵ_d in the concrete struts is derived in Fig. 7.4 (b). This derivation is quite different from that in Fig. 7.3 (c) and, therefore, requires some explanation. In Fig. 7.3 (c) we assume that the shortening of the diagonal CB occurs at one end, B, while the other end, C, remains in place. In Fig. 7.4 (b), however, we shall assume that the shortening of the diagonal CB occurs at both ends with respect to the point O. Point O is the end point of line AO, which is perpendicular to the diagonal CB. Since the length of line CO is $\sin \alpha$, the shortening CG at the end C is $\epsilon_d \sin \alpha$. At the other end, B, the shortening BE is $\epsilon_d \cos \alpha \cot \alpha$ because the length of line OB is $\cos \alpha \cot \alpha$. To maintain geometric compatibility after deformation, the shortened diagonal GE must move to the new position HF. Point F is the intersection of displacement BF (which is perpendicular to the original edge AB) and the displacement

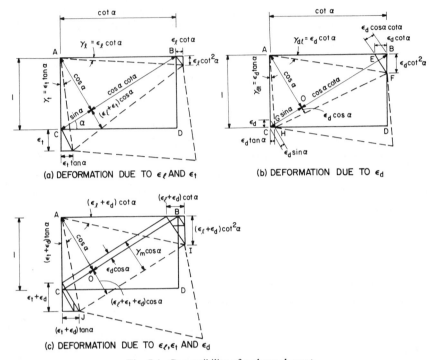

(a) DEFORMATION DUE TO ϵ_ℓ AND ϵ_t

(b) DEFORMATION DUE TO ϵ_d

(c) DEFORMATION DUE TO $\epsilon_\ell, \epsilon_t$ AND ϵ_d

Fig. 7.4. Compatibility of a shear element.

EF (which is perpendicular to the diagonal *GE*). Similarly, point *H* is the intersection of displacements *CH* and *GH*.

The distortion due to ϵ_d, as illustrated in Fig. 7.4 (b), has two components, the longitudinal component $\gamma_{d\ell}$ and the transverse component γ_{dt}:

$$\gamma_{d\ell} = \frac{BF}{AB} = \epsilon_d \cot \alpha \qquad (7\text{–}20)$$

$$\gamma_{dt} = \frac{CH}{AC} = \epsilon_d \tan \alpha \qquad (7\text{–}21)$$

When Eqs. (7–20) and (7–21) are added, the combined distortion due to ϵ_d is $\gamma_d = \gamma_{d\ell} + \gamma_{dt} = \epsilon_d(\tan \alpha + \cot \alpha)$. This is identical to Eq. (7–17), which was derived from Fig. 7.3 (c). The difference between the derivation in Fig. 7.3 (c) and that in Fig. 7.4 (b) is now apparent. The latter allows us to obtain the two components of the distortion due to ϵ_d, whereas the former gives us only a combined value.

Let us now add the distortion due to ϵ_d in Fig. 7.4 (b) to the distortion due to ϵ_l and ϵ_t in Fig. 7.4 (a). The resulting distortion due to all three strains, ϵ_l, ϵ_t, and ϵ_d, is shown in Fig. 7.4 (c). The longitudinal and the transverse components of the combined distortion are:

$$\gamma_l + \gamma_{dl} = (\epsilon_l + \epsilon_d) \cot \alpha \qquad (7\text{–}22)$$

$$\gamma_t + \gamma_{dt} = (\epsilon_t + \epsilon_d) \tan \alpha \qquad (7\text{–}23)$$

If we equate the total longitudinal components of distortion to the total transverse components of distortion, then:

$$(\epsilon_l + \epsilon_d) \cot \alpha = (\epsilon_t + \epsilon_d) \tan \alpha \qquad (7\text{–}24)$$

and:

$$\tan^2 \alpha = \frac{\epsilon_l + \epsilon_d}{\epsilon_t + \epsilon_d} \qquad (7\text{–}25)$$

Note that Eq. (7–25) is identical to Eq. (7–19). This equation, therefore, has a physical meaning. It means that the distortion due to ϵ_l, ϵ_t, and ϵ_d should produce a longitudinal component and a transverse component, which are equal in magnitude. As a corollary of this statement, it can be shown by trigonometry and geometry that the diagonal IJ after deformation should be parallel to the diagonal CB before deformation (see Fig. 7.4c). Furthermore, the strain perpendicular to the diagonal CB should be $\epsilon_l + \epsilon_t + \epsilon_d$. The corresponding displacement should be $(\epsilon_l + \epsilon_t + \epsilon_d) \cos \alpha$, since the length AO is equal to $\cos \alpha$.

7.2.2.3 Mohr's Circle for Strains.
In view of Eq. (7–24), the first term in Eq. (7–18) is equal to the second term. Hence, each of the two terms in Eq. (7–18) must be equal to $\gamma/2$:

$$\frac{\gamma}{2} = (\epsilon_l + \epsilon_d) \cot \alpha \qquad (7\text{–}26)$$

$$\frac{\gamma}{2} = (\epsilon_t + \epsilon_d) \tan \alpha \qquad (7\text{–}27)$$

Equations (7–26) and (7–27) can also be written as:

$$\frac{\gamma}{2} \tan \alpha = \epsilon_l + \epsilon_d \qquad (7\text{–}28)$$

$$\frac{\gamma}{2} \cot \alpha = \epsilon_t + \epsilon_d \tag{7-29}$$

Summing Eqs. (7–28) and (7–29) gives:

$$\frac{\gamma}{2} \frac{1}{\sin \alpha \cos \alpha} = \epsilon_l + \epsilon_t + 2\epsilon_d \tag{7-30}$$

Let's define:

$$\gamma_m = \epsilon_l + \epsilon_t + 2\epsilon_d \tag{7-31}$$

Then Eq. (7–30) becomes:

$$\frac{\gamma}{2} = \frac{\gamma_m}{2} \sin 2\alpha \tag{7-32}$$

Substituting Eq. (7–32) into Eqs. (7–28) and (7–29), we have:

$$\epsilon_l + \epsilon_d = \gamma_m \sin^2 \alpha \tag{7-33}$$

$$\epsilon_t + \epsilon_d = \gamma_m \cos^2 \alpha \tag{7-34}$$

From trigonometry Equations (7–33) and (7–34) can be written as:

$$\epsilon_l = \left(\frac{\gamma_m}{2} - \epsilon_d\right) - \frac{\gamma_m}{2} \cos 2\alpha \tag{7-35}$$

$$\epsilon_t = \left(\frac{\gamma_m}{2} - \epsilon_d\right) + \frac{\gamma_m}{2} \cos 2\alpha \tag{7-36}$$

Equations (7–32), (7–35), and (7–36) can be expressed graphically by Mohr's strain circle, as shown in Fig. 7.5. In Mohr's strain circle, the horizontal and vertical axes represent the normal strain, ϵ, and one-half of the shear strain, $\gamma/2$, respectively. A point on the circle gives the strain condition in a specific direction. The strain condition is represented by the coordinates of the point, while the subtending angle gives twice the angle between the direction of the strain and a reference direction. The reference direction can be taken as the longitudinal direction, along which the given strains ϵ_l and $\gamma/2$ are acting. This reference direction is represented by point A on the circle. On the opposite side of the circle (180° away), point B provides the given strains ϵ_t and $\gamma/2$ in the transverse direction. Point C, which has a

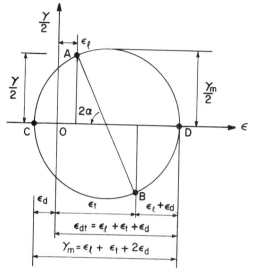

Fig. 7.5. Mohr's circle for average strains in panel.

subtending angle of 2α, represents the direction of the diagonal concrete struts. In this direction only the strain ϵ_d is acting. The direction perpendicular to the diagonal concrete strut is represented by point D, which is 180° away from point C. Along this direction the principal tensile strain is:

$$\epsilon_{dt} = \epsilon_l + \epsilon_t + \epsilon_d \qquad (7\text{-}37)$$

This relationship was previously observed by examining the deformation in Fig. 7.4 (c). Finally, it should be noted that the diameter of Mohr's circle for strain is equal to γ_m, which is the maximum total shear distortion in the element. The relationship between the maximum distortion γ_m and the strains ϵ_l, ϵ_t, and ϵ_d was given in Eq. (7–31).

7.2.3 Compression Stress–Strain Relationship of Concrete in a Shear Element

7.2.3.1 Stress-Strain Relationship. Seventeen shear panels similar to that shown in Fig. 7.1 (a) were tested at the University of Toronto.[13] Three average strains of this panel, ϵ_l, ϵ_t, and ϵ_{45}, were measured. The strain ϵ_{45} was in the inclined 45° direction. (Actually, two mutually perpendicular strains in the 45° and −45° directions were measured. The additional one was used as a check.) From these strains, Mohr's strain circle as explained in Fig. 7.5 could be established. Based on this measured Mohr's circle, the angle α, as well as the principal compressive strain, ϵ_d, could be determined.

From the measured strains ϵ_l and ϵ_t, the stresses in the longitudinal and transverse steel bars, f_l and f_v, could also be calculated using the stress–strain relationship of the reinforcement, i.e., $f_l = E_s \epsilon_l$ and $f_v = E_s \epsilon_t$. The stresses in the concrete, σ_l and σ_t, could then be calculated by $\sigma_l = r_l f_l$ and $\sigma_t = r_v f_v$, where r_l and r_v are the reinforcement ratios in the longitudinal and transverse directions, respectively. Using these calculated concrete stresses, σ_l and σ_t, together with the applied shear stress, τ, Mohr's circle for stresses (explained in Fig. 7.2) could be established. This measured Mohr's stress circle should provide the principal compressive stress in the concrete, σ_d. This principal compressive stress of concrete, σ_d, and the corresponding principal compressive strain of concrete, ϵ_d (obtained from Mohr's strain circle), provide one point on the stress–strain curve of concrete. By applying the shear stress on the panel in increments and measuring the strains at each load stage, a complete stress–strain curve of concrete was achieved.

It is interesting to point out that the measured Mohr's circle for average concrete stresses was not exactly the same as indicated in Fig. 7.2. It was observed that the measured Mohr's circle had, in general, shifted to the right along the σ-axis by an amount $\bar{\sigma}_t$, as shown in Fig. 7.6. In other words, the stresses in the steel, f_l and f_t, and, therefore, the stresses in the concrete, σ_l and σ_t, were smaller than predicted by the truss model presented in Section 7.2.1. The reason for this discrepancy is quite simple. In the truss model presented, we neglected the contribution of the tensile strength of concrete. Consequently, $\bar{\sigma}_t$ represented an average uniform tensile stress of concrete, which is assumed to be constant in all directions. The quantity

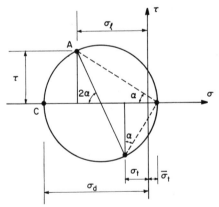

Fig. 7.6. Mohr's circle for average stress in concrete (tensile strength of concrete considered).

$\bar{\sigma}_t$ is also called the principal tensile stress of concrete, σ_{dt}, because it happens to be the only stress perpendicular to the direction of the principal compressive stress. Since the tensile strength of concrete is usually very small in comparison to the compressive strength of concrete, it is often acceptable to neglect the tensile strength of concrete in the design of reinforced concrete structures. However, the tensile strength of concrete must be taken into account in research to predict the behavior of a reinforced concrete shear panel.

Based on the testing of 17 shear panels,[13] expressions for the stress–strain curve of concrete in the direction of the principal compression (σ_d vs. ϵ_d) were proposed as follows:

$$\sigma_d = f'_c \left[2 \left(\frac{\epsilon_d}{\epsilon_0} \right) - \lambda \left(\frac{\epsilon_d}{\epsilon_0} \right)^2 \right] \tag{7-38}$$

where f'_c = standard 6 in. by 12 in. cylinder compressive strength of concrete; ϵ_0 = strain corresponding to the peak stress f'_c in a stress–strain curve for a standard 6 in. by 12 in. cylinder; and λ is an empirical coefficient given as a function of γ_m / ϵ_d:

$$\lambda = \sqrt{\frac{\gamma_m}{\epsilon_d}} - \nu \tag{7-39}$$

where ν is Poisson's ratio taken as 0.3, a value commonly observed near failure of a concrete cylinder.

Equation (7–38) is a modification of the commonly used parabolic stress–strain curve, except that the second term, $(\epsilon_d/\epsilon_0)^2$, in the bracket is modified by an empirical coefficient λ. This equation is plotted in a nondimensional form (σ_d/f'_c vs. ϵ_d/ϵ_0) in Fig. 7.7 for the ascending portion of the curves to the left of the dotted line. Four curves, with the parameter $1/\lambda = 1$, 0.75, 0.50, and 0.25, have been included. It can be seen that the peak strength of these four curves decreases in direct proportion to $1/\lambda$. Let's define σ_p as the peak compressive strength of concrete; then:

$$\sigma_p = \frac{1}{\lambda} f'_c \tag{7-40}$$

Similarly, the strain corresponding to the peak strength is defined as ϵ_p, where:

$$\epsilon_p = \frac{1}{\lambda} \epsilon_0 \tag{7-41}$$

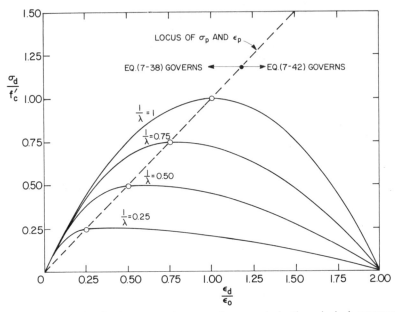

Fig. 7.7. Nondimensionalized stress–strain curves for concrete in the principal compression direction of a shear panel (Vecchio and Collins).

When a straight line is drawn through the peak strengths of the four curves, we obtain the locus of σ_p and ϵ_p as shown by the 45° dotted line in Fig. 7.7. Equation (7–38) applies only to the left of the locus line for the ascending portion of the stress–strain curves. In other words, it is applicable only when $\epsilon_d < \epsilon_p$.

Beyond the peak strength and in the descending portion of the stress–strain curve, (i.e., $\epsilon_d > \epsilon_p$), the curves are expressed by another equation:

$$\sigma_d = f_p \left(1 - \eta^2\right) \qquad (7\text{–}42)$$

where:

$$\eta = \frac{\epsilon_d - \epsilon_p}{2\,\epsilon_0 - \epsilon_p} \qquad (7\text{–}43)$$

Equation (7–42) is also a second-order parabolic curve that connects the point at peak strength to a fixed point at $\epsilon_d/\epsilon_o = 2$ and $\sigma_d/f'_c = 0$.

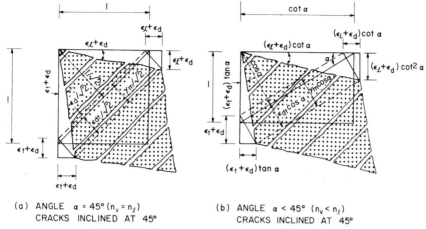

(a) ANGLE α = 45° (n$_v$ = n$_\ell$)
CRACKS INCLINED AT 45°

(b) ANGLE α < 45° (n$_v$ < n$_\ell$)
CRACKS INCLINED AT 45°

Fig. 7.8. Deformation and crack of a shear element.

7.2.3.2 Softening Coefficient 1/λ. The empirical coefficient λ, or its reciprocal 1/λ, deserves some discussion. The coefficient 1/λ can be considered to provide a "softening effect" that reduces the strength of the concrete from the standard strength of f'_c to $(1/\lambda)f'_c$. This softening effect is apparently caused by the diagonal shear cracking of concrete. Consequently, 1/λ must be a function of a parameter that measures the severity of the cracking.

To find a parameter that describes cracking, let us examine Fig. 7.8 (a). Figure 7.8 (a) shows the deformation of a square shear element for an α angle of 45°. This α angle which represents the direction of the principal compression *after* cracking, will occur when the steel forces per unit length in the longitudinal and transverse directions are equal (or $\eta_v = \eta_l$). On the other hand, the direction of the crack should also be 45°. This is so because the direction of the cracks is always perpendicular to the direction of the principal tensile stress *before* cracking, and the latter should be −45° when subjected to pure shear. Since the angle of cracks is the same as the α angle, crack widths will open exactly in the direction of the principal tensile stress. Consequently, the principal tensile strain ϵ_{dt} ($= \epsilon_l + \epsilon_t + \epsilon_d$), or the maximum total distortion γ_m ($= \epsilon_l + \epsilon_t + 2\epsilon_d$), can be taken as a measure of the severity of cracking. The corresponding elongations, $\epsilon_{dt}/\sqrt{2}$ and $\gamma_m/\sqrt{2}$ for α = 45°, are indicated in Fig. 7.8 (a).

The nondimensionalized γ_m/ϵ_d ratio has been chosen by Vecchio and Collins to be the parameter for the empirical coefficient λ, resulting in Eq. (7–39). For an uncracked concrete cylinder subjected to axial load $\epsilon_l = -\epsilon_d$, $\epsilon_t/\epsilon_d = \nu = 0.3$, and $\gamma_m/\epsilon_d = 1.3$. Substituting this value into

Eq. (7–39), we obtain $\lambda = 1$ for the stress–strain curve of a standard cylinder. This degeneration of λ into unity shows that λ is applicable for the whole range from a severely cracked case to an uncracked case.

Let us also examine Fig. 7.8 (b), which shows a case where the α angle is less than 45°. This α angle could occur when the transverse steel force per unit length, n_v, is less than the longitudinal steel force per unit length, n_l. However, since the angle of the cracks is determined by the direction of the principal tensile stress before cracking, it will remain 45°. So we have a case in which the direction of the principal compressive stress is not parallel to the direction of the cracks. That is, the principal compression must cross the cracks. This situation will produce shear stresses in the 45° concrete struts as well as shear stresses due to aggregate interlock in the cracks. These shear stresses certainly will weaken the compressive strength of concrete in the direction of the principal compression stress. Consequently, the angle α should also be a parameter for the softening coefficient $1/\lambda$. It is hoped that this refinement of the $1/\lambda$ coefficient could be made in the future.

The softening of concrete must be considered in the analysis of the strength and the behavior of a reinforced concrete member subjected to shear or torsion. This softening effect has been theoretically taken into account in the construction of a torque-twist curve for a torsional member in Example 7.2 (Section 7.8). In design practice the softening effect is considered by specifying a so-called effective strength of concrete.

7.3 TRUSS MODEL FOR BEAMS

7.3.1 Beam Element Subjected to Shear and Bending

7.3.1.1 Equilibrium of a Beam Element. In Section 7.2.1, the equilibrium of a reinforced concrete panel subjected to pure shear was presented. The principles used will now be extended to study a reinforced concrete beam subjected to combined shear and bending. Figure 7.9 (a) shows a simple beam subjected to a concentrated load at midspan. In this truss model, it will be assumed that all the longitudinal steel in the beam is concentrated in the top and bottom stringers. The distance between these two stringers will be designated d_v. This distance d_v can also be considered as the length of the vertical bars.

If two cuts are made on the beam at Sections I–I and II–II, we can isolate a beam element A of length $d_v \cot \alpha$. This beam element A is subjected to a shear force V and a bending moment M on the left face. On the right face, the shear force remains the same, but the bending moment has received an increment of $V d_v \cot \alpha$.

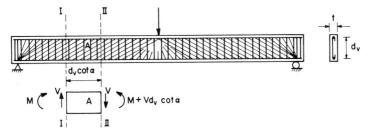

(a) A SIMPLE BEAM UNDER A CONCENTRATED LOAD

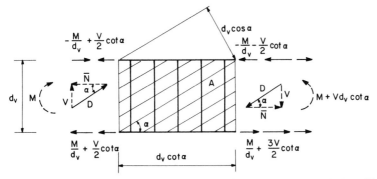

(b) EQUILIBRIUM OF BEAM ELEMENT A (ONLY FORCES WITH SOLID LINES ARE REAL FORCES FOR EQUILIBRIUM OF FREE BODY)

Fig. 7.9. Equilibrium of beam element in a simple beam under midspan load.

The equilibrium of beam element A is illustrated in Fig. 7.9 (b). The shear force V acting on the left face can be resolved into two components, $\overline{N}$ and D, as shown by the force triangle. The component $\overline{N}$ should be taken by all the longitudinal steel in Section I–I. From the force triangle:

$$\overline{N} = V \cot \alpha \qquad (7\text{-}44)$$

The other component, diagonal force D, is:

$$D = \frac{V}{\sin \alpha} \qquad (7\text{-}45)$$

The stresses in the diagonal concrete struts, σ_d, can be obtained from Eq. (7–45) if we observe that D is acting on a length of $d_v \cos \alpha$:

$$\sigma_d = \frac{D}{t d_v \cos \alpha} = \frac{V}{t d_v \sin \alpha \cos \alpha} \qquad (7\text{-}46)$$

Equations (7–44) and (7–45) are the same as Eqs. (7–1) and (7–2) for a panel subjected to pure shear. Equation (7–46) is also identical to Eq. (7–8), if the shear force V is considered to be uniformly distributed on the cross-sectional area of concrete, td_v.

The component $\overline{N}$ is shown in Fig. 7.9 (b) and is assumed to be divided equally between the top steel and the bottom steel, each part being $(V \cot \alpha)/2$. In addition to this force contributed by shear force V, the forces in the longitudinal steel on the left face are also contributed by the moment M. The force in the top and bottom steel due to moment is M/d_v. Summing these two contributions, the forces in the top and bottom stringers, N_t and N_b, are:

$$N_t = -\frac{M}{d_v} + \frac{V}{2} \cot \alpha \qquad (7\text{–}47)$$

$$N_b = \frac{M}{d_v} + \frac{V}{2} \cot \alpha \qquad (7\text{–}48)$$

On the right face of the beam element, the moment is $M + Vd_v \cot \alpha$. Consequently, the forces in the top and bottom stringers on the right face are:

$$N_t = -\frac{M + Vd_v \cot \alpha}{d_v} + \frac{V}{2} \cot \alpha = -\frac{M}{d_v} - \frac{V}{2} \cot \alpha \qquad (7\text{–}49)$$

$$N_b = +\frac{M + Vd_v \cot \alpha}{d_v} + \frac{V}{2} \cot \alpha = +\frac{M}{d_v} + \frac{3V}{2} \cot \alpha \qquad (7\text{–}50)$$

The beam element should be in equilibrium under these four stringer forces, Eqs. (7–47) through (7–50), plus the diagonal forces, D, on the left and right faces, Eq. (7–45).

The forces in the transverse steel can be obtained by making a horizontal cut through the beam element at an arbitrary distance y from the bottom edge so as to isolate a rectangular free body as shown in Fig. 7.10 (a). To maintain the horizontal equilibrium of the free body, there should be a horizontal shear force $V \cot \alpha$ on the cut surface. This horizontal shear force can be resolved into two components, $n_v d_v \cot \alpha$ and D, as shown by the force triangle. The term $n_v d_v \cot \alpha$ is the transverse steel force per unit length, n_v, acting on a longitudinal length of $d_v \cot \alpha$. This transverse force is resisted by the transverse steel bars, as indicated, in a uniform manner. D is the diagonal force acting on the concrete struts. From this force triangle, we have:

$$n_v d_v = V \tan \alpha \qquad (7\text{--}51)$$

Inserting the definition of $n_v = A_v f_v / s$ into Eq. (7–51), we can write:

$$A_v f_v = V \frac{s}{d_v} \tan \alpha \qquad (7\text{--}52)$$

Equations (7–51) and (7–52) are the same as Eqs. (7–3) and (7–5), respectively, if V is considered to be uniformly distributed along the depth, d_v.

Another way to find the transverse steel force, which is commonly used, is to make a diagonal cut at an angle α through the beam element, so as to isolate a triangular free body as shown in Fig. 7.10 (b). The forces acting on this free body are also indicated. Equations (7–51) and (7–52) will result directly from the equilibrium of forces in the transverse direction. To maintain

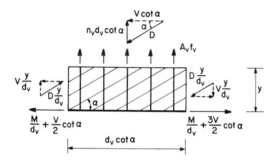

(a) RECTANGULAR FREE BODY

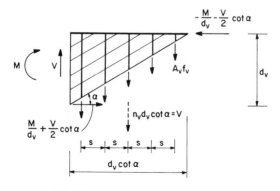

(b) TRIANGULAR FREE BODY

Fig. 7.10. Equilibrium of free bodies.

moment equilibrium the transverse force $n_v d_v \cot \alpha$ should be distributed uniformly among the transverse steel bars.

7.3.1.2 Shear–Bending Interaction Curve.

Let us now study the interaction of shear and moment at yield load. Failure may occur in two modes. In the first mode, failure is caused by the yielding in the bottom stringer and in the transverse steel. Consequently, Eqs. (7–48) and (7–52) become:

$$A_{lb}f_{ly} = \frac{M_y}{d_v} + \frac{V_y}{2}\cot\alpha \qquad (7\text{--}53)$$

$$A_v f_{vy} = V_y \frac{s}{d_v}\tan\alpha \qquad (7\text{--}54)$$

In the above two equations the subscripts y for the force, moment, and stress (V_y, M_y, and f_{vy}) indicate yielding; and N_b has been substituted by the area of the bottom stringer times the yield strength of longitudinal steel, $A_{lb}f_{ly}$.

Solving for $\tan \alpha$ in Eq. (7–54) and substituting it into Eq. (7–53) give:

$$\frac{M_y}{A_{lb}f_{ly}d_v} + \frac{V_y^2}{2(A_{lb}f_{ly})(A_v f_{vy})\dfrac{d_v}{s}} = 1 \qquad (7\text{--}55)$$

Let's define:

$$M_{0y} = A_{lb}f_{ly}d_v \qquad (7\text{--}56)$$

$$V_{0y} = \sqrt{2(A_{lb}f_{ly})(A_v f_{vy})\frac{d_v}{s}} \qquad (7\text{--}57)$$

where

M_{0y} = pure bending moment ($V_y = 0$) at yielding of bottom stringer

V_{0y} = pure shear force ($M_y = 0$) at yielding of bottom stringer and transverse steel

Then, Eq. (7–55) can be written as:

$$\frac{M_y}{M_{0y}} + \left(\frac{V_y}{V_{0y}}\right)^2 = 1 \qquad (7\text{--}58)$$

Equation (7–58) is the nondimensionalized shear–moment interaction curve for the first mode of failure when the bottom stringer and the transverse steel yield.

The second mode of failure is caused by the yielding of the top stringer and the transverse steel. Equation (7–47) becomes:

$$A_{1t}f_{1y} = -\frac{M_y}{d_v} + \frac{V_y}{2}\cot\alpha \qquad (7\text{--}59)$$

where N_t has been taken as the area of the top stringer times the yield strength of longitudinal steel, $A_{1t}f_{1y}$. Substituting $\tan\alpha$ from Eq. (7–54) into Eq. (7–59) results in:

$$-\frac{M_y}{A_{1t}f_{1y}d_v} + \frac{V_y^2}{2(A_{1t}f_{1y})(A_vf_{vy})\dfrac{d_v}{s}} = 1 \qquad (7\text{--}60)$$

In addition to the definitions in Eqs. (7–56) and (7–57), let us define R as the ratio of top steel area to bottom steel area:

$$R = \frac{A_{1t}}{A_{1b}} \qquad (7\text{--}61)$$

Then, Eq. (7–60) becomes:

$$-\frac{M_y}{M_{0y}} + \left(\frac{V_y}{V_{0y}}\right)^2 = R \qquad (7\text{--}62)$$

Equation (7–62) is the nondimensionalized shear–moment interaction curve for the second mode of failure when the top stringer and the transverse steel yield. This mode of failure will occur only when the bending moment is small and the area of the top stringer is considerably less than the area of the bottom stringer.

The two shear–moment interaction curves derived in Eqs. (7–58) and (7–62) are plotted in Fig. 7.11 using the steel area ratio R as a parameter. The former is drawn as a solid curve, while the latter is given as a series of dotted curves. In this figure a positive bending moment is defined as the moment that produces tension in the bottom stringer, while a negative bending moment is one that creates tension in the top stringer.

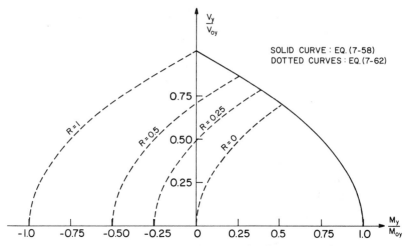

Fig. 7.11. Nondimensionalized interaction curves for shear and bending moment.

7.3.2 Effect of Concentrated Loads and the "Shift Rule"

In Section 7.3.1 and in Fig. 7.9 we studied the equilibrium of a beam element A. This element is cut out of a beam sufficiently far from the concentrated loads at the end and at the midspan to avoid the local effects of concentrated loads. The equilibrium condition of beam element A will remain valid at the end of the beam if the end surface is furnished with a force system as shown in Fig. 7.12 (a). This force system includes uniformly distributed shear and compressive stresses on the concrete, plus tensile forces in the top and bottom stringers.

To achieve this particular force system at the end surface, it is possible to provide a steel plate at the end surface and to fasten it to the top and bottom longitudinal bars as shown in Fig. 7.12 (b). Such a measure is, of course, uneconomical and is used here only to illustrate a possible hypothetical remedy. A more practical case, however, is that of a beam supported at the end by an intersecting beam, as shown in Fig. 7.12 (c). It would be reasonable to assume that the intersecting beam will furnish the required force system.

If the end reaction of a beam is supplied by a concentrated load on the bottom surface, then an interesting local effect will occur to modify the force system in the vicinity of the end reaction, such that a new equilibrium condition (other than that in beam element A) is reached. This local effect is shown in Fig. 7.13 as the "fanning" of the compression struts from the application point of the concentrated load. It is assumed here that the angle

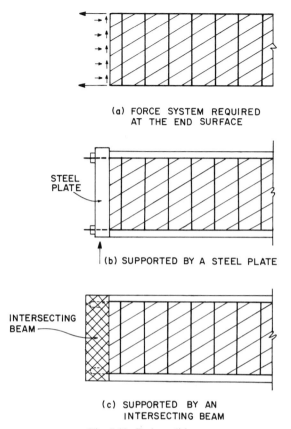

(a) FORCE SYSTEM REQUIRED
AT THE END SURFACE

(b) SUPPORTED BY A STEEL PLATE

(c) SUPPORTED BY AN
INTERSECTING BEAM

Fig. 7.12. End conditions.

of principal compression stresses coincides with the angle of the concrete
struts, and this angle varies from 90° to α. Based on this model, compression
forces are assumed to radiate upward from the application point of the concentrated load and to form truss action with the forces in the top stringer and
in the transverse bars. Similarly, Fig. 7.13 also shows the "fanning" of the
compression struts under the concentrated load at midspan. Since the application point of the concentrated load is on the top surface, the compression
forces should radiate downward and form truss action with the forces in
the bottom stringer and in the transverse bars.

Let us now examine the variation of forces along the beam in the bottom
stringer, particularly near the midspan where the angle of the compression
struts changes. The variation is expressed by the diagram shown in Fig.
7.13. To plot this diagram, we shall assume that the primary angle of inclina-

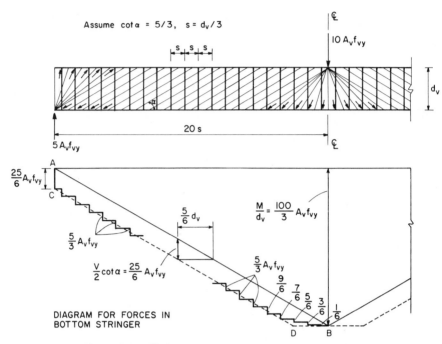

Fig. 7.13. Equilibrium at support and under concentrated load.

tion of the principal compression is given by $\cot \alpha = 5/3$. This angle is the smallest permitted by the CEB-FIP Model Code. To simplify the calculation, we will also assume that the spacing of the transverse bars is one-third of the depth ($s = d_v/3$). Consequently, a crack inclined at an α angle will intersect five transverse bars; and the end reaction and the midspan concentrated load can be taken as $5A_v f_{vy}$ and $10A_v f_{vy}$, respectively ($A_v f_{vy}$ being the yield force in one transverse bar). If we further assume that the half-span of the beam is 20 times the spacing of the transverse bars, then the moment at midspan $M = 100A_v f_{vy}s$, and the force in the bottom stringer at midspan is $M/d_v = (100/3)A_v f_{vy}$. Since the force in the bottom stringer due to the bending moment is zero at support, the diagram for the bottom stringer force due to moments has a triangular shape, as indicated by the solid line AB in Fig. 7.13.

In addition to the contribution of moment M, the force in the bottom stringer is also contributed by the shear force V. Equation (7–48) and Fig. 7.9 (b) show that this contribution is $(V/2) \cot \alpha = (25/6)A_v f_{vy}$. This additional force is applicable throughout the beam ($V = $ constant) and is shown by the dotted line CD in Fig. 7.13. In actuality, of course, the transverse steel is not uniformly distributed, but is concentrated at discrete points. Therefore, the bottom stringer force contributed by shear force should change at each transverse bar and should have a stepped shape as indicated. Each step of change should introduce a bottom stringer force of $A_v f_{vy}$ cot α. For most of the beam length, cot $\alpha = 5/3$, and each step is $(5/3)A_v f_{vy}$. When the midspan is approached, however, cot α gradually decreases, and the steps decrease accordingly in the following sequence: $(5/3)A_v f_{vy}$, $(9/6)A_v f_{vy}$, $(7/6)A_v f_{vy}$, $(5/6)A_v f_{vy}$, $(3/6)A_v f_{vy}$, and $(1/6)A_v f_{vy}$. This stepped curve near the midspan can be approximated conservatively by a horizontal dotted line DB, which is commonly used in design.

As shown in Fig. 7.13, the dotted line CD in the diagram for bottom stringer forces is displaced downward by an amount $(25/6)A_v f_{vy}$ from the solid line AB. It is also possible to view the dotted line CD as having been displaced leftward by a distance $(5/6)d_v$ from the solid line AB. This distance can be determined by dividing the vertical displacement, $(25/6)A_v f_{vy}$, by the slope of the solid line, $V/d_v = 5A_v f_{vy}/d_v$. This shifting of the diagram has been recognized in many codes throughout the years and is known as the "shift rule." The ACI Codes and the CEB Codes require that the design moment diagram be shifted by a distance d (effective depth) toward the support. This is an indirect recognition of the additional longitudinal tensile steel area demanded by the applied shear force. This provision is obviously conservative because the theoretically required shifting of $(5/6)d_v$ has been calculated from the smallest α angle (and, therefore, the largest cot α) generally accepted.

7.3.3 Effect of Distributed Loads[14]

In Sections 7.3.1 and 7.3.2 we studied the equilibrium of a simple beam subjected to a concentrated load at midspan (Fig. 7.9). In such a beam the shear force V is constant for a shear span between the support reaction and the midspan load. Consequently, the shear stresses are distributed uniformly throughout the shear span. The stresses in the transverse steel bars are also uniform, as shown by the free bodies in Fig. 7.10. Now let us examine the equilibrium of a simple beam subjected to a uniformly distributed load w, as shown in Fig. 7.14 (a). In this case the shear force varies linearly from a maximum at the support to zero at midspan. In a beam with such varying shear force the stresses in the beam and in the transverse steel bars are no longer uniform. This nonuniform stress distribution is the direct result of the effect of the distributed loads.

In Fig. 7.14 (a) a beam element B of length d_v cot α is isolated. This beam element is cut out from a location sufficiently far from the support and the midspan to avoid the local effect discussed in Section 7.3.2. It is subjected on the left face to a shear force V and a bending moment M.

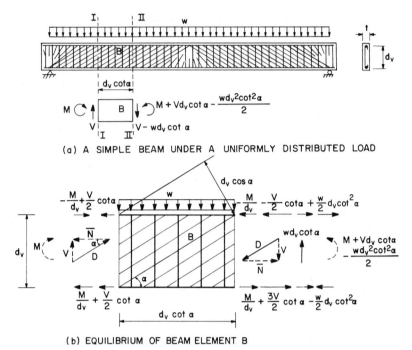

(a) A SIMPLE BEAM UNDER A UNIFORMLY DISTRIBUTED LOAD

(b) EQUILIBRIUM OF BEAM ELEMENT B

Fig. 7.14. Equilibrium of beam element in a simple beam under uniformly distributed load.

On the right face, however, the shear force is reduced by the increment $wd_v \cot \alpha$, while the bending moment is increased by the amount $Vd_v \cot \alpha - (w/2)d_v^2 \cot^2 \alpha$.

The equilibrium of beam element B is shown in Fig. 7.14 (b). The derivation of the forces on the left face is identical to that discussed in Section 7.3.1.1 in connection with Fig. 7.9 (b). The force in the concrete struts D is given by Eq. (7–45), and the forces in the top and bottom stringers are derived in Eqs. (7–47) and (7–48), respectively. On the right face, however, the force system in Fig. 7.14 (b) is different from that shown in Fig. 7.9 (b) because two additional terms due to the uniformly distributed load w [a shear of $-wd_v \cot \alpha$ and a moment of $(w/2)d_v^2 \cot^2 \alpha$] are involved. Incorporating the second new term, the forces in the top and bottom stringers are:

$$
\begin{aligned}
N_t &= -\frac{M + Vd_v \cot \alpha - \dfrac{w}{2} d_v^2 \cot^2 \alpha}{d_v} + \frac{1}{2} V \cot \alpha \\
&= -\frac{M}{d_v} - \frac{V}{2} \cot \alpha + \frac{w}{2} d_v \cot^2 \alpha
\end{aligned}
\tag{7–63}
$$

$$N_b = + \frac{M + V d_v \cot \alpha - \frac{w}{2} d_v^2 \cot^2 \alpha}{d_v} + \frac{1}{2} V \cot \alpha$$

$$= \frac{M}{d_v} + \frac{3V}{2} \cot \alpha - \frac{w}{2} d_v \cot^2 \alpha$$

(7-64)

It can be seen that Eqs. (7-63) and (7-64) are identical to Eqs. (7-49) and (7-50), respectively, except that a new force, $(w/2)d_v \cot^2 \alpha$, has been created in the top and bottom stringers.

If the whole equilibrium force system in Fig. 7.14 (b) is compared to that in Fig. 7.9 (b), it can be seen that the equilibrium force system in Fig. 7.14 (b) can be resolved into two equilibrium force systems: the first one identical to that in Fig. 7.9 (b), and the other as shown in Fig. 7.15. This new force system on beam element B (Fig. 7.15) involves only those three forces that are related to the uniformly distributed load w. They are: (1) the load w, acting on the top face of element B; (2) the shear force $w d_v \cot \alpha$, acting on the right face (this shear force is assumed to be uniformly distributed along the depth); and (3) the moment $(w/2)d_v^2 \cot^2 \alpha$, also acting on the right face. This moment can be replaced by a couple of forces acting on the top and bottom stringers, each force being $(w/2)d_v \cot^2 \alpha$. The statically admissible stress field within beam element B that is consistent with this external force system must be added to the uniform stress field that corresponds to the external force system in Fig. 7.9 (b). In short, the stress field

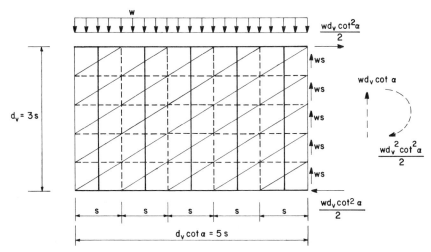

Fig. 7.15. Equilibrium of beam element B under force system created by uniformly distributed load w (to be added to Fig. 7.9).

in Fig. 7.15 should represent the special effect of a uniformly distributed load w on the beam element B.

To find a statically admissible stress field within the beam element B in Fig. 7.15, we shall make two assumptions. First, it was previously assumed that the shear stresses must be *uniformly* distributed along the depth of the beam element B at each cross section. Second, in view of the uniformly distributed load, it is logical to assume that the shear force varies *linearly* along the length of the beam from zero at the left surface to $wd_v \cot \alpha$ at the right surface. To illustrate this stress field, we shall divide and separate beam element B into 25 equal subelements, each containing a vertical steel bar at its center (see Fig. 7.16). The shear forces acting on each subelement are indicated in the figure, and they vary from subelement to subelement in accordance with the two assumptions.

Take the upper right corner subelement, for example. It is subjected to a shear force of ws on the right face and a shear force of $0.8ws$ on the left face. This change is in accordance with the second assumption. On the top

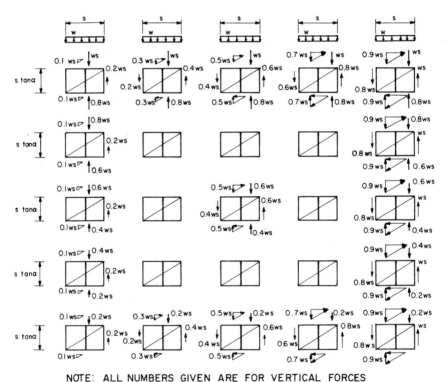

NOTE: ALL NUMBERS GIVEN ARE FOR VERTICAL FORCES

Fig. 7.16. Equilibrium within a beam element B (uniformly distributed load on top surface).

face we have a vertical force ws due to external load acting along the center line of the subelement. (This force is drawn to the right of the center line for clarity). To maintain vertical equilibrium the vertical force on the bottom face must be $0.8ws$ acting along the center line (also drawn to the right of the center line). This compressive force must be carried by the vertical steel bar because vertical stresses cannot pass through the diagonal cracks except through the connecting steel bars. To maintain moment equilibrium and horizontal force equilibrium the shear forces on the top and bottom faces must both be $(0.9ws)$ cot α and act in the opposite direction. This equality of the top and bottom shear forces is in accordance with the first assumption. Each of these two shear forces can be resolved into a diagonal tensile force in the concrete struts and a vertical compression force of $0.9ws$ in the vertical steel. Summing the two vertical compressive forces in the steel bar on the top face gives $1.9ws$. On the bottom face, however, the sum of the two vertical compressive steel forces is $1.7ws$, which is different from that on the top face. If we observe the equilibrium of the five subelements of the extreme right column in Fig. 7.16, it can be seen that the compressive steel forces on the top faces decrease linearly downward in the following sequence: $1.9ws$, $1.7ws$, $1.5ws$, $1.3ws$, and $1.1ws$. If we also look at the top faces of the top five subelements, the compressive steel forces decrease in this same sequence from right to left. Similarly, the sequence of compressive steel forces, $0.9ws$, $0.7ws$, $0.5ws$, $0.3ws$, and $0.1ws$, can be observed on the bottom face of the five extreme left subelements from top to bottom, as well as on the bottom faces of the five bottom subelements from right to left.

The compressive forces in the transverse steel bars as illustrated in Fig. 7.16 can now be subtracted from the uniform tensile steel forces for the force system in Fig. 7.9 (b). For the latter force system each transverse steel bar should carry a uniform force of $V/5$ (see Fig. 7.10). The algebraic summation of the two force systems is given in Fig. 7.17. This figure shows clearly that the tensile forces in the transverse steel bars increase linearly downward, and that the maximum forces are located at the bottom of the bars. The forces at these lowest locations vary linearly along the length of the beam according to the conventional shear diagram. Since the design of the transverse steel bars must be based on the maximum force in each bar, the calculation of transverse steel area based on the conventional shear diagram should be considered correct.

A triangular free body is also shown in Fig. 7.18 so as to expose the forces in the transverse steel bars along a diagonal crack. These forces are labeled according to the distribution pattern of Fig. 7.17. It is quite obvious that these forces cannot be considered as uniform. An assumption of uniform transverse steel forces along a diagonal crack, which is made in the so called staggering concept of shear design, will lead to an erroneous and unconservative design.[14]

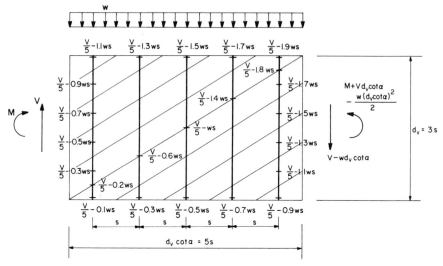

Fig. 7.17. Variation of stirrup forces in compatibility truss model.

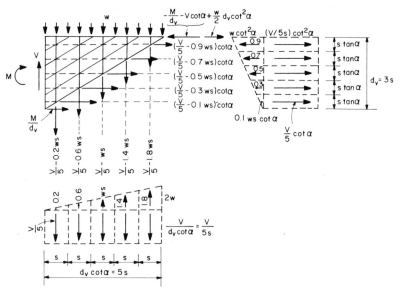

Fig. 7.18. Stirrup and longitudinal steel forces along a diagonal crack—compatibility truss model.

Distribution of the horizontal forces in the region between the top and bottom stringers can also be derived from the equilibrium of the subelements in Fig. 7.16. In this figure an upward shear force ws is shown to act on the right face of each of the five subelements in the far right column. As shown in Fig. 7.19, each of these forces can be resolved into a longitudinal compressive force, $ws \cot \alpha$, and a diagonal tensile force resisted by the concrete struts. To develop a truss action, these five longitudinal compressive forces would demand five longitudinal steel bars to be distributed in the region between the top and bottom stringers. If such longitudinal steel bars are available, the distribution of forces in these steel bars due to w is shown in Fig. 7.19, varying linearly in the longitudinal direction. This variation of longitudinal steel forces must be subtracted from a uniform tensile force of $(V/5) \cot \alpha$, caused by a uniform shear force V along the length of the beam.

Distribution of the five longitudinal forces along a diagonal crack is plotted in Fig. 7.18. It can be seen that the five longitudinal steel forces in the region between the top and bottom stringers vary linearly from $(V/5 - 0.9ws)$ $\cot \alpha$ on the top to $(V/5 - 0.1ws) \cot \alpha$ at the bottom. One can easily verify that all equilibrium conditions are satisfied for this triangular free body.

If the five longitudinal bars between the top and bottom stringers are not available, a redistribution of forces will occur. Most of the longitudinal

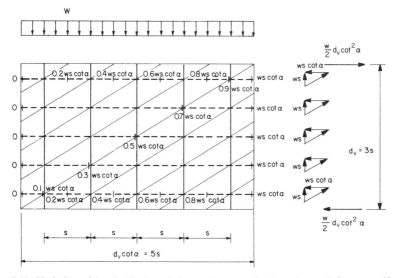

Fig. 7.19. Variation of longitudinal steel forces due to w (to be subtracted from a uniform tensile force of $(V/5) \cot \alpha$).

tensile forces will have to be resisted by the top and bottom stringers. However, some longitudinal forces are still required in the region between the top and bottom stringers to ensure moment equilibrium of the triangular free body. These forces could be supplied from three sources: (1) the dowel action of the transverse steel bars, (2) the shear resistance of the concrete struts in connection with the aggregate interlock in the cracks, and (3) the angle change of a concrete strut at each stirrup so as to produce a longitudinal component of the strut forces. A detailed study of the redistribution of longitudinal forces is beyond the scope of this book. Suffice it to say that the redistribution of forces should be in such a manner that the moment equilibrium of the triangular free body is satisfied.

Let us also consider a simple beam subjected to a uniformly distributed load on the bottom face. Similar analysis of a typical beam element results in a variation pattern of the transverse steel forces as shown in Fig. 7.20. This figure again shows that the transverse steel forces increase downward in a linear fashion. The maximum forces at the bottom of the bars also vary linearly along the beam. A comparison of Fig. 7.20 (load on bottom surface) to Fig. 7.17 (load on top surface) shows clearly that the former force pattern in the transverse bars is simply the latter force pattern plus a uniform tensile force of ws. This additional tensile force must be taken into account in design by providing additional areas for transverse steel. The

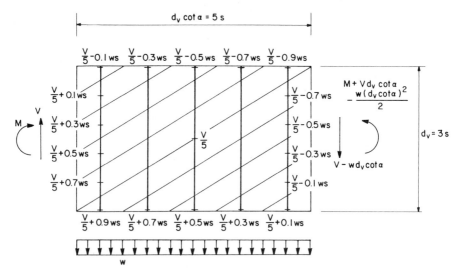

Fig. 7.20. Variation of stirrup forces in compatibility truss model (uniformly distributed load on bottom surface).

additional area of each transverse bar should carry a load of ws. In simpler language, these additional areas of transverse steel are required to "hang up" the loads by transmitting them from the bottom surface to the top surface.

7.3.4 Beam Element Subjected to Axial Tension, Shear, and Bending

In Section 7.3.1 we studied the equilibrium of a beam element subjected to shear and bending (Fig. 7.9). Now let us introduce an axial tensile force and study its interaction with shear force and bending moment. In the truss model, it can be assumed that the axial load is resisted only by the longitudinal steel bars. Consequently, the axial load does not destroy the internal equilibrium of the beam truss action under shear and bending. The only addition to the equilibrium condition is the forces in the top and bottom stringer. Hence, Eqs. (7–47) and (7–48) should include a simple new term due to the axial tensile force P:

$$N_t = \frac{P}{2} - \frac{M}{d_v} + \frac{V}{2} \cot \alpha \qquad (7\text{–}65)$$

$$N_b = \frac{P}{2} + \frac{M}{d_v} + \frac{V}{2} \cot \alpha \qquad (7\text{–}66)$$

The other equations in Section 7.3.1 remain valid for forces in the concrete struts, Eq. (7–45), and for forces in the transverse steel bars, Eq. (7–52).

The interaction equations for axial tension, shear, and bending can easily be derived using the logic of Section 7.3.1.2, except that the new term involving P must be included. For the first mode of failure, where yielding occurs in both the bottom stringer and in the transverse steel, the interaction equation is:

$$\frac{P_y}{P_{0y}} + \frac{M_y}{M_{0y}} + \left(\frac{V_y}{V_{0y}}\right)^2 = 1 \qquad (7\text{–}67)$$

where P_y = the yield axial tensile force under combined loading; $P_{0y} = 2A_{lb}f_{ly}$ = two times the yield force of the bottom stringer; and M_{0y} and V_{0y} have been given previously, in Eqs. (7–56) and (7–57), respectively.

For the second mode of failure, yielding should occur both in the top stringer and in the transverse steel bars. The interaction equation so derived is:

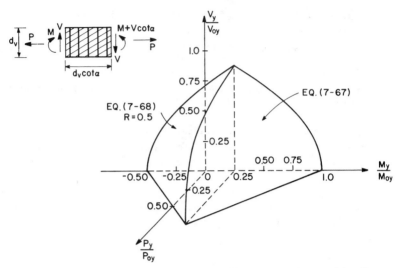

Fig. 7.21. Nondimensional interaction surface for axial load, shear, and bending.

$$\frac{P_y}{P_{0y}} - \frac{M_y}{M_{0y}} + \left(\frac{V_y}{V_{0y}}\right)^2 = R \qquad (7\text{--}68)$$

where $R = A_{lt}/A_{lb}$ in Eq. (7–61).

Equations (7–67) and (7–68) are plotted in Fig. 7.21 in a non-dimensionalized form using (P_y/P_{0y}), (M_y/M_{0y}), and (V_y/V_{0y}) as axes. Equation (7–67) is represented by the interaction surface on the right half, and Eq. (7–68) by the interaction surface on the left half assuming $R = 0.5$.

7.3.5 Effect of Prestressing

Longitudinal prestressing of a reinforced concrete member introduces a self-equilibrium axial force and/or a self-equilibrium bending moment into the member. The tensile force of the self-equilibrium force system is resisted by the prestressing steel, and the compressive force is carried by the concrete. This compressive stress in the concrete improves the cracking behavior of a member under service load. This is the main purpose of prestressing.

When an external tensile force is applied to a prestressed member, the compressive stresses in the concrete due to prestressing will decrease. If the external tensile force system has a magnitude equal to the self-equilibrium force system induced by prestress, then the compressive stresses in the concrete will vanish and the prestressing steel will carry all the external force system. This is known as the "decompression of concrete." Beyond the "decompression of concrete" the member will behave like an ordinary reinforced concrete beam. The strains and cracks can be predicted as in Section 7.2.2 except that a "strain at decompression" must be added to the prestressing steel. If the strain at decompression is designed to be sufficiently high, the high-strength prestressing steel and the mild steel can reach yielding at the same time. Under this condition, all the equilibrium equations derived previously will remain valid if the yield force of the combined reinforcement (prestressing steel plus ordinary reinforcement) is introduced.

To ensure the simultaneous yielding of both the high-strength prestressing steel and the mild reinforcing steel, the level of prestress strain must be taken as the difference between the yield strain of the prestressing steel and that of the mild steel:

$$\epsilon_{dec} = \epsilon_{py} - \epsilon_y \qquad (7\text{–}69a)$$

where ϵ_{dec} = strain at decompression in the prestressing steel
ϵ_{py} = yield strain of prestressing steel
ϵ_y = yield strain of mild steel reinforcement

The corresponding level of decompression prestress σ_{dec} should be:

$$\sigma_{dec} = \Psi(\epsilon_d) \qquad (7\text{–}69b)$$

where the function Ψ describes the shape of the stress–strain curve of the prestressing steel. The decompression stress σ_{dec} and the decompression strain ϵ_{dec} can be illustrated by the superposition of the stress–strain curve of the mild steel onto the stress-strain curve of the prestressing steel as shown in Fig. 7.22.

The unified treatment of nonprestressed and prestressed concrete members discussed here is one of the primary features of truss model analysis. This analysis is, of course, based on the assumption of full truss action. This assumption is reasonable if the member is subjected to large shear stresses and is reinforced with a high percentage of shear steel. It is not sufficiently accurate if the member is subjected to relatively small shear stresses. In this latter circumstance, the "contribution of concrete" to the shear resistance of the member must be taken into account.

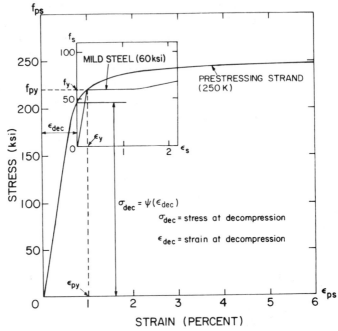

Fig. 7.22. Superposition of stress–strain diagrams for prestressing and mild steels.

7.4 CEB-FIP MODEL CODE FOR SHEAR[9,10]

The 1978 CEB-FIP Model Code includes two design methods. The first is the "standard" method, which is based on the traditional 45° truss model, modified by the addition of an empirical term, attributed to the "contribution of concrete." This method follows basically the same philosophy used by the ACI Code. It is intended for the rapid design of small or medium-size structural members, especially those used in buildings. The second, "accurate," method, however, is based on the variable-angle truss model. It is intended for the design of large girders, particularly those employed in bridges and guideways. In this section, we shall emphasize the accurate method, but will also discuss the standard method as a simplification of the accurate method.

7.4.1 Accurate Method

Three strength criteria must be considered in the design of a reinforced concrete beam. First, the web reinforcement must have enough strength to resist the applied shear force. Second, the longitudinal force caused by the applied shear force must be determined. This longitudinal force should be added to the longitudinal forces caused by the bending and the axial load in the calculation of the longitudinal steel area. Third, the diagonal concrete struts should have sufficient strength to resist the compressive stress caused by the applied shear, until the yielding of the reinforcing steel. This purpose is achieved by specifying a maximum shear strength in connection with an effective compressive stress in the concrete struts.

7.4.1.1 Design of Web Reinforcement.

In Section 7.3.1, we discussed the equilibrium of a beam element subjected to shear and bending. In this beam element, the transverse steel bars were assumed to be perpendicular to the longitudinal axis of the beam. In the CEB-FIP Model Code, however, a more general equation is given for the design of inclined web reinforcement. A portion of the beam with inclined web reinforcement is shown in Fig. 7.23. This beam is identical to that in Fig. 7.9, except that the angle β between the web reinforcement and the longitudinal bars is less than 90°.

The forces in the reinforcement and in the concrete can be obtained from the equilibrium of a beam element of length $d_v \cot \alpha$ as shown in Fig. 7.24 (a). To find the forces in the inclined web reinforcement, a horizontal cut is applied at an arbitrary distance y from the bottom stringer so as to isolate a rectangular free body as shown in Fig. 7.24 (b). A horizontal shear force $V \cot \alpha$ must exist on the cut surface to maintain equilibrium of the horizontal forces. This shear force can be resolved into two components. One component, D, is parallel to the compression field and is resisted by the concrete struts; the other component, $A_v f_v d_v \cot \alpha / s$, is the force in each inclined bar, $A_v f_v$, times the number of bars $(d_v \cot \alpha)/s$. From the geometry of this force triangle:

$$A_v f_v \frac{d_v \cot \alpha}{s} = \frac{z}{\sin \beta} \tag{7-70}$$

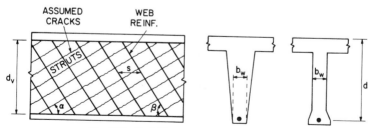

Fig. 7.23. Definitions of symbols for beams.

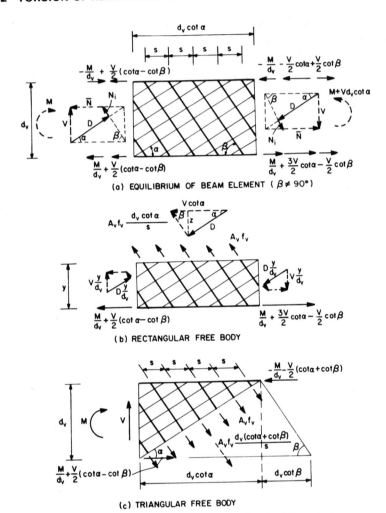

Fig. 7.24. Beam element with inclined web reinforcement.

and:

$$V \cot \alpha = z (\cot \alpha + \cot \beta) \qquad (7\text{–}71)$$

Eliminating z from Eqs. (7–70) and (7–71) gives:

$$V = \frac{A_v f_v}{s} d_v \sin \beta (\cot \alpha + \cot \beta) \qquad (7\text{–}72)$$

Equation (7–72) can also be derived easily by isolating the triangular free body as shown in Fig. 7.24 (c) and taking the equilibrium of the vertical forces. It can also be seen that Eq. (7–72) degenerates into Eq. (7–52) when $\beta = 90°$.

At yielding of the inclined web reinforcement $f_v = f_{vy}$ and $V = V_n$. Also d_v can be taken as $0.9d$ in design, where d is the effective depth, measured from the centroid of the bottom steel to the top extreme compression fiber. Making these substitutions in Eq. (7–72) produces:

$$V_n = \frac{A_v f_{vy}}{s} 0.9d \sin \beta(\cot \alpha + \cot \beta) \tag{7–73}$$

Equation (7–73) is the form given in the CEB-FIP Model Code. It is applicable to the case of high shear when $V_n \geq 3\phi V_c$. The quantity ϕ is the material reduction factor, taken as $1/1.5$ in the CEB-FIP Model Code; and V_c is an empirical shear resistance given by:

$$V_c = 2.5\tau_R b_w d \tag{7–74}$$

In Eq. (7–74) the term b_w is the minimum web thickness along the effective depth d as shown in Fig. 7.23; and $\tau_R = f_t'/4$. The tensile strength of concrete, f_t', is taken as $0.214(f_c')^{2/3}$, where f_t' and f_c' must be in the SI unit of MPa.

For the case of low shear, i.e., $V_u < 3\phi V_c$, an empirical shear resistance V_{cv} must be added to Eq. (7–73), resulting in:

$$V_n = V_{cv} + \frac{A_v f_{vy}}{s} 0.9d \sin \beta(\cot \alpha + \cot \beta) \tag{7–75}$$

V_{cv} is known as the resistance contributed by concrete. However, it actually takes into account all the effects that have been neglected in the truss model analysis, such as the shear resistance of the concrete struts, the dowel action of the reinforcement, the residual concrete tensile strength, etc. In Eq. (7–75):

$$V_{cv} = V_c \quad \text{for } V_u \leq \phi V_c \tag{7–76}$$
$$V_{cv} = 0 \quad \text{for } V_u \geq 3\phi V_c \tag{7–77}$$

For values between Eqs. (7–76) and (7–77) a straight-line interpolation is recommended.

7.4.1.2 Longitudinal Force Due to Shear. The longitudinal force due to shear can be found from the force polygon for the vertical face of the beam element as shown in Fig. 7.24 (a). It can be seen that the applied vertical shear, V, can be resolved into three components, D, N_i, and $\overline{N}$. D is the diagonal force resisted by concrete struts, N_i is a force resisted by the inclined steel bars, and $\overline{N}$ is the longitudinal force that must be carried by the longitudinal steel. From geometry:

$$\overline{N} = V(\cot \alpha - \cot \beta) \tag{7-78}$$

In the CEB-FIP Model Code, $\overline{N}$ is assumed to be resisted equally by the top and bottom stringers. Hence, the longitudinal force for the bottom stringer $\overline{N}_b$ at ultimate load is:

$$\overline{N}_b = \frac{1}{2} V_u (\cot \alpha - \cot \beta) \tag{7-79}$$

in which V_u is the applied design shear, and $\cot \alpha$ is taken from Eq. (7-72), except that d_v and V are replaced by d and V_u, respectively:

$$\cot \alpha = \frac{V_u s}{A_v f_{vy} d \sin \beta} - \cot \beta \tag{7-80}$$

Inserting Eq. (7-80) into Eq. (7-79) gives the CEB-FIP equation:

$$\overline{N}_b = A_{1b} f_{1y} = \frac{V_u^2 s}{2 A_v f_{vy} d \sin \beta} - V_u \cot \beta \tag{7-81}$$

This formulation permits an increase of the longitudinal reinforcement, A_{1b}, when the transverse reinforcement, A_v, is reduced owing to the inclusion of the empirical shear resistance, V_{cv}. The calculated bottom longitudinal reinforcement, A_{1b}, must be added to those required by the bending moment and the axial load. It is interesting to point out that the effect of the longitudinal force $\overline{N}$ on the top compression stringer is not recognized by the CEB-FIP Code. In other words, the effect of shear has been neglected in the design of longitudinal steel in the flexural compression zone.

7.4.1.3 Maximum Shear Strength Due to Web Crushing. The force triangle for the horizontal cut surface in Fig. 7.24 (b) also shows that:

$$D = \frac{z}{\sin \alpha} \qquad (7\text{-}82)$$

Eliminating z from Eqs. (7–82) and (7–71) gives:

$$V = D \frac{\sin^2 \alpha}{\cos \alpha} (\cot \alpha + \cot \beta) \qquad (7\text{-}83)$$

Notice that D is acting on an area $b_w d_v \cos \alpha$, and the stress in the concrete struts $\sigma_d = D/b_w d_v \cos \alpha$. Hence, Equation (7–83) can be written as:

$$V = \sigma_d b_w d_v \sin^2 \alpha (\cot \alpha + \cot \beta) \qquad (7\text{-}84)$$

In the CEB-FIP Code the effective strength of concrete at failure is taken as $\sigma_d = 0.6 f'_c$, while the value of d_v is approximated by d. The maximum shear strength, $V_{n,\,\text{max}}$, then becomes:

$$V_{n,\,\text{max}} = 0.6 f'_c b_w d \sin^2 \alpha (\cot \alpha + \cot \beta) \qquad (7\text{-}85)$$

with the limitation of β such that:

$$V_{n,\,\text{max}} \not> 0.45 f'_c b_w d \sin 2\alpha \qquad (7\text{-}86)$$

For vertical web reinforcement $\beta = 90°$ and Eq. (7–85) degenerates into:

$$V_{n,\,\text{max}} = 0.3 f'_c b_w d \sin 2\alpha \qquad (7\text{-}87)$$

Equations (7–85) through (7–87) are given in the CEB-FIP Code.

7.4.2 Standard Method

The strength design equations for the standard method can be easily derived by simplifications of those equations for the accurate method. For the design of web reinforcement, we can substitute $\alpha = 45°$ and $V_{cv} = V_c$ into Eq. (7–75) to get:

$$V_n = V_c + \frac{A_v f_{vy}}{s} 0.9d \sin \beta (1 + \cot \beta) \qquad (7\text{-}88)$$

where V_c has been defined in Eq. (7–74).

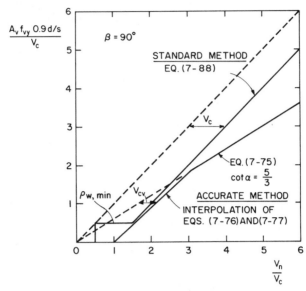

Fig. 7.25. Comparison of standard method and accurate method in 1978 CEB-FIP model code.

A comparison of the standard method, Eq. (7–88), and the accurate method, Eqs. (7–75) through (7–77), for the case of vertical web reinforcement, $\beta = 90°$, is shown in Fig. 7.25. In plotting Fig. 7.25 all the equations have been nondimensionalized by V_c. Equation (7–88) is shown to be parallel to the diagonal 45° line of equality and is displaced to the right by a distance indicated by V_c (not nondimensionalized for convenience). It is cut off at the lower portion by the minimum reinforcement ($\rho_{w, min}$), assumed to correspond to $V_c/2$. Also, Eq. (7–75) is plotted using cot $\alpha = 5/3$, the maximum value permitted by the CEB-FIP Code. This equation is valid when $V_n/V_c > 3$. In the region $1 \leq V_n/V_c \leq 3$, a straight line interpolation is shown according to Eqs. (7–76) and (7–77). It can be seen that the two methods are quite close when $V_n \leq 3V_c$. In fact, the accurate method in this region has been tailored purposely to approximate the standard method because tests and experience have consistently shown that the standard method is reasonable in this region. In the high shear region of $V_n > 3V_c$, however, tests have shown that the accurate method is quite safe, whereas the standard method is overly conservative.

To design the longitudinal steel area required by shear, the standard method does not ask a designer to calculate the longitudinal force due to shear. It simply utilizes the shift rule discussed in Section 7.3.2. This rule requires that the applied moment diagram be shifted toward the support by a distance

d in the design of tensile longitudinal steel. The increased tensile longitudinal steel area due to the shifting should account for the steel area required by the applied shear. This rule has been shown to be both simple and reasonably conservative.

The equations for diagonal web crushing in the standard method can be obtained by simply substituting $\alpha = 45°$ into Eqs. (7–85) through (7–87):

$$V_{n,\max} = 0.3 f'_c b_w d (1 + \cot \beta) \qquad (7\text{–}89)$$

but:

$$V_{n,\max} \not> 0.45 f'_c b_w d \qquad (7\text{–}90)$$

For vertical web reinforcement, when $\beta = 90°$:

$$V_{n,\max} = 0.3 f'_c b_w d \qquad (7\text{–}91)$$

7.4.3 Serviceability Requirements

The CEB-FIP Model Code provides two serviceability criteria, as discussed in the following paragraphs.

7.4.3.1 Inclination of α Angle.
In design the inclination of the α angle must be limited to a certain range so that the transverse reinforcement and the longitudinal reinforcement will both yield at failure, and so that the crack openings are tolerable under service load. To achieve these purposes, let us recall Section 7.2.3.2, in which the principal tensile strain, ϵ_{dt}, was shown to be capable of serving as a parameter for the severity of diagonal cracking. For simplicity, the strain in the diagonal concrete struts, ϵ_d, is neglected in Eq. (7–37):

$$\epsilon_{dt} = \epsilon_l + \epsilon_t \qquad (7\text{–}92)$$

From Eq. (7–25), the compatibility equation can also be simplified as:

$$\tan^2 \alpha = \frac{\epsilon_l}{\epsilon_t} \qquad (7\text{–}93)$$

Inserting Eq. (7–93) into Eq. (7–92) gives:

$$\epsilon_{dt} = \epsilon_l \left(1 + \cot^2 \alpha\right) \qquad (7\text{–}94)$$

or:

$$\epsilon_{dt} = \epsilon_t \left(1 + \tan^2 \alpha\right) \tag{7-95}$$

At the yielding of longitudinal steel, $\epsilon_l = \epsilon_y$, Eq. (7-94) becomes:

$$\frac{\epsilon_{dt}}{\epsilon_y} = \left(1 + \cot^2 \alpha\right) \tag{7-96}$$

At the yielding of transverse steel, $\epsilon_t = \epsilon_y$, Eq. (7-95) is:

$$\frac{\epsilon_{dt}}{\epsilon_y} = \left(1 + \tan^2 \alpha\right) \tag{7-97}$$

Equations (7-96) and (7-97) are plotted in Fig. 7.26. It can be seen that the least cracking condition is obtained when $\alpha = 45°$. At this angle the

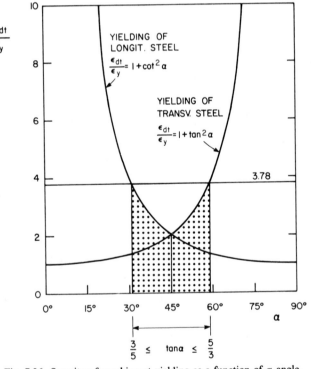

Fig. 7.26. Severity of cracking at yielding as a function of α angle.

transverse steel and the longitudinal steel will yield simultaneously. When the α angle is decreased below $45°$, the transverse steel will first yield, causing the cracks to develop very rapidly until the yielding of the longitudinal steel. On the other hand, when the α angle is increased beyond $45°$, the longitudinal steel will first yield, and will stimulate the rapid development of cracks until the yielding of the transverse steel. In order to control cracking and to ensure the yielding of both types of steel, the CEB-FIP Code limits the range of the α angle to:

$$\frac{3}{5} \leq \tan \alpha \leq \frac{5}{3} \qquad (7\text{-}98)$$

This range of α is also indicated in Fig. 7.26. A measure of the severity of cracking at the specified extremes (tan $\alpha = 3/5$ and $5/3$) is $\epsilon_{dt}/\epsilon_y = 3.78$. This is approximately twice the value of $\epsilon_{dt}/\epsilon_y = 2$ when $\alpha = 45°$.

7.4.3.2 Checking Crack Width.
The CEB-FIP Code also provides a method for checking the crack width of shear cracks as follows:

$$w_k = 1.7 k_w \epsilon_{tm} s_{crm} \qquad (7\text{-}99)$$

where

$w_k =$ characteristic crack width
$\epsilon_{tm} =$ mean strain in transverse steel
$s_{crm} =$ mean spacing of cracks
$k_w = 1.2$ for vertical web reinforcement
$\quad = 0.8$ for inclined web reinforcement with $\beta = 45°\text{-}60°$

In Eq. (7–99) ϵ_{tm} can be calculated from:

$$\epsilon_{tm} = \frac{f_y}{E_s}\left[1 - \left(\frac{V_c}{V_{se}}\right)^2\right] \geq 0.4\frac{f_v}{E_s} \qquad (7\text{-}100)$$

$$f_v = \frac{(V_{se} - V_c)s}{A_v d(\sin \beta + \cos \beta)} \qquad (7\text{-}101)$$

where $V_{se} =$ shear force at service load and V_c is given by Eq. (7–74). In Eq. (7–100) the factor $[1 - (V_c/V_{se})^2]$ accounts for the stiffening effect due to the tensile strength of concrete; and Eq. (7–101) is obtained from Eq. (7–72) by taking $\alpha = 45°$, $d_v = d$, and $V = V_{se} - V_c$.

Also, in Eq. (7–99) s_{crm} is computed by:

$$s_{crm} = 2\left(c + \frac{s}{10}\right) + k_1 k_2 \frac{d_b}{\rho_r} \leq \frac{d - x}{\sin \beta} \qquad (7\text{--}102)$$

where

$c = $ concrete cover

$s = $ spacing of steel bars; use $s = 15\ d_b$ when $s > 15\ d_b$

$d_b = $ diameter of a steel bar

$\rho_r = A_s/A_{c,ef}$

$A_{c,ef} = $ cross section of the effective concrete embedment zone defined in Fig. 15.1 of the 1978 CEB-FIP Code.

$A_s = $ area of reinforcement contained in $A_{c,ef}$

$k_1 = 0.4$ for high bond bars; 0.8 for plain bars

$k_2 = 0.125$ in bending; 0.25 for pure tension

$x = $ height of compression zone in the cracked section

The crack width w_k calculated by Eq. (7–99) should not exceed a specified value given in Table 15.1 of the 1978 CEB-FIP Code. This specified value varies from 0.1 mm to 0.4 mm, depending on the condition of exposure, the combination of actions, the sensitivity of reinforcement to corrosion, and the limit state. In view of the complexity of checking crack widths, the code also states that the checking of crack widths can be waived if the spacing of web reinforcement is less than a specified value. This specified value varies from 3 in. (75 mm) to 12 in. (300 mm), depending on the condition of exposure, the yield strength of steel, the bond property of reinforcement, and the applied shear force. It is given in Table 15.3 of the 1978 CEB-FIP Model Code.

7.4.4 Detailing of Reinforcement

7.4.4.1 Spacing of Web Reinforcement. The maximum spacing for web reinforcement is limited to:

$$s_{\max} = 0.5d \leq 12 \text{ in.} \qquad \text{for } V_u \leq \frac{2}{3}\,\phi V_{n,\max} \qquad (7\text{--}103)$$

$$s_{\max} = 0.3d \leq 8 \text{ in.} \qquad \text{for } V_u > \frac{2}{3}\,\phi V_{n,\max} \qquad (7\text{--}104)$$

in which $V_{n,\max}$ is defined in Eqs. (7–85) and (7–86) for the accurate method and in Eqs. (7–89) and (7–90) for the standard method. Physically, Eqs. (7–103) and (7–104) means that the effective strength of concrete, σ_d, in Eqs. (7–85) has been reduced from $0.6f'_c$ to $0.4f'_c$, when the spacing of web reinforcement is increased from $0.3d$ to $0.5d$. It is also interesting to point out that the maximum spacing is not a function of the angle α. This format, in fact, is quite similar to that employed by the ACI Code.

7.4.4.2 Minimum Reinforcement.

The web reinforcement should also satisfy a specified minimum reinforcement ratio, $\rho_{w,\,min}$, as given in Table 7.1.

Table 7.1 Minimum shear reinforcement ratio, $\rho_{w,\,min}$

f_c', MPa (psi) \ f_y, MPa (ksi)	S200 (31.4)	S400 (58.0)	S500 (72.5)
C12 to C20 (1,740–2,900)	0.0016	0.0009	0.0007
C25 to C35 (3,625–5,075)	0.0024	0.0013	0.0011
C40 to C50 (5,800–7,250)	0.0030	0.0016	0.0013

The ratio ρ_w is calculated by:

$$\rho_w = \frac{A_v}{b_w s \sin \beta} \tag{7–105}$$

7.4.4.3 Shear Near Supports.

The CEB-FIP Code states that the shear stresses within a distance d from the face of a direct support need not be checked, but the web reinforcement calculated at the distance d should be continued up to the support.

7.4.4.4 "Staggering Concept" for Shear Design.

The CEB-FIP Code permits the design of transverse reinforcement using the average shear over a length l, where l is defined as:

$$l = 1.5d \quad \text{for } V_u \leq \frac{2}{3}\phi V_{n,\,max} \tag{7–106}$$

$$l = d \quad \text{for } V_u > \frac{2}{3}\phi V_{n,\,max} \tag{7–107}$$

This "staggering concept" is based on the assumption that web reinforcement will yield throughout the depth of beam and throughout the length l. This assumption has been shown in Section 7.3.3 to be incorrect for a uniformly distributed load. Consequently, the author suggests that this "staggering concept" be abandoned. The designer should design the web reinforcement according to the conventional shear diagram.[14]

7.5 BOX SECTIONS SUBJECTED TO PURE TORSION

7.5.1 Equilibrium of Box Sections Under Torsion

In Section 3.2.1 we introduced Rausch's space truss analogy. This theory postulates that when a torsional moment is applied to a reinforced concrete member and causes the member to crack, the concrete will be separated by the diagonal cracks into a series of inclined concrete struts. These concrete struts will interact with the longitudinal and hoop reinforcement to form a space truss, which could resist the external torque. Rausch further assumed that the concrete struts are inclined at a 45° angle. This assumption of 45° results in the requirement of a definite ratio of longitudinal steel to lateral steel. For the case of pure torsion, this definite ratio amounts to an equal volume of longitudinal steel and hoop steel.

It has been shown by tests, however, that both the longitudinal steel and the lateral steel may yield within a wide range of the ratios of longitudinal to lateral steel. This phenomenon could be explained by assuming that the concrete struts are inclined at a variable angle, α. This angle α is governed by the actual volume ratios of longitudinal to lateral steel.

A variable-angle space truss model for a rectangular box member subjected to pure torsion is shown in Fig. 7.27. We shall study only rectangular box sections for three reasons. First, rectangular sections are the most commonly used cross section. Second, rectangular sections can illustrate most clearly the interaction of torsion, shear, and bending without the unnecessary geometric complexity involved in other types of cross sections. Third, it has been shown by the author[15] that the behavior and strength of a solid section near failure are identical to that of a hollow section with the same overall dimension, material, and steel arrangement. In other words, the concrete core is not effective when failure is approached, and the torsional moment is carried mainly by the outer shell.

A box section, shown in Fig. 7.27 (a) and (b), can be considered as a tube in resisting torsional moment, so that Bredt's thin-tube theory, discussed in Section 1.3, is applicable. Under torsion, a constant circulatory shear flow, q, will develop in the wall of the box section. From Eq. (1–72):

$$q = \frac{T}{2A_0} \qquad (7-108)$$

where A_0 = area enclosed by the center line of the shear flow.

A shear force acting on one straight wall of a box section is identical to

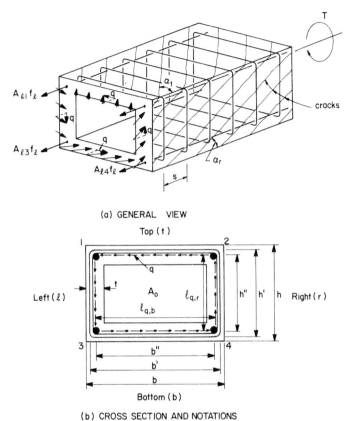

(a) GENERAL VIEW

(b) CROSS SECTION AND NOTATIONS

Fig. 7.27. Box section subjected to pure torsion.

a shear force acting on a beam element, as discussed in Section 7.3.1.1 and Fig. 7.9 (b), except that the transverse length of the beam element becomes ℓ_q. The length ℓ_q is the straight portion of the center line of a shear flow q. The shear force in one straight wall becomes $V = q\ell_q$. Hence, the longitudinal steel force in one wall can be obtained from Eq. (7–44):

$$\overline{N} = q\ell_q \cot \alpha \qquad (7\text{–}109)$$

If the total force of the longitudinal bars of the box section is defined as ΔN, then:

$$\Delta N = \Sigma \overline{N} = q p_0 \cot \alpha \qquad (7\text{–}110)$$

where $p_0 = \Sigma l_q = $ the perimeter of the center line of shear flow. Substituting q from Eq. (7–108) into Eq. (7–110), we arrive at:

$$\Delta N = \frac{Tp_0}{2A_0} \cot \alpha \qquad (7\text{–}111)$$

The force in each hoop bar can similarly be derived from Eq. (7–52), if we recognize that $q = V/d_v$:

$$A_t f_t = qs \tan \alpha \qquad (7\text{–}112)$$

where A_t and f_t are the area and stress of one hoop steel bar. Substituting q from Eq. (7–108) into Eq. (7–112) results in:

$$A_t f_t = \frac{Ts}{2A_0} \tan \alpha \qquad (7\text{–}113)$$

The stress in the diagonal concrete struts can be obtained from Eq. (7–46):

$$\sigma_d = \frac{q}{t \sin \alpha \cos \alpha} \qquad (7\text{–}114)$$

Inserting q from Eq. (7–108) into Eq. (7–114) results in:

$$\sigma_d = \frac{T}{2A_0 t \sin \alpha \cos \alpha} \qquad (7\text{–}115)$$

Equations (7–108), (7–111), (7–113), and (7–115) are the four basic equilibrium equations for torsion in the theory of the variable-angle truss model.

At yielding of both longitudinal and hoop reinforcement, i.e., $\Delta N = \Delta N_y$, $f_t = f_{ty}$, and $T = T_y$, Eqs. (7–111) and (7–113) become:

$$\Delta N_y = \frac{T_y p_0}{2A_0} \cot \alpha \qquad (7\text{–}116)$$

$$A_t f_{ty} = \frac{T_y s}{2A_0} \tan \alpha \qquad (7\text{–}117)$$

Eliminating T_y from the above two equations gives:

$$\tan \alpha = \sqrt{\frac{A_t f_{ty} p_0}{\Delta N_y s}} \qquad (7\text{--}118)$$

Similarly, eliminating α from Eqs. (7–116) and (7–117) produces:

$$T_y = 2A_0 \sqrt{\frac{\Delta N_y (A_t f_{ty})}{p_0 s}} \qquad (7\text{--}119)$$

We shall now study the resultant longitudinal force due to torsion, ΔN_y, in Eq. (7–119). First, let us consider the case of a box section with symmetric reinforcement. In this symmetrical case, Fig. 7.27 (b) consists of:

areas: $A_{11} = A_{12} = A_{13} = A_{14} = A_1$
yield forces: $N_{1y} = N_{2y} = N_{3y} = N_{4y} = A_1 f_{1y}$

where the subscripts 1, 2, 3 and 4 refers to the corners indicated in Fig. 7.27 (b). Therefore, in Eq. (7–119):

$$\Delta N_y = 4A_1 f_{1y} \qquad (7\text{--}120)$$

In the case of box sections with unsymmetric "bending-type" reinforcement:

Top reinforcement:

area: $A_{11} = A_{12} = A_{1t}$
yield force: $N_{1y} = N_{2y} = A_{1t} f_{1y}$

Bottom reinforcement:

area: $A_{13} = A_{14} = A_{1b} > A_{1t}$
yield force: $N_{3y} = N_{4y} = A_{1b} f_{1y} > A_{1t} f_{1y}$

ΔN_y should be the minimum value of the four resultants obtained by taking moments of the longitudinal steel forces about all four sides of the box. For the "bending-type" reinforcement where $A_{1b} > A_{1t}$, it is obvious that the resultant ΔN_y would be a minimum when moment is taken about an axis through the centroids of the bottom bars:

$$\Delta N_y = \frac{N_{1y}h'' + N_{2y}h''}{\left(\dfrac{h''}{2}\right)} = 4A_{1t}f_{1y} \tag{7-121}$$

In Eq. (7–121), $(h''/2)$ in the denominator is the distance from the centroid of a perimeter connecting the longitudinal bars to this moment axis.

7.5.2 Bending of Concrete Struts

When a box section is subjected to torsion, the concrete struts not only will receive axial forces but also will be subjected to bending. Therefore, the stresses and strains induced by bending must be superimposed on the stresses and strains produced by axial force. This interesting phenomenon was first observed by Lampert and Thurlimann.[5]

The geometry that gives rise to the bending of a concrete strut is shown in Fig. 7.28. Let us look at the top wall of a box section under torsion in Fig. 7.28 (a). A plane surface $OABC$ in the top wall through the center line of the shear flow is indicated. A coordinate system is imposed with the x-axis in the longitudinal direction and the y-axis in the transverse direction. The origin O is fixed at the far left corner. A concrete strut with an angle of inclination α is represented by the diagonal line OB. Since the transverse length OA of the plane is l_q, the longitudinal length OC will be $l_q \cot \alpha$.

The plane surface $OABC$ is isolated in Fig. 7.28 (b). When the box beam is subjected to an angle of twist, θ (angle per unit length), the edge CB will rotate to the position CD through an angle $\theta l_q \cot \alpha$. The new surface $OADC$ becomes a hyperbolic paraboloid, and the concrete strut OD becomes curved. To derive the curvature of the concrete struts, we shall first express the hyperbolic parabolic surface by an equation:

$$w = \theta xy \tag{7-122}$$

where w is the displacement perpendicular to the x–y plane. Let us impose another s-axis along the direction of the diagonal concrete strut. By differentiating Eq. (7–122) with respect to s, we can find the slope of the concrete strut, dw/ds, with the help of the chain rule:

$$\frac{dw}{ds} = \frac{\partial w}{\partial x}\frac{dx}{ds} + \frac{\partial w}{\partial y}\frac{dy}{ds}$$

$$= (\theta y)\cos\alpha + (\theta x)\sin\alpha \qquad (7\text{-}123)$$

The curvature of the concrete strut, ψ, is the second derivative of w with respect to s:

$$\frac{d^2 w}{ds^2} = \frac{\partial\left(\dfrac{dw}{ds}\right)}{\partial x}\frac{dx}{ds} + \frac{\partial\left(\dfrac{dw}{ds}\right)}{\partial y}\frac{dy}{ds}$$

$$= (\theta\sin\alpha)\cos\alpha + (\theta\cos\alpha)\sin\alpha$$

$$\psi = \theta\sin 2\alpha \qquad (7\text{-}124)$$

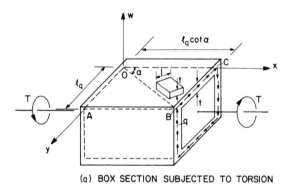

(a) BOX SECTION SUBJECTED TO TORSION

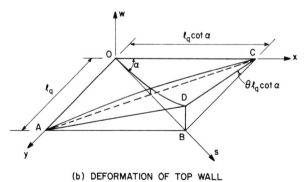

(b) DEFORMATION OF TOP WALL

Fig. 7.28. Bending of a concrete strut in the wall of a box section subjected to torsion.

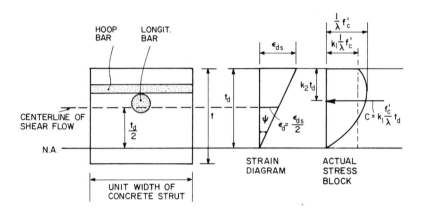

1/λ = EMPIRICAL COEFFICIENT REPRESENTING THE "SOFTENING EFFECT" OF CONCRETE
DUE TO SHEAR CRACKING. SEE EQ.(7-39) AND FIG. 7.7

$f'_c/λ$ = PEAK CONCRETE STRENGTH. SEE EQ.(7-40)

k_1 = RATIO OF AVERAGE STRESS TO PEAK STRESS

$k_1 f'_c/λ$ = AVERAGE STRESS OF CONCRETE STRESS BLOCK

t_d = DEPTH OF COMPRESSION STRESS BLOCK

k_2 = RATIO OF THE DISTANCE FROM RESULTANT C TO EXTREME FIBRE TO THE
DEPTH OF COMPRESSION STRESS BLOCK t_d

Fig. 7.29. Strain and stress distribution in a wall of a box section subjected to torsion.

Equation (7–124) relates the curvature of the concrete strut, ψ, to the angle of twist of the box beam, θ.

A unit width of concrete struts as indicated in Fig. 7.28 (a) is isolated from the top wall and is shown in Fig. 7.29. The total thickness of the wall is designated t. However, owing to the bending of the concrete struts, a certain portion of the area near the bottom edge may be in tension. This area should be neglected according to the basic assumption 4 in Section 7.1.3. The portion of the area that is in compression will be considered as effective. The depth of this effective area is designated t_d. If the strain distribution within the effective depth t_d is assumed to be linear, then the maximum compressive strain ϵ_{ds} at the surface can be written as:

$$\epsilon_{ds} = \psi t_d \qquad (7–125)$$

Equation (7–125) relates the maximum compressive strain, ϵ_{ds}, to the curvature of the concrete strut, ψ. The value of t_d will be determined from equilibrium conditions and the properties of the stress block in the next section.

Equations (7–124) and (7–125) are the two basic compatibility equations for the bending of concrete struts due to torsion.

7.5.3 Compressive Stress Block and the Effective Wall Thickness

The stress diagram within the effective depth t_d is also shown in Fig. 7.29. The actual stress distribution must, of course, be obtained from tests. Let us define the peak concrete strength of the stress block as f'_c/λ where λ is an empirical coefficient. As discussed in Section 7.2.3, the reciprocal $1/\lambda$ represents the "softening effect" of concrete due to diagonal shear cracking. The coefficient λ was given in Eq. (7–39), and the effect of $1/\lambda$ on the stress–strain relationship of concrete was illustrated in Fig. 7.7. In Eq. (7–39) the value of λ is not sensitive to the Poisson's ratio, ν. If ν is neglected, then Eq. (7–39) becomes:

$$\lambda = \sqrt{\frac{\gamma_m}{\epsilon_d}} \qquad (7\text{–}126)$$

A comparison of $1/\lambda$ values computed by Eqs. (7–39) and (7–126) is given in Table 7.2.

Table 7.2 Comparison of $1/\lambda$ values

γ_m/ϵ_d	2	4	10	15
$1/\lambda$, Eq. (7–39)	0.77	0.52	0.32	0.26
$1/\lambda$, Eq. (7–126)	0.71	0.50	0.32	0.26

Recalling that $\gamma_m = \epsilon_l + \epsilon_t + 2\epsilon_d$ (Fig. 7.5), and $\tan^2 \alpha = (\epsilon_l + \epsilon_d)/(\epsilon_t + \epsilon_d)$ from Eq. (7–19), Eq. (7–126) can be expressed as:

$$\lambda = \frac{\sqrt{\epsilon_t + \epsilon_d}}{\sqrt{\epsilon_d}\,\cos \alpha} \qquad (7\text{–}127)$$

This expression of λ will be used in the next Section 7.5.4 to determine the strains in the reinforcement.

The average stress of the stress block can be expressed as $k_1 f'_c/\lambda$, where k_1 is defined as:

$$k_1 = \frac{\text{Average stress}}{\text{Peak stress}} \qquad (7\text{-}128)$$

Hence, the resultant of the stress block C for a unit width of concrete strut is:

$$C = k_1 \frac{f'_c}{\lambda} t_d \qquad (7\text{-}129)$$

The location of the resultant C can be defined by a distance $k_2 t_d$ from the extreme compression fiber, where:

$$k_2 = \frac{\text{Distance from resultant } C \text{ to extreme compression fiber}}{\text{Effective depth of compression stress block } (t_d)} \qquad (7\text{-}130)$$

The coefficients k_1 and k_2 have been determined by Hognestad et al.[16] from flexural tests of nonsoftened concrete. They are shown to be a function of f'_c (psi):

$$k_1 = 0.94 - \frac{f'_c}{26{,}000} \qquad (7\text{-}131)$$

$$k_2 = 0.50 - \frac{f'_c}{80{,}000} \qquad (7\text{-}132)$$

Typical values of k_1 and k_2 are listed in Table 7.3.

Table 7.3 Coefficient k_1 and k_2 for nonsoftened concrete

f'_c (PSI)	3,000	4,000	5,000	6,000	7,000	8,000
k_1	0.82	0.79	0.75	0.71	0.67	0.63
k_2	0.46	0.45	0.44	0.42	0.41	0.40

The coefficient k_1 for softened concrete can be calculated by integrating Eq. (7–38) and (7–42). These two equations describe the shape of the stress-strain curve of softening concrete as shown in Fig. 7.7. When $\epsilon_{ds} < \epsilon_0/\lambda$

$$k_1 \frac{1}{\lambda} f_c' = \frac{1}{\epsilon_{ds}} \underbrace{\int_0^{\epsilon_{ds}} f_c' \left[2 \left(\frac{\epsilon_d}{\epsilon_0} \right) - \lambda \left(\frac{\epsilon_d}{\epsilon_0} \right)^2 \right] d\epsilon_d}_{\text{Eq. (7–38)}} \qquad (7\text{–}133)$$

$$k_1 = \lambda \frac{\epsilon_{ds}}{\epsilon_0} - \frac{\lambda^2}{3} \left(\frac{\epsilon_{ds}}{\epsilon_0} \right)^2 \qquad (7\text{–}134)$$

When $\epsilon_{ds} > \epsilon_0/\lambda$

$$k_1 \frac{1}{\lambda} f_c' = \frac{1}{\epsilon_{ds}} \underbrace{\int_0^{\epsilon_0/\lambda} f_c' \left[2 \frac{\epsilon_d}{\epsilon_0} - \lambda \left(\frac{\epsilon_d}{\epsilon_0} \right)^2 \right] d\epsilon_d}_{\text{Eq. (7–38)}}$$

$$+ \frac{1}{\epsilon_{ds}} \underbrace{\int_{\epsilon_0/\lambda}^{\epsilon_{ds}} \frac{1}{\lambda} f_c' \left[1 - \left(\frac{\epsilon_d - \epsilon_p}{2\epsilon_0 - \epsilon_p} \right)^2 \right] d\epsilon_d}_{\text{Eq. (7–42)}} \qquad (7\text{–}135)$$

$$k_1 = \left(1 - \frac{1}{(2\lambda - 1)^2} \right) \left(1 - \frac{1}{3} \frac{\epsilon_p}{\epsilon_{ds}} \right) + \frac{1}{(2\lambda - 1)^2} \frac{\epsilon_{ds}}{\epsilon_p} \left(1 - \frac{1}{3} \frac{\epsilon_{ds}}{\epsilon_p} \right) \qquad (7\text{–}136)$$

where $\epsilon_p = \epsilon_0/\lambda$. The coefficient k_1 has been tabulated in Table 7.4 as a function of $1/\lambda$ and ϵ_{ds}, assuming $\epsilon_0 = 0.002$.

Table 7.4 k_1 as a function of $1/\lambda$ and ϵ_{ds} for softened concrete ($\epsilon_0 = 0.002$)

$1/\lambda$ \ ϵ_{ds}	0.005	0.001	0.0015	0.002	0.0025	0.003	0.0035
0.10	0.8654	0.9215	0.9218	0.8994	0.8610	0.8089	0.7439
0.20	0.7333	0.8611	0.8883	0.8806	0.8513	0.8048	0.7429
0.30	0.6018	0.7980	0.8526	0.8604	0.8409	0.8005	0.7419
0.40	0.4948	0.7331	0.8147	0.8385	0.8294	0.7956	0.7407
0.50	0.4167	0.6667	0.7747	0.8148	0.8167	0.7901	0.7394
0.60	0.3588	0.6019	0.7325	0.7891	0.8026	0.7840	0.7379
0.70	0.3146	0.5442	0.6889	0.7613	0.7870	0.7771	0.7362
0.80	0.2799	0.4948	0.6445	0.7314	0.7698	0.7693	0.7342
0.90	0.2521	0.4527	0.6018	0.6997	0.7506	0.7603	0.7319
1.00	0.2292	0.4167	0.5625	0.6667	0.7292	0.7500	0.7292

For a great majority of torsional members, $1/\lambda$ varies between 0.4 and 0.6 and the maximum compressive strain, ϵ_{ds}, at failure varies between 0.002 and 0.003. For these ranges of $1/\lambda$ and ϵ_{ds}, a reasonable k_1-value of 0.80 is indicated in Table 7.4. This k_1-value will be assumed for design purposes.

Since the k_1-value for nonsoftened concrete is a function of the concrete strength, f'_c, (Table 7.3), it would seem logical that the k_1-value for softened concrete should also be affected by f'_c. However, no tests are available at present to make such refinement. It should also be noted that $k_1 = 0.80$ is also specified in Table 7.3 for a nonsoftened concrete with f'_c about 4,000 psi. This concrete has a k_2-value of 0.45. Therefore, a k_2-value of 0.45 for softened concrete would appear to be appropriate for design purposes.

At failure of the beam the depth of the compression stress block t_d can be obtained from Eq. (7–115), noting that $t = t_d$, $\sigma_d = k_1 f'_c / \lambda$ and $T = T_n$:

$$t_d = \frac{T_n}{2A_0(k_1 f'_c/\lambda) \sin \alpha \cos \alpha} \tag{7–137}$$

The center line of the shear flow is assumed to lie midway in the depth of the stress block. Consequently, the perimeter of the center line of shear flow, p_0, and the area bounded by the center line of shear flow, A_0, can be approximated for a box section by:

$$p_0 = p_c - 4t_d \tag{7–138}$$

$$A_0 = A_c - \frac{t_d}{2} p_c \tag{7–139}$$

where p_c = outer perimeter of the concrete cross section
   A_c = cross-sectional area within the outer perimeter of concrete

Substituting Eq. (7–139) into Eq. (7–137) and solving for t_d result in:

$$t_d = \frac{A_c}{p_c} \left[1 - \sqrt{1 - \left(\frac{T_n p_c}{A_c^2} \right) \frac{1}{(k_1 f'_c/\lambda) \sin \alpha \cos \alpha}} \right] \tag{7–140}$$

In Eq. (7–140) $k_1 f'_c / \lambda$ is the effective strength of concrete in the wall of a box section subjected to torsion. This effective strength takes into account the softening of concrete due to diagonal shear cracking as well as the bending of the concrete struts. From Section 7.4.1.3 we recall that the effective strength of concrete is taken as $0.6 f'_c$ in the CEB-FIP Code for the maximum shear strength of a member. For box sections subjected to torsion, the CEB-FIP Code suggested a value of $0.5f'_c$ for the maximum torsional strength (see Section 7.7.2.3 later). For underreinforced torsional members, however, recent research[17] shows that the effective strength of concrete varies between $0.30f'_c$ and $0.50f'_c$ with an average value of about $0.4f'_c$. Inserting $k_1f'_c/\lambda = 0.4f'_c$ into Eq. (7–140) and defining:

$$\tau_n = \frac{T_n p_c}{A_c^2} \tag{7-141}$$

we arrive at a simplified expression for t_d:

$$t_d = \frac{A_c}{p_c}\left[1 - \sqrt{1 - \left(\frac{\tau_n}{0.20f'_c}\right)\left(\frac{1}{\sin 2\alpha}\right)}\right] \tag{7-142a}$$

It can be seen that the parameter for effective wall thickness, $t_d p_c/A_c$ is a function of two variables, namely, the torsional stress ratio τ_n/f'_c and the angle α. Hence, $t_d p_c/A_c$ is plotted in Fig. 7.30a as a function of τ_n/f'_c, using α as a parameter. The two dotted curves provide the boundaries of Eq. (7–142a) when α varies from 45° to arctan (3/5) or arctan (5/3). The latter two angles are the minimum and maximum angles permitted by the CEB-FIP Model Code.

The assumption of $k_1f'_c/\lambda = 0.4f'_c$, however, are found to be unsatisfactory,

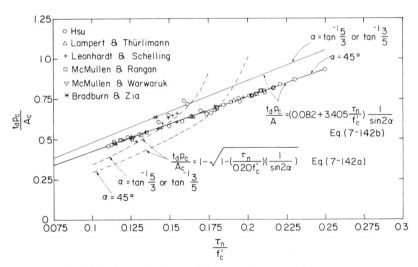

Fig. 7.30a. Determination of effective depth t_d at maximum torque.

because the effective wall thickness, t_d, is quite sensitive to $k_1 f'_c / \lambda$. A rigorous analysis of k_1 and $1/\lambda$ using a complex trial and error method for 61 eligible specimen available in literature has been made in Ref. 17. The theoretical $t_d p_c / A_c$ values for these specimen are also plotted in Fig. 7.30a. It can be seen that Eq. (7–142a) does not fit the specimen points in a satisfactory manner. The t_d values are considerably underestimated when τ_n / f'_c is low and considerably overestimated when τ_n / f'_c is high.

A more logical expression for t_d can be obtained directly from the analysis of the specimen points. A linear regression analysis gives the following simple expression:

$$t_d = \frac{A_c}{p_c} \left(0.082 + 3.405 \frac{\tau_n}{f'_c} \right) \frac{1}{\sin 2\alpha} \qquad (7\text{–}142b)$$

This expression is conservative within the range of $(3/5) \leq \tan \alpha \leq (5/3)$. Eq. (7–142b) is plotted in Fig. 7.30a as two straight solid lines with $\tan \alpha = 1$ and $\tan \alpha = (3/5)$ or $(5/3)$.

When the concrete cover is too thick as compared to the effective wall thickness, the cover may spall before the maximum torque is reached. A non-dimensional parameter $\bar{c}/t_d$ has been used to represent the concrete cover. The quantity $\bar{c}$ is defined as the distance from the surface of the member to the inner face of the transverse hoop bars. This inner face of the transverse hoop bars is taken as the centroidal line of the steel cage, including both longitudinal and transverse steel. It was found in Ref. 17 that spalling of the concrete cover may occur when $\bar{c}/t_d$ exceeds 0.75. If this is the case, then the concrete cover should be deleted in the calculation of the torsional strength.

It is interesting to mention here that Collins and Mitchell[12] have treated the compression stress block in a different way. First, they have neglected the concrete cover; second, the effective thickness of the wall is assumed to be the depth of an equivalent rectangular stress block; third, the center line of the shear flow is assumed to coincide with the center line of the compression resultant; and fourth, the softening of the concrete due to diagonal shear cracking has been neglected.

7.5.4 Strains in Transverse and Longitudinal Steel, and Provisions to Prevent Overreinforcement

The strains in the transverse and longitudinal steel can be derived by using five compatibility equations. These five compatibility equations are summarized as follows:

Eq. (7–26): $\dfrac{\gamma}{2} = (\epsilon_l + \epsilon_d)\cot\alpha$

Eq. (7–27): $\dfrac{\gamma}{2} = (\epsilon_t + \epsilon_d)\tan\alpha$

Eq. (1–77): $\theta = \dfrac{p_0}{2A_0}\gamma$

Eq. (7–124): $\psi = \theta\sin 2\alpha$

Eq. (7–125): $\epsilon_{ds} = \psi t_d$

It should be noted in Eq. (1–77) that γ is a constant and $p_0 = \oint ds$ is the perimeter of the center line of the shear flow.

Substituting Eq. (7–27) into Eq. (1–77) gives:

$$\theta = \frac{p_0}{A_0}(\epsilon_t + \epsilon_d)\tan\alpha \qquad (7\text{–}143)$$

Inserting this θ into Eq. (7–124), we have:

$$\psi = \frac{p_0}{A_0}(\epsilon_t + \epsilon_d)\tan\alpha \sin 2\alpha \qquad (7\text{–}144)$$

Inserting the above ψ into Eq. (7–125) results in:

$$\epsilon_{ds} = \frac{2p_0 t_d}{A_0}(\epsilon_t + \epsilon_d)\tan\alpha \sin\alpha \cos\alpha \qquad (7\text{–}145)$$

Substituting t_d from Eq. (7–137) into Eq. (7–145) gives:

$$\epsilon_{ds} = \frac{p_0 T_n \tan\alpha}{A_0^2(k_1 f_c'/\lambda)}(\epsilon_t + \epsilon_d) \qquad (7\text{–}146)$$

Inserting $k_1 = 0.80$, λ from Eq. (7–127), $T_n = \tau_n A_c^2/p_c$ from Eq. (7–141), and $\epsilon_d = \epsilon_{ds}/2$ (see Fig. 7.29) into Eq. (7–146), we have:

$$\epsilon_t = \left[\left(\frac{0.57 f_c' A_0^2 p_c \cos\alpha}{\tau_n A_c^2 p_0 \tan\alpha}\right)^{2/3} - \frac{1}{2}\right]\epsilon_{ds} \qquad (7\text{–}147)$$

Assuming $p_c/p_0 = 1.2$ and recalling from Eq. (7–139) that $A_0/A_c = 1 - t_d p_c/2A_c$, Eq. (7–147) can be expressed as:

$$\epsilon_t = \left\{ \left[\left(\frac{0.68 f'_c}{\tau_n} \right) \left(\frac{\cos^2 \alpha}{\sin \alpha} \right) \left(1 - \frac{t_d p_c}{2 A_c} \right)^2 \right]^{2/3} - \frac{1}{2} \right\} \epsilon_{ds} \qquad (7\text{-}148)$$

Substituting t_d from Eq. (7-142b) into Eq. (7-148) we arrive at:

$$\epsilon_t = \left\{ \left[\left(\frac{0.68 f'_c}{\tau_n} \right) \left(\frac{\cos^2 \alpha}{\sin \alpha} \right) \right. \right.$$
$$\left. \left. \left(1 - \left(0.041 + 1.703 \frac{\tau_n}{f'_c} \right) \frac{1}{\sin 2\alpha} \right)^2 \right]^{2/3} - \frac{1}{2} \right\} \epsilon_{ds} \qquad (7\text{-}149)$$

Similarly, the strain in the longitudinal steel has been derived as:

$$\epsilon_l = \left\{ \left[\left(\frac{0.68 f'_c}{\tau_n} \right) \left(\frac{\sin^2 \alpha}{\cos \alpha} \right) \right. \right.$$
$$\left. \left. \left(1 - \left(0.041 + 1.703 \frac{\tau_n}{f'_c} \right) \frac{1}{\sin 2\alpha} \right)^2 \right]^{2/3} - \frac{1}{2} \right\} \epsilon_{ds} \qquad (7\text{-}150)$$

At failure it would appear reasonable to assume $\epsilon_{ds} = 0.003$ in the above two equations.[17] This value has been specified by the ACI Code for flexure of nonsoftened concrete and is found to be applicable also to softened concrete in torsion. Equations (7-149) and (7-150) also show that the steel strains, ϵ_t and ϵ_l, are a function of only two variables, namely, the torsional stress ratio, τ_n/f'_c, and the angle α. To facilitate the design process, these two equations are plotted in Fig. 7.30b using τ_n/f'_c and α as coordinates. For a given angle α and a given ratio of τ_n/f'_c, the longitudinal and transverse steel strains can readily be obtained from Fig. 7.30b.

The design chart, Fig. 7.30b, can also be thought of as defining the limit of α for a given level of stress and given yield strains of steel ϵ_{ty} and ϵ_{ly}. For the usable range of steel yield strains between 0.001 and 0.003 the family of solid curves in Fig. 7.30b can be approximated by a family of dotted straight lines as shown. This family of dotted straight lines has the following expression:

$$12° + \frac{(\tau_n/f'_c)}{(0.27 - 45\epsilon_{ly})} 33° \leq \alpha \leq 78° - \frac{(\tau_n/f'_c)}{(0.27 - 45 \epsilon_{ty})} 33° \qquad (7\text{-}151)$$

Expression (7-151) gives the range of α which ensures the yielding of both the longitudinal and transverse steel. It can be seen that with higher level of stress this range of α becomes narrower.

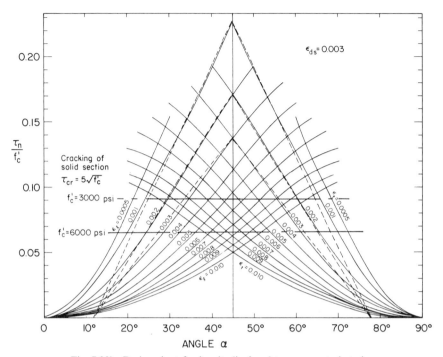

Fig. 7.30b. Design chart for longitudinal and transverse steel strains.

A similar expression has been proposed for design by Collins and Mitchell.[12] Their expression was derived without considering the softening of concrete and by neglecting the concrete cover. Although the effects of these two measures tend to cancel each other, Collins and Mitchell's expression was found to be unconservative when compared to many of the PCA tests.[15]

The equations for equilibrium, compatibility and stress-strain curve of concrete derived in Section 7.5 can also be used to construct a theoretical torque-twist curve for a member subjected to pure torsion. This treatment is given in Example 7.2 (Section 7.8).

7.5.5. RECOMMENDATIONS FOR TORSION DESIGN[17]

The basic equations for design are summarized as follows:

For transverse hoop steel

Eq. (7–117)
$$T_n = \frac{2A_0 A_t f_{ty}}{s} \cot \alpha \qquad \text{(a)}$$

where

Eq. (7–139)
$$A_0 = A_c - \frac{t_d}{2} p_c + t_d^2 \qquad \text{(b)*}$$

Eq. (7–142b)
$$t_d = \frac{A_c}{p_c} \left(0.082 + 3.405 \frac{\tau_n}{f'_c} \right) \frac{1}{\sin 2\alpha} \qquad \text{(c)}$$

Eq. (7–141)
$$\tau_n = \frac{T_n p_c}{A_c^2} \qquad \text{(d)}$$

For longitudinal steel

Eq. (7–118)
$$\Delta N_y = A_1 f_{1y} = \frac{A_t f_{ty} p_0}{s} \cot^2 \alpha \qquad \text{(e)}$$

where

Eq. (7–138)
$$p_0 = p_c - 4 t_d \qquad \text{(f)}$$

Check minimum reinforcement (or maximum member size)

$$\tau_n \geq \frac{p_c h}{A_c} (5 \sqrt{f'_c}) \qquad h = \frac{A_c}{p_c} \text{ for solid sections} \qquad \text{(g)}$$

Check overreinforcement (or minimum member size)

Eq. (7–151)
$$12° + \frac{(\tau_n/f'_c)}{(0.27 - 45 \, \epsilon_{1y})} 33° \leq \alpha \leq 78° - \frac{(\tau_n/f'_c)}{(0.27 - 45 \, \epsilon_{ty})} 33° \qquad \text{(h)}$$

Check serviceability (excessive cracking)

Eq. (7–98)
$$\frac{3}{5} \leq \tan \alpha \leq \frac{5}{3} \qquad \text{(i)}$$

Check concrete cover

If
$$\frac{\bar{c}}{t_d} \leq 0.75 \qquad \text{O.K.} \qquad \text{(j)}$$

---

* For rectangular sections Eq. (b) is simply $A_0 = (x - t_d)(y - t_d)$. Eq. (b) is a more accurate expression than Eq. (7–139).

If $\qquad \dfrac{\bar{c}}{t_d} > 0.75 \qquad$ Neglect concrete cover $\hspace{2cm}$ (k)

Check stirrup spacing

Eq. (4–73) $\hspace{3cm} s \leq \dfrac{u}{8}$ and 12 in. $\hspace{2cm}$ (ℓ)

For a given torque (T_n), and the material properties (f'_c, f_{ly}, f_{ty}), the procedures to design the size of the member, the transverse hoop steel and the longitudinal steel are as follows:

(1) Select a cross section and calculate A_c and p_c.
(2) Calculate torsional stress τ_n from Eq. (d) and check the maximum size of the member by Eq. (g).
(3) Select an α angle which satisfies Eqs. (h) and (i). If no range of α angle is indicated in Eq. (h), the size of the member must be increased. If a range of α angle is available, the smallest α will usually give the most economical design.
(4) Determine the effective thickness of diagonal concrete struts t_d from Eq. (c) and check the concrete cover by Eqs. (j) or (k). The $\bar{c}$ value is often governed by the fire-proof requirement.
(5) Calculate A_0 from Eq. (b) and p_0 from Eq. (f).
(6) Calculate the transverse steel area per unit length, A_t/s, from Eq. (a). Select a bar size and then determine the spacing s. Check transverse hoop bar spacing by Eq. (ℓ)
(7) Determine the total area of longitudinal steel by Eq. (e) and distribute this steel around the perimeter such that the spacing does not exceed 12 in.

7.6   BOX SECTIONS SUBJECTED TO COMBINED LOADINGS

7.6.1   Interaction of Torsion and Bending

The interaction of torsion and bending for a box section will now be examined. The forces in the top and bottom stringers are assumed to be simply the superposition of the longitudinal forces induced by torsion and bending, as shown in Fig. 7.31. It can be seen that the forces in the bottom stringer due to torsion and bending are additive, whereas the forces in the top stringer are subtractive. Assuming that the force induced by the bending moment is M/d_v and noticing that the force in either the top or the bottom stringer

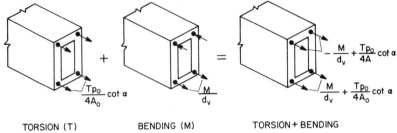

TORSION (T)   BENDING (M)   TORSION + BENDING

Fig. 7.31. Superposition of stringer forces due to torsion and bending.

due to torsion is one-half of that given in Eq. (7–111), we can express the forces in the top and bottom stringers as follows:

$$N_t = -\frac{M}{d_v} + \frac{Tp_0}{4A_0}\cot\alpha \qquad (7\text{–}152a)$$

$$N_b = \frac{M}{d_v} + \frac{Tp_0}{4A_0}\cot\alpha \qquad (7\text{–}152b)$$

The force in the hoop reinforcement is induced by torsion alone, so Eqs. (7–113) and (7–117) remain valid for the case of combined torsion and bending.

Failure of a box section subjected to torsion and bending may occur in two modes. The first mode is caused by the yielding of the bottom stringer and the hoop steel. Eq. (7–152b) can, therefore, be written as:

$$A_{lb}f_{ly} = \frac{M_y}{d_v} + \frac{T_y p_0}{4A_0}\frac{1}{\tan\alpha} \qquad (7\text{–}153)$$

Substituting $\tan\alpha$ from Eq. (7–117) into Eq. (7–153) gives:

$$\frac{M_y}{A_{lb}f_{ly}d_v} + \frac{T_y^2}{4A_0^2\left(\dfrac{2A_{lb}f_{ly}}{p_0}\right)\left(\dfrac{A_t h_{ty}}{s}\right)} = 1 \qquad (7\text{–}154)$$

Notice that $M_{0y} = A_{lb}f_{ly}d_v$ and:

$$T_{0y} = 2A_0\sqrt{\left(\frac{2A_{lb}f_{ly}}{p_0}\right)\left(\frac{A_t f_{ty}}{s}\right)} \qquad (7\text{–}155)$$

where T_{0y} = pure torsional moment ($M_y = 0$) at the yielding of the bottom stringer and hoop steel. Comparing Eq. (7–155) to Eq. (7–119), we see that they are the same because ΔN_y in the latter equation is indeed equal to $2A_{1b}f_{1y}$ in the former equation.

Inserting M_{0y} and T_{0y} into Eq. (7–154) results in:

$$\frac{M_y}{M_{0y}} + \left(\frac{T_y}{T_{0y}}\right)^2 = 1 \qquad (7\text{–}156)$$

Equation (7–156) is the nondimensionalized torsion–moment interaction curve for the first mode of failure.

The second mode of failure is caused by the yielding of the top stringer and the hoop steel. Equation (7–152a) then becomes:

$$A_{1t}f_{1y} = -\frac{M_y}{d_v} + \frac{T_y p_0}{4A_0}\frac{1}{\tan\alpha} \qquad (7\text{–}157)$$

Substituting $\tan\alpha$ from Eq. (7–117) into Eq. (7–157) gives:

$$-\frac{M_y}{A_{1t}f_{1y}d_v} + \frac{T_y^2}{4A_0^2\left(\dfrac{2A_{1t}f_{1y}}{p_0}\right)\left(\dfrac{A_t f_{ty}}{s}\right)} = 1 \qquad (7\text{–}158)$$

Recalling the definition of $R = A_{1t}/A_{1b}$, and utilizing the definitions of M_{0y} and T_{0y}, Eq. (7–158) is modified into a nondimensionalized form:

$$-\frac{M_y}{M_{0y}} + \left(\frac{T_y}{T_{0y}}\right)^2 = R \qquad (7\text{–}159)$$

Equations (7–156) and (7–159) for combined torsion and bending are identical to Eqs. (7–58) and (7–62), respectively, if T_y/T_{0y} is replaced by V_y/V_{0y}. Therefore, the nondimensionalized interaction curve in Fig. 7.11 is also applicable to combined torsion and bending, if the vertical axis of V_y/V_{0y} is replaced by T_y/T_{0y}.

7.6.2   Interaction of Torsion, Shear, and Bending

Now that the torsion–bending interaction and the shear–bending interaction have been separately established, we shall examine the combined interaction of torsion, shear, and bending. The interaction of torsion and shear turns

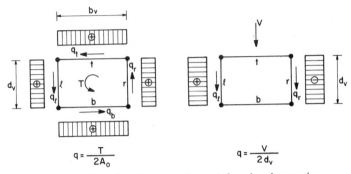

Fig. 7.32. Shear flow due to torsion and shear in a box section.

out to be more complex than the other interactions, and the solution was first worked out by Elfgren.[6]

When both shear and torsion are present, the shear flows q on the four walls of a box section will differ as shown in Fig. 7.32. It can be seen that the shear flows due to torsion and shear are additive on the left wall and are subtractive on the right wall. On the top and bottom walls, however, only shear flow due to torsion exists. Utilizing the subscripts l, r, t, and b to indicate left, right, top, and bottom walls, the shear flows in the four walls are:

$$q_l = \frac{T}{2A_0} + \frac{V}{2d_v} \qquad (7\text{--}160)$$

$$q_r = \frac{T}{2A_0} - \frac{V}{2d_v} \qquad (7\text{--}161)$$

$$q_t = q_b = \frac{T}{2A_0} \qquad (7\text{--}162)$$

These shear flows are also illustrated in Fig. 7.33. It is assumed that the center line of the shear flow coincides with both the center line of the hoop bar and the line connecting the centroids of the longitudinal bars.

The angle α of the concrete struts will also be different in the four walls. The value of α can be expressed either by Eq. (7–5) for shear or by Eq. (7–112) for torsion:

$$\cot \alpha = \frac{qs}{A_w f_s} \qquad (7\text{--}163)$$

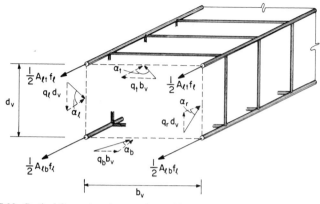

Fig. 7.33. Statical forces in a box section subjected to torsion, shear, and bending.

where A_w and f_s are the area and the stress, respectively, of the web reinforcement required by combined torsion and shear. Substituting q from Eqs. (7–160) through (7–162) into Eq. (7–163), we have:

$$\cot \alpha_1 = \frac{s}{A_w f_s} \left(\frac{T}{2A_0} + \frac{V}{2d_v} \right)$$

(7–164)

$$\cot \alpha_r = \frac{s}{A_w f_s} \left(\frac{T}{2A_0} - \frac{V}{2d_v} \right)$$

(7–165)

$$\cot \alpha_t = \cot \alpha_b = \frac{s}{A_w f_s} \left(\frac{T}{2A_0} \right)$$

(7–166)

These four α angles are also indicated in Fig. 7.33.

A rectangular box section subjected to torsion, shear, and bending may fail in three distinct ductile modes. These three modes have been called Modes 1, 2, and 3:

Mode 1: failure is caused by yielding in the *bottom* stringer and in the *web* reinforcement on the side where shear flows due to torsion and shear are *additive*. The equilibrium equation can be derived by taking moment about the *top* side.

Mode 3: failure is caused by yielding in the *top* stringer and in the *web* reinforcement on the side where shear flows due to torsion and shear are *additive*. The equilibrium equation can be derived by taking moment about the *bottom* side.

Mode 2: failure is caused by yielding in the *top* bar, in the *bottom* bar, and in the *web* reinforcement, all on the side where shear flows due to torsion and shear are *additive*. The equilibrium equation is obtained by taking moment about the side where shear flows due to torsion and shear are *subtractive*.

7.6.2.1 Mode 1 Failure. Referring to Fig. 7.33 and taking moment about the top side will furnish the following equilibrium equation:

$$M = A_{1b}f_{1y}d_v - q_1 d_v \cot \alpha_1 \frac{d_v}{2} - q_r d_v \cot \alpha_r \frac{d_v}{2} - q_b b_v \cot \alpha_b \, d_v \quad (7\text{-}167)$$

where

A_{1b} = total area of bottom longitudinal bars
f_{1y} = yield stress of longitudinal bars
M = bending moment taken as positive when the bottom fiber is in tension.

Substituting q_1, q_r, q_b from Eqs. (7–160) through (7–162) and $\cot \alpha_1$, $\cot \alpha_r$, $\cot \alpha_b$ from Eqs. (7–164) through (7–166) into Eq. (7–167) and simplifying give:

$$\frac{M}{A_{1b}f_{1y}d_v} + \left(\frac{T}{2A_0}\right)^2 \frac{s}{A_w f_{sy}} \frac{(d_v + b_v)}{A_{1b}f_{1y}} + \left(\frac{V}{2d_v}\right)^2 \frac{s}{A_w f_{sy}} \frac{d_v}{A_{1b}f_{1y}} = 1 \quad (7\text{-}168)$$

Let us define:

$$M_0 = A_{1b}f_{1y}d_v \qquad (7\text{-}169)$$

$$T_0 = 2A_0 \sqrt{\frac{A_{1t}f_{1y}}{d_v + b_v} \frac{A_w f_{sy}}{s}} \qquad (7\text{-}170)$$

$$V_0 = 2d_v \sqrt{\frac{A_{1t}f_{1y}}{d_v} \frac{A_w f_{sy}}{s}} \qquad (7\text{-}171)$$

$$R = \frac{A_{1t}}{A_{1b}} \qquad (7\text{-}172)$$

It should be noted that definitions for M_0, T_0, and V_0 are chosen as the *lowest positive values* from the three modes of failure. (See Section 6.5.3 in Chapter 6.) Utilizing these definitions, Eq. (7–168) becomes:

$$\frac{M}{M_0} + \left(\frac{T}{T_0}\right)^2 R + \left(\frac{V}{V_0}\right)^2 R = 1 \qquad (7\text{-}173)$$

This is the nondimensionalized interaction equation for mode 1 failure.

7.6.2.2 Mode 3 Failure. Again referring to Fig. 7.33 and taking moment about the bottom side will furnish the following equilibrium equation:

$$M = - A_{lt}f_{ly}d_v + q_l d_v \cot \alpha_l \frac{d_v}{2} + q_r d_v \cot \alpha_r \frac{d_v}{2} + q_t b_v \cot \alpha_t \, d_v \qquad (7\text{-}174)$$

where A_{lt} = total area of the top longitudinal bars. As in the derivation of the equation for mode 1 failure, we shall substitute the following quantities into Eq. (7–174): q_l, q_r, q_t from Eqs. (7–160) through (7–162) and cot α_l, cot α_r, cot α_t from Eqs. (7–164) through (7–166). After simplification we should obtain:

$$-\frac{M}{M_0}\frac{1}{R} + \left(\frac{T}{T_0}\right)^2 + \left(\frac{V}{V_0}\right)^2 = 1 \qquad (7\text{-}175)$$

This is the nondimensionalized interaction equation for mode 3 failure.

7.6.2.3 Mode 2 Failure. The equilibrium equation for mode 2 failure is obtained by taking moment about the right side of the truss model in Fig. 7.33. It is obvious that moments with respect to this vertical axis must be equal to zero:

$$0 = \frac{1}{2}(A_{lb} + A_{lt})f_{ly}b_v - q_l d_v \cot \alpha_l b_v - q_t b_v \cot \alpha_t \frac{b_v}{2}$$
$$- q_b b_v \cot \alpha_b \frac{b_v}{2} \qquad (7\text{-}176)$$

Substituting q_l, q_t, q_b from Eqs. (7–160) and (7–162) and cot α_l, cot α_t, cot α_b from Eqs. (7–164) and (7–166), and R from Eq. (7–172) into Eq. (7–176), we have:

$$1 = \left(\frac{T}{2A_0}\right)^2 \frac{s}{A_w f_{sy}} \frac{(d_v + b_v)}{A_{lt}f_{ly}} \frac{2R}{(R+1)} + \left(\frac{V}{2d_v}\right)^2 \frac{s}{A_w f_{sy}} \frac{d_v}{A_{lt}f_{ly}} \frac{2R}{(R+1)}$$
$$+ \left(\frac{T}{2A_0}\right)\left(\frac{V}{2d_v}\right) \frac{s}{A_w f_{sy}} \frac{2d_v}{A_{lt}f_{ly}} \frac{2R}{(R+1)} \qquad (7\text{-}177)$$

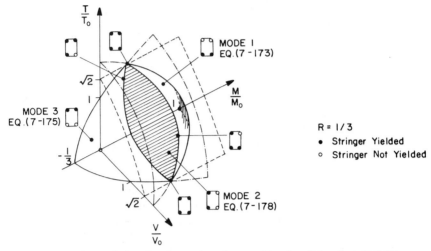

Fig. 7.34. Interaction surface for torsion, shear, and bending (taken from Ref. 11).

Utilizing the definition of T_0 and V_0 in Eqs. (7–170) and (7–171), Eq. (7–177) can be expressed in a nondimensionalized form.

$$\left(\frac{T}{T_0}\right)^2 \frac{2R}{R+1} + \left(\frac{V}{V_0}\right)^2 \frac{2R}{R+1} + \left(\frac{TV}{T_0 V_0}\right) \frac{2R}{(R+1)} \frac{2}{\sqrt{1+b_v/d_v}} = 1 \quad (7\text{–}178)$$

The three nondimensionalized interaction equations, Eqs. (7–173), (7–175), and (7–178), for failure modes 1, 3, and 2, respectively, are identical to Eqs. (6–171), (6–172), and (6–173) in Chapter 6. The latter three equations were derived from the skew failure mechanism. An interaction surface formed by the three equations is shown in Fig. 7.34 for an R value of ⅓. The different modes of failure are also indicated.

7.7 CEB-FIP MODEL CODE FOR TORSION

7.7.1 Types of Torsion

7.7.1.1 Equilibrium Torsion versus Compatibility Torsion. The
1978 CEB-FIP Model Code separates torsion into two types, namely, equilibrium torsion and compatibility torsion. Equilibrium torsion occurs when torsional moment is required for a structure to be in equilibrium. Such torsional moment *cannot* be reduced by redistribution of internal forces within the structure. Compatibility torsion arises when torsional moment is induced from twisting of the member in order to maintain compatibility of deformation. Such torsional moment *can* be reduced by redistribution of internal

forces within the structure. A detailed discussion of compatibility torsion and equilibrium torsion is given in Section 9.1 (Chapter 9). For equilibrium torsion, the CEB-FIP Model Code requires that the full torsional moment should be used for the design of a member. For compatibility torsion, however, the code permits a designer to neglect completely the torsional moment in the ultimate strength design of a member. Such treatment of compatibility torsion, in the author's opinion, is dangerous and should not be followed. A correct design method for a member subjected to compatibility torsion is the torsional limit design method presented in Section 9.3 (Chapter 9).

7.7.1.2 Circulatory Torsion versus Warping Torsion.

The terms "circulatory torsion" and "warping torsion" appear in the CEB-FIP Model Code to distinguish two types of torsional resistance in a member. Circulatory torsion is also known as St. Venant torsion; in it the torsional resistance is generated by the shear stresses flowing in a circulatory manner on the cross section of a member. In contrast, warping torsion furnishes the torsional resistance from the differential in-plane bending and shear in the component walls of a member.

In general, both circulatory torsion resistance and warping torsion resistance occur side by side in a member subjected to torsion. In solid or hollow members with bulky cross section, circulatory torsion should predominate, and warping torsion is often neglected. Since these types of members are most commonly used in reinforced concrete, most codes for reinforced concrete provide detailed design procedures for circulatory torsion while ignoring warping torsion. It should be realized, however, that warping torsion may predominate in members with a thin-walled open section. In the design of such members, warping torsion resistance must be taken into account. Calling attention to this need, the CEB-FIP Model Code provides brief instructions for the design of open sections having three walls in separate planes.

7.7.2 Strength Design for Circulatory Torsion

In the 1978 CEB-FIP Model Code, simplifications were made in defining the center line of the shear flow and the effective wall thickness in a solid or hollow section. The center line of the shear flow is assumed to coincide with the perimeter connecting the centroids of the corner longitudinal bars. This assumed shear flow perimeter is shown by a dotted polygon for an arbitrary cross section in Fig. 7.35 (a). From this dotted polygon the area, A_0, and the perimeter of this area, p_0, can be determined.

The effective wall thickness is defined in the following manner. Draw the

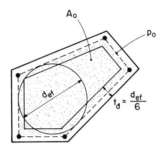

(a) ARBITRARY CROSS SECTION

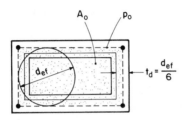

(b) RECTANGULAR BOX SECTION

Fig. 7.35. Definitions of A_0, p_0, and t_d in CEB-FIP model code.

largest circle that can be contained within the effective perimeter, p_0, and denote the diameter of this circle as d_{ef}. Then the effective wall thickness, t_d, is assumed to be one-sixth of this diameter (i.e., $t_d = d_{ef}/6$).

The definitions of A_0, p_0, and t_d in the code are also illustrated in Fig. 7.35 (b) for a rectangular box section. It is assumed that the effective wall thickness, t_d, is less than the actual wall thickness. If the opposite is true, then the actual wall thickness should be taken as the effective wall thickness.

As in shear design, three strength criteria must be considered in the design of a member subjected to torsion. First, the hoop steel must have sufficient strength to resist the applied torsional moment. Second, the longitudinal force (or steel area) due to torsion must be determined and added to the longitudinal forces (or steel area) caused by other actions. Third, a maximum torsional strength should be established by specifying an effective compressive stress in the diagonal concrete struts such that the hoop and longitudinal reinforcement will yield before the crushing of the concrete struts.

7.7.2.1 Design of Web Reinforcement. In the CEB-FIP Code, the design of web reinforcement to resist torsion is based on Eq. (7–117), which gives:

$$T_n = \frac{A_t f_{ty}}{s} 2A_0 \cot \alpha \qquad (7\text{--}179)$$

This equation is applicable to the case of high torsion when $T_u \geq 3\phi T_c$. The quantity ϕ is the material reduction factor, taken as $1/1.5$ in the code, and T_c is an empirical torsional resistance given by:

$$T_c = 2.5\tau_R t_d 2A_0 \qquad (7\text{--}180)$$

where $\tau_R = f'_t/4$. The tensile strength of concrete, f'_t, is taken as $0.214(f'_c)^{2/3}$, where f'_t and f'_c must be in the SI unit of MPa.

For the case of low torsion, i.e., $T_u < 3\phi T_c$, Eq. (7–179) should be modified by adding an empirical torsional resistance, T_{cv}:

$$T_n = T_{cv} + \frac{A_t f_{ty}}{s} 2A_0 \cot \alpha \qquad (7\text{--}181)$$

The term T_{cv} is known as the resistance contributed by concrete, but actually includes all the effects that are neglected in the truss model, such as shear resistance of concrete struts, dowel action of reinforcement, aggregate interlock, etc. In the above equation:

$$T_{cv} = T_c \qquad \text{for } T_u \leq \phi T_c \qquad (7\text{--}182)$$
$$T_{cv} = 0 \qquad \text{for } T_u \geq 3\phi T_c \qquad (7\text{--}183)$$

Intermediate values can be determined by linear interpolation.

In combined torsion and shear the torsional web reinforcement calculated by Eq. (7–179) or (7–181) should be added to the web reinforcement required by shear. On the side of the wall where shear flows due to torsion and shear are additive, a simple addition of the steel areas would suffice. However, on the side of the wall where shear flows due to torsion and shear are subtractive, the code allows the steel area required by torsion to be reduced by the area required by shear, or vice versa.

It should also be noted that the CEB-FIP Code permits a conservative design in combined torsion and shear by taking $T_c = V_c = 0$.

7.7.2.2 Design of Longitudinal Steel.

In the CEB-FIP Code the design of longitudinal steel to resist torsion is based on Eq. (7–116), in which the total longitudinal force due to torsion, ΔN_y, is replaced by $A_l f_{ly}$:

$$T_n = \frac{A_l f_{ly}}{p_0} 2A_0 \tan \alpha \qquad (7\text{--}184)$$

Notice that the above equation gives the total area of longitudinal steel due to torsion alone. In combined torsion and bending, a part of this longitudinal steel area in the flexural tension zone is additive to the flexural tensile steel. The other part of the longitudinal steel area in the flexural compression zone, however, should be subtractive to the flexural compression steel.

For combined torsion and shear the CEB-FIP Code requires that the longitudinal steel due to torsion should be added to the longitudinal steel due to shear. This provision is, of course, very conservative because on one side of the cross section the shear flow due to torsion and that due to shear are actually in opposite directions.

A more logical way to treat the combination of torsion and shear is as follows. First, the shear flows in each wall of a cross section are determined individually for shear and for torsion; second, the shear flows in each wall due to torsion and shear are combined; and finally, each wall is designed according to the combined shear flow. This method, though logical, is apparently quite tedious.

It should be pointed out that the CEB-FIP design format for torsion is quite different from that for shear in Eq. (7–81). For example, the formulation of Eq. (7–184) does not permit an increase of longitudinal reinforcement when the hoop steel is reduced. This reduction of hoop steel is achieved by the inclusion of the empirical terms, T_{cv}, in the torsional resistance, Eq. (7–181).

A consistent design format for torsion and shear has been proposed by Collins and Mitchell.[12] First, the longitudinal steel force due to torsion alone is calculated from Eq. (7–116):

$$\Delta N_u = \frac{T_u p_0}{2 A_0} \cot \alpha \qquad (7\text{–}185)$$

In combined torsion and shear, an equivalent tensile force $(\Delta N_u)_{eq}$ can be assumed as the square root of the sum of the squares of the individually calculated tensile forces:

$$(\Delta N_u)_{eq} = \cot \alpha \sqrt{V_u^2 + \left(\frac{T_u p_0}{2 A_0}\right)^2} \qquad (7\text{–}186)$$

In the flexural tension zone the longitudinal steel area can be determined by $M_u/d_v + (\Delta N_u)_{eq}/2$, and in the flexural compression zone the longitudinal steel is found from $-M_u/d_v + (\Delta N_u)_{eq}/2$. In regions of high moment, when $M_u/d_v > (\Delta N_u)_{eq}/2$, no reinforcement is required in the flexural compression zone.

7.7.2.3 Maximum Torsional Strength Due to Concrete Crushing.

The compressive stress in the concrete struts, σ_d, was derived in Eq. (7–115). At failure the effective strength of concrete was chosen as $0.5f'_c$. Inserting $\sigma_d = 0.5f'_c$, $t = t_d$, and $T = T_{n,\max}$ into Eq. (7–115), we have:

$$T_{n,\max} = 0.5f'_c A_0 t_d \sin 2\alpha \qquad (7\text{–}187)$$

Equation (7–187) is given in the CEB-FIP Code. This equation was derived originally from large-size box sections used in bridges and was calibrated for such structures. However, it was found to be unreasonably conservative for smaller-size solid sections used in buildings. The source of the difficulty stems from the definitions of the area, A_0, and the effective wall thickness, t_d, given in Fig. 7.35. A_0 and t_d are related to the dotted polygon constructed through the centers of longitudinal corner bars. For large box sections, where the concrete cover and the size of the steel bars are small in comparison to the overall dimensions, the dotted polygon represents the center line of the shear flow with reasonable accuracy. However, for smaller-size solid sections the concrete cover and the size of the steel bars are quite significant with respect to the overall dimensions. Therefore, the area within the dotted polygon, A_0, becomes considerably smaller than the area within the outer concrete perimeter, and the dotted polygon (representing the center line of shear flow) may lie completely outside the effective wall thickness, t_d. Furthermore, the maximum torque calculated by Eq. (7–187) may become smaller than the cracking torque. These awkward situations are, of course, unacceptable, and will be illustrated in Example 7.1.

To overcome this difficulty, the author would like to propose the following new equation, which is a modification of Eq. (7–187). This new equation has been checked by the PCA tests[15] with specimen sizes varying from 6 in. by 12 in. to 10 in. by 20 in. Therefore, this equation is applicable not only to large box sections used in bridges, but also to smaller members normally employed in buildings.

$$T_{n,\max} = 0.5f'_c A_c t_c \sin 2\alpha \qquad (7\text{–}188a)$$

where:

$$t_c = 0.45 \frac{A_c}{p_c} \qquad (7\text{–}188b)$$

In these two equations:

A_c = cross-sectional area within the outer perimeter of concrete
p_c = outer perimeter of concrete
t_c = wall thickness defined by Eq. (7–188b)

The thickness t_c in Eq. (7–188b) has the same parameter A_c/p_c as the effective thickness t_d in Eq. (7–140), if the value within the bracket in Eq. (7–140) is taken to be a constant. Based on test results this constant was evaluated to be 0.45 for t_c. Inserting t_c from Eq. (7–188b) into Eq. (7–188a) gives:

$$\tau_{n,\,\max} = \frac{T_{n,\,\max}\, p_c}{A_c^2} = 0.225 f_c' \sin 2\alpha \qquad (7\text{–}189)$$

In combined torsion and shear, the code simply specifies a straight-line interaction between the two, resulting in:

$$\frac{T_u}{\phi T_{n,\,\max}} + \frac{V_u}{\phi V_{n,\,\max}} \leq 1 \qquad (7\text{–}190)$$

where $T_{n,\,\max}$ and $V_{n,\,\max}$ are given in Eqs. (7–187) and (7–85), respectively.

An interesting provision is also given in the code for the case of torsion combined with a large bending moment. Such a combination may cause critical principal compressive stress in the compression zone, particularly in box sections. The principal compressive stress can be calculated from the mean longitudinal compression due to flexure and from the tangential stress due to torsion, taken as being equal to $T_u/2A_0 t_d$. The combined stress so obtained should not exceed $0.85 f_c'$.

7.7.3 Serviceability Requirements and Detailing of Reinforcement

The two serviceability requirements for torsion are identical to those for shear, which were discussed in Section 7.4.3. First, the angle α is limited to $3/5 \leq \tan \alpha \leq 5/3$, and, second, the crack width due to diagonal cracks must be limited to a specified size. The crack width can be calculated by means of Eqs. (7–99) through (7–102).

The detailing of torsional reinforcement are also similar to those for shear reinforcement in Section 7.4.4, except that the maximum hoop spacing is limited to:

$$s_{\max} = \frac{p_0}{8} \qquad (7\text{–}191)$$

This spacing is basically the same as that required in the ACI Building Code, Eq. (4–73).

7.8 EXAMPLES 7.1 AND 7.2

Example 7.1 Design of a Beam According to CEB-FIP Code

In order to compare the CEB-FIP Code with the ACI Code, we shall redesign the beam in Example 4.4 (see Section 4.5) using the 1978 CEB-FIP Code. The frame and loading are given in Fig. 4.24 (a), and the cross section of the beam is shown in Fig. 4.24 (b). Material properties remain $f'_c = 4,000$ psi and $f_y = 60,000$ psi.

Solution

Load and Material Factors:
The limit state design in the CEB-FIP Code employs load factors and material factors that are quite different from those used in the ACI Code. The factors that are necessary in this example problem are summarized below:

Load Factors		*ϕ-Factor*	
Permanent load	1.35	Concrete $\phi_c = 1/1.5 = 0.667$	
Variable load	1.50	Steel $\phi_s = 1/1.15 = 0.87$	

Flexural Moments, Shears, and Torques (Fig. 7.36)
Beam dead weight:

$$w_d = 1.35 \, (150) \frac{18(28)}{144} = 709 \text{ plf}$$

Superimposed concentrated load:

Dead load	1.35(42) =	56.7 k
Live load	1.5(36) =	54.0 k
	Design load =	110.7 k

Bending moment at face of column (assumed fixed ends):

$$M_u = \frac{P_u \ell}{8} + \frac{w_d \ell^2}{12} = \frac{110.7(30)(12)}{8} + \frac{0.790(30)^2(12)}{12} = 5,620 \text{ in.-k}$$

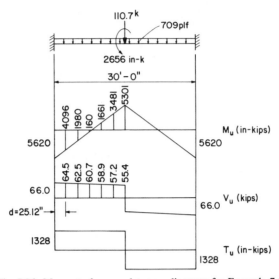

Fig. 7.36. Moment, shear, and torque diagrams for Example 7.1.

Bending moment at midspan:

$$M_u = \frac{P_u \ell}{8} + \frac{w_d \ell^2}{24} = \frac{110.7(30)(12)}{8} + \frac{0.709(30)^2(12)}{24} = 5{,}301 \text{ in.-k}$$

Shear at face of column:

$$V_u = \frac{P_u}{2} + \frac{w_d \ell}{2} = \frac{110.7}{2} + \frac{0.709(30)}{2} = 66.0 \text{ k}$$

Shear at a distance d from face of column (assume $d = 25.12$ in.):

$$V_u = \frac{P_u}{2} + w_d \left(\frac{\ell}{2} - d\right) = \frac{110.7}{2} + 0.709 \left(15 - \frac{25.12}{12}\right) = 64.5 \text{ k}$$

Shear at midspan:

$$V_u = \frac{P_u}{2} = \frac{110.7}{2} = 55.4 \text{ k}$$

Torsional moment throughout beam:

$$T_u = \frac{110.7(2)(12)}{2} = 1,328 \text{ in.-k}$$

Cross-sectional properties:

Assume No. 6 stirrups and No. 10 longitudinal bars
$b_w = 18$ in.
$d = 28 - 1.5 - 0.75 - 1.27/2 = 25.12$ in.
$x_2 = 18 - 2(1.5 + 0.75 + 1.27/2) = 12.23$ in.
$y_2 = 28 - 2(1.5 + 0.75 + 1.27/2) = 22.23$ in.
$d_{ef} = x_2 = 12.23$ in.
$t_d = d_{ef}/6 = 12.23/6 = 2.04$ in.
$A_0 = x_2 y_2 = 12.23(22.23) = 271.9$ in.^{2}
$p_0 = 2(x_2 + y_2) = 2(12.23 + 22.23) = 68.92$ in.
Assume $\alpha = 45°$, $\beta = 90°$

Material properties:

$f'_c = 4,000$ psi
$f_y = 60,000$ psi

Check Maximum Shear and Torsional Strengths:

Eq. (7–91): $\phi_c V_{n,\text{max}} = \phi_c(0.3 f'_c) b_w d$
$= 0.667(0.3)(4,000)(18)(25.12) = 362,000$ lb

Eq. (7–187): $\phi_c T_{n,\text{max}} = \phi_c(0.5 f'_c) A_0 t_d \sin 2\alpha$
$= 0.667(0.5)(4,000)(271.9)(2.04)(1)$
$= 740,000$ in.-lb

At a distance d from the face of the column, $V_u = 64.5$ and $T_u = 1,328$ in.-k.

Eq. (7–190): $\dfrac{T_u}{\phi_c T_{n,\text{max}}} + \dfrac{V_u}{\phi_c V_{n,\text{max}}} = \dfrac{1,328}{740} + \dfrac{64.5}{362} = 1.97$ N.G.

This means that the cross section is insufficient even though the selected beam size has been found to be quite ample in Example 4.4 using the ACI Code. The difficulty arises from the calculation of $T_{n,\text{max}}$ using Eq. (7–187). As discussed in Section 7.7.2.3 this equation is unacceptably conservative for smaller solid members commonly used in buildings. It is interesting to point out that the effective wall thickness $t_d = 2.04$ in. lies well outside the area A_0. The maximum torsional strength $\phi T_{n,\,max}$ is also smaller than the cracking torque T_{cr}, which is calculated conservatively by the skew-bending theory using Eq. (2–14):

$$T_{cr} \approx T_{np} = 6(x^2 + 10)y \sqrt[3]{f_c'} = 6(18^2 + 10)(28) \sqrt[3]{4,000}$$
$$= 891,000 \text{ in.-lb} > 740,000 \text{ in.-lb} \ (\phi T_{n,\,max})$$

or by the plastic theory using Eq. (2–7) and assuming $f_t' = 5\sqrt{f_c'}$:

$$T_{cr} \approx T_p = \left(0.5 - \frac{x}{6y}\right) x^2 y (5\sqrt{f_c'})$$

$$= \left(0.5 - \frac{18}{6(28)}\right) (18)^2 (28)(5\sqrt{4000})$$

$$= 1,127,000 \text{ in.-lb} > 740,000 \text{ in.-lb} \ (\phi T_{n,\,max})$$

Since Eq. (7–187) is unreasonable for this beam, we shall use Eq. (7–189), which is applicable to members normally used in buildings:

$$A_c = 18(28) = 504 \text{ in.}^2$$
$$p_c = 2(18 + 28) = 92 \text{ in.}^2$$

Eq. (7–189): $\phi_c T_{n,\,max} = \phi_c (0.225 f_c')(\sin 2\alpha)(A_c^2/p_c)$
$$= 0.667(0.225)(4,000)(1)(504^2/92) = 1,657,000 \text{ in.-lb}$$

Eq. (7–190): $\dfrac{T_n}{\phi_c T_{n,\,max}} + \dfrac{V_u}{\phi_c V_{n,\,max}} = \dfrac{1,328}{1,657} + \dfrac{64.5}{362} = 0.98 < 1 \qquad$ O.K.

Shear and Torsional Web Reinforcement:

$$f_c' = 4,000 \text{ psi} = 27.59 \text{ MPa}$$
$$f_t' = 0.214\ (f_c')^{2/3} = 0.214(27.59)^{2/3}$$
$$= 1.954 \text{ MPa} = 283 \text{ psi}$$
$$\tau_R = \frac{f_t'}{4} = \frac{283}{4} = 70.8 \text{ psi}$$

Eq. (7–74): $V_c = 2.5\ \tau_R b_w d = 2.5(70.8)(18)(25.12) = 80,000 \text{ lb}$
Eq. (7–76): $\phi_c V_c = 0.667(80.0) = 53.3 \text{ k} < 64.5 \text{ k} \ (V_u)$

Eq. (7–77):
$$3\phi_c V_c = 3(0.667)(80.0) = 160 \text{ k} > 64.5 \text{ k } (V_u)$$
$$V_{cv} = \frac{160 - 64.5}{160 - 53.3} V_c = 0.895(80.0) = 71.6 \text{ k}$$

Eq. (7–180):
$$T_c = 2.5 \tau_R t_d 2A_0 = 2.5(70.8)(2.04)(2)(271.9)$$
$$= 196.000 \text{ in.-lb}$$

Eq. (7–182):
$$\phi_c T_c = 0.667(196) = 130.7 \text{ in.-k} < 1{,}328 \text{ in.-k } (T_u)$$

Eq. (7–183):
$$3\phi_c T_c = 3(0.667)(196) = 392.2 \text{ in.-k} < 1{,}328 \text{ in.-k } (T_u)$$
$$T_{cv} = 0$$

Eq. (7–75):
$$\frac{A_v}{s} = \frac{V_u - \phi_c V_{cv}}{\phi_s f_{vy} 0.9d \sin \beta (\cot \alpha + \cot \beta)}$$
$$= \frac{64.5 - 0.667(71.6)}{0.87(60)(0.9)(25.12)(1)(1 + 0)} = 0.0142 \text{ in.}^2/\text{in.}$$

Eq. (7–181):
$$\frac{A_t}{s} = \frac{T_u - \phi_c T_{cv}}{\phi_s f_{ty} 2A_0 \cot \alpha} = \frac{1{,}328 - 0.667(0)}{0.87(60)(2)(271.9)(1)}$$
$$= 0.0468 \text{ in.}^2/\text{in.}$$
$$\frac{A_t}{s} + \frac{1}{2}\frac{A_v}{s} = 0.0468 + \frac{0.0142}{2} = 0.0539 \text{ in.}^2/\text{in.}$$

Use No. 6 closed stirrups, $s = \dfrac{0.44}{0.0539} = 8.16 \text{ in.}$

It should be mentioned that the CEB-FIP Code allows the shear steel to be subtracted from the torsional steel for the vertical side where shear and torsion are subtractive. This would be economical for large box sections but impractical for this member in which continuous closed stirrups are used. Now let us check the maximum spacing.

Eq. (7–191):
$$s_{max} = \frac{p_0}{8} = \frac{68.92}{8} = 8.6 \text{ in.} > 8.16 \text{ in.} \qquad \text{O.K.}$$
$$\frac{2}{3}\phi_c V_{n, max} = \frac{2}{3} 362 = 241 \text{ kips} > 64.5 \text{ k}$$

Eq. (7–103):
$$s_{max} = 0.5d = 0.5(25.12) = 12.56 \text{ in.} > 8.16 \text{ in.} \qquad \text{O.K.}$$
$$s_{max} = 12 \text{ in.} > 8.16 \text{ in.} \qquad \text{O.K.}$$

The minimum web reinforcement is checked as follows:

Eq. (7–105):
$$\rho_w = \frac{A_v}{b_w s \sin \beta} = \frac{0.0539(2)}{18(1)} = 0.00600$$

ρ_w in Eq. (7–105) is given in the CEB-FIP Code for shear reinforcement alone. It is assumed here to be applicable to shear plus torsional reinforcement. From Table 7.1 ($f'_c = 4,000$ psi and $f_y = 60$ ksi):

$$\rho_{w, \min} = 0.0012 < 0.00600 \qquad \text{O.K.}$$

Use No. 6 closed stirrups at 8-in. spacing throughout the beam.

Shear Longitudinal Reinforcement:
At a distance d from column face: $V_u = 64.5$ k. The shear longitudinal steel area in the flexural tensile zone is:

Eq. (7–81):
$$A_{lv} = \frac{V_u^2}{2 \left(\dfrac{A_v}{s} \right) f_{vy} f_{ly} d \sin \beta} - \frac{V_u \cot \beta}{f_{ly}}$$

$$= \frac{64.5^2}{2(0.0142)(60)^2(25.12)(1)} - \frac{64.5(0)}{60} = 1.62 \text{ in.}^2$$

Torsional Longitudinal Reinforcement:
The total longitudinal steel area is:

Eq. (7–184):
$$A_l = \frac{T_u p_0}{\phi_s f_y 2 A_0 \tan \alpha} = \frac{1,328(68.92)}{0.87(60)(2)(271.9)(1)} = 3.22 \text{ in.}^2$$

The CEB-FIP Code does not specify a maximum spacing for longitudinal bars. However, to control crack width it would be advisable to limit this spacing to 12 in. as required by the ACI Code. To satisfy this requirement we shall provide three layers of steel with equal distribution, each layer having an area of:

$$\frac{A_l}{3} = \frac{3.22}{3} = 1.07 \text{ in.}^2$$

Using two No. 7 bars at mid-height will give an area of $2(0.60) = 1.20$ in.^{2} ≈ 1.07 in.^{2} Then the top and bottom longitudinal bar areas are:

$$A_{l,top} = A_{l,bot} = \frac{1}{2}(3.22 - 1.20) = 1.01 \text{ in.}^2$$

Flexural Longitudinal Reinforcement:
At face of column: $M_u = 5,620$ in.-k.

$$M_{max} = 937bd^2 = 937(18)(25.12)^2 = 10,651 \text{ in.-k.}$$
$$> 5,620 \text{ in.-k.} \quad \text{Singly reinforced}$$

Assume $a = 4.5$ in.

$$A_s = \frac{M_u}{\phi_s f_y \left(d - \dfrac{a}{2}\right)} = \frac{5,620}{0.87(60)\left(25.12 - \dfrac{4.5}{2}\right)} = 4.71 \text{ in.}^2$$

$$a = \frac{A_s f_y}{0.85 f'_c b} = \frac{4.71(60)}{0.85(4)(18)} = 4.61 \text{ in.} \approx 4.5 \text{ in.} \quad \text{O.K.}$$

At a distance d from column face: $M_u = 4,096$ in.-k. Assume $a = 3.3$ in.

$$A_s = \frac{4,096}{0.87(60)\left(25.12 - \dfrac{3.3}{2}\right)} = 3.34 \text{ in.}^2$$

$$a = \frac{3.34(60)}{0.84(4)(18)} = 3.27 \text{ in.} \approx 3.3 \text{ in.} \quad \text{O.K.}$$

At midspan: $M_u = 5,301$ in.-k. Assume $a = 4.2$ in.

$$A_s = \frac{5,301}{0.87(60)\left(25.12 - \dfrac{4.2}{2}\right)} = 4.41 \text{ in.}^2$$

$$a = \frac{4.41(60)}{0.85(4)(18)} = 4.32 \text{ in.} \approx 4.2 \text{ in.} \quad \text{O.K.}$$

Arrangement of Longitudinal Bars (see Fig. 7.37):
At face of column:

Top bars:

$$A_s + A_{l,top} = 4.71 + 1.01 = 5.72 \text{ in.}^2$$
Use 5 No. 10 $5(1.27) = 6.35 \text{ in.}^2 > 5.72 \text{ in.}^2$

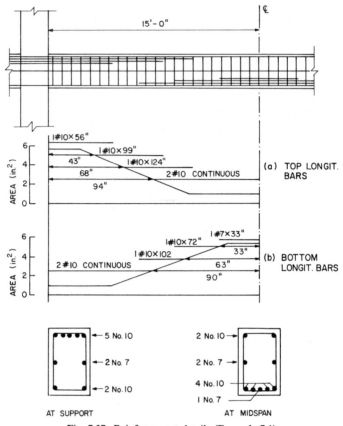

Fig. 7.37. Reinforcement details (Example 7.1).

Bottom bars:

$$-A'_s + A_{l,\text{bot}} = 0 + 1.01 = 1.01 \text{ in.}^2$$
$$\text{Use 2 No. 10} \qquad 2(1.27) = 2.54 > 1.01 \text{ in.}^2$$

At a distance d from column face:

Top bars:

$$A_s + A_{lv} + A_{l,\text{top}} = 3.34 + 1.62 + 1.10 = 5.97 \text{ in.}^2$$
$$\text{Since } 5.97 \text{ in.}^2 > 5.72 \text{ in.}^2 \text{ at column face, use } 5.72 \text{ in.}^2$$

Bottom bars:

$$-A'_s + A_{l,\text{bot}} = 0 + 1.01 = 1.01 \text{ in.}^2$$

At midspan:

Top bars:

$$-A'_s + A_{1,\text{top}} = 0 + 1.01 = 1.01 \text{ in.}^2$$
Use 2 No. 10 $2(1.27) = 2.54 > 1.01 \text{ in.}^2$

Bottom bars:

$$A'_s + A_{1,\text{bot}} = 4.41 + 1.01 = 5.42 \text{ in.}^2$$
Use 4 No. 10 + 1 No. 7 $4(1.27) + 0.60 = 5.68 \text{ in.}^2 > 5.42 \text{ in.}^2$

Calculations for reinforcement areas at other sections along the member are given in Table 7.5.

Cut-Off Points for Negative Bars (see Fig. 7.37):
The development length l_d is specified in the CEB-FIP Code as:

$$l_d = \frac{d_b}{4} \frac{(\phi_s f_y)}{(\phi_c f_b)}$$

where $f_b = 0.482\,(f'_c)^{2/3}$, and f_b and f'_c must be in MPa. For $f'_c = 4,000$ psi $= 27.59$ MPa:

$$f_b = 0.482(27.59)^{2/3} = 4.40 \text{ MPa} = 638 \text{ psi}$$

For bottom No. 10 bar:

$$l_d = \frac{1.27(0.87)(60,000)}{4(0.667)(638)} = 39.0 \text{ in.}$$

For top No. 10 bar:

$$l_d = \frac{39.0}{0.7} = 55.7 \text{ in.} \approx 56 \text{ in.}$$

For bottom No. 7 bar:

$$l_d = \frac{0.875(0.87)(60,000)}{4(0.667)(638)} = 26.8 \text{ in.} \approx 27 \text{ in.}$$

Negative moment region:
First cut-off point (one No. 10 bar):

    Steel area requirement = 43 in. from graph
    Development length requirement = $l_d = 56$ in.   governs

Table 7.5.   Summary of reinforcement areas (Example 7.1)

ITEM	UNITS	AT COLUMN FACE	AT DISTANCE d FROM COL. FACE	DISTANCE FROM CENTER OF SPAN, FT				
				10.0	7.5	5.0	2.5	0
Top Longitudinal Bars								
$-M_u$	in.-kips	5,620	4,096	1,980	160	—	—	—
A_s	in.²	4.71	3.34	1.56	0.12	—	—	—
V_u	kips	—	64.5	—	—	—	—	—
A_{1v}	in.²	—	1.62	1.62[a]	1.62[a]	0[b]	0[b]	0[b]
T_u	in.-kips	1,328	1,328	1,328	1,328	1,328	1,328	1,328
$A_{1,top}$	in.²	1.01	1.01	1.01	1.01	1.01	1.01	1.01
$A_s + A_{1v} + A_{1,top}$	in.²	5.72	5.72[c]	4.19	2.75	1.01	1.01	1.01
Use		5 No. 10						2 No. 10
Bottom Longitudinal Bars								
$+M_u$	in.-kips	—	—	—	—	1,661	3,481	5,301
A_s	in.²	—	—	—	—	1.30	2.79	4.41
V_u	kips	—	64.5	—	—	—	—	—
A_{1v}	in.²	—	0[b]	0[b]	0[b]	1.62[a]	1.62[a]	—
T_u	in.-kips	1,328	1,328	1,328	1,328	1,328	1,328	1,328
$A_{1,bot}$	in.²	1.01	1.01	1.01	1.01	1.01	1.01	1.01
$A_s + A_{1v} + A_{1,bot}$	in.²	1.01	1.01	1.01	1.01	3.93	5.42	5.42
Use		2 No. 10						4 No. 10 + 1 No. 7

Note: The values underlined have been calculated in the text.

[a] CEB-FIP Code requires all shear longitudinal steel in the flexural tension zone to be calculated for the section of maximum shear.

[b] CEP-FIP Code neglects the shear longitudinal steel in the flexural compression zone.

[c] Since the calculated value 5.97 in.² > 5.72 in.² at column face, use 5.72 in.²

Second cut-off point (one No. 10 bar):

68 in. from graph
43 in. from graph + l_d = 43 + 56 = 99 in.   governs

Third cut-off point (one No. 10 bar):

94 in. from graph
68 in. from graph + l_d = 68 + 56 = 124 in.   governs

Positive moment region:
First cut-off point (one No. 7 bar):

33 in. from graph   governs
l_d = 27 in.

Second cut-off point (one No. 10 bar):

63 in. from graph
33 in. from graph + l_d = 33 + 39 = 72 in.   governs

Third cut-off point (one No. 10 bar):

90 in. from graph
63 in. from graph + l_d = 63 + 39 = 102 in.   governs

All the cut-off points for longitudinal bars are plotted in Fig. 7.37.

Comparison:
When the reinforcement designed by the CEB-FIP Code (Fig. 7.37) is compared to that designed by the ACI Code (Fig. 4.25), it can be seen that the former is slightly less conservative than the latter. This difference, however, is caused mainly by the lower load factors inherent in the CEB-FIP Code. As far as the design formulas are concerned, both codes will give results that are reasonably close.

Example 7.2   Construction of a Torque-Twist Curve

This example is aimed to illustrate the calculation and the plotting of a torque-twist curve for a member subjected to pure torsion, using the truss model theory explained in this chapter. The member selected is Beam M2

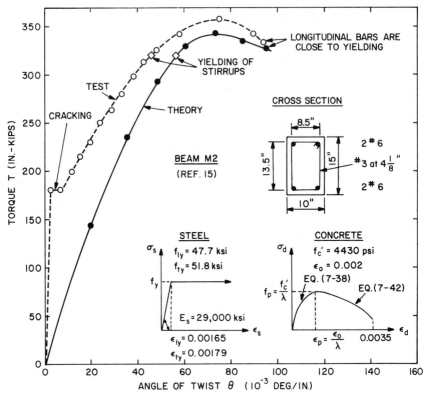

Fig. 7.38. Comparison of theoretical and experimental torque-twist curves for beam M2.

given in Ref. 15. Both the experimental and theoretical torque-twist curves for this beam are plotted in Fig. 7.38. The inset of Fig. 7.38 also gives the cross-sectional dimensions and the reinforcement details of the beam, as well as the assumed stress-strain curves of steel and concrete. The stress-strain curve of concrete takes into account the "softening effect" due to the diagonal shear cracking and, therefore, is a function of the coefficient λ (see Section 7.2.3 and Fig. 7.7). The ascending and descending portions of the stress-strain curve for concrete are expressed by Eqs. (7–38) and (7–42), respectively.

The construction of a torque-twist curve requires three equilibrium equations to determine the torque T, the effective depth t_d and the angle α. The torque T at any load level is calculated by Eq. (7–115) in which t is taken as the effective thickness, t_d (see Fig. 7.29):

$$T = 2A_0 \, t_d \, \sigma_d \sin \alpha \cos \alpha \qquad (7\text{--}192)$$

Substituting T from Eq. (7–192) into Eqs. (7–111) and (7–113) we obtained the angle α in two forms:

$$\cos^2 \alpha = \frac{A_l f_l}{p_0 \, \sigma_d t_d} \tag{7–193a}$$

$$\sin^2 \alpha = \frac{A_t f_t}{s \, \sigma_d \, t_d} \tag{7–193b}$$

In Eq. (7–193a) the total force of longitudinal steel ΔN has been taken as $A_l f_l$ for non-prestressed concrete. By summing Eqs. (7–193a) and (7–193b) the effective depth t_d can be expressed as:

$$t_d = \frac{A_l f_l}{p_0 \, \sigma_d} + \frac{A_t f_t}{s \, \sigma_d} \tag{7–194}$$

Notice that Eqs. (7–192) to (7–194) are all expressed in terms of σ_d, the stress in the diagonal concrete struts.

Three additional equations are also required to determine the strains in the transverse and longitudinal steel, ϵ_t and ϵ_l, as well as the angle of twist θ. Substituting t_d from Eq. (7–192) into Eq. (7–145) and taking $\epsilon_d = \epsilon_{ds}/2$ gives the strain ϵ_t:

$$\epsilon_t = \left(\frac{A_0^2 \, \sigma_d}{p_0 \, T \tan \alpha} - \frac{1}{2} \right) \epsilon_{ds} \tag{7–195}$$

Similarly, the conjugate equation for the strain ϵ_l can be derived as:

$$\epsilon_l = \left(\frac{A_0^2 \, \sigma_d}{p_0 \, T \cot \alpha} - \frac{1}{2} \right) \epsilon_{ds} \tag{7–196}$$

Finally, eliminating ψ from Eqs. (7–124) and (7–125), the angle of twist θ can be expressed as:

$$\theta = \frac{\epsilon_{ds}}{2 t_d \sin \alpha \cos \alpha} \tag{7–197}$$

Notice that Eqs. (7–195) through (7–197) are all expressed in terms of ϵ_{ds}, the maximum concrete strain at the surface. Equations (7–195) and (7–196) are also functions of σ_d.

The stress in the diagonal concrete struts σ_d is defined as the average stress for a non-uniform compression stress block in the concrete strut of a torsional member (see Section 7.5.3 and Fig. 7.29):

$$\sigma_d = k_1 \frac{1}{\lambda} f'_c \qquad (7\text{-}198)$$

where

$$\lambda = \sqrt{\frac{\epsilon_l + \epsilon_t + 2(\epsilon_{ds}/2)}{(\epsilon_{ds}/2)} - 0.3} \qquad (7\text{-}199)$$

The coefficient k_1 in Eq. (7–198) has been given by Eqs. (7–134) and (7–136), and has been tabulated in Table 7.4 assuming $\epsilon_0 = 0.002$.

The trial-and-error procedure for constructing a torque-twist curve is as follows:

1. Select ϵ_{ds} and assume t_d, α and λ. Find k_1 from Table 7.4, and σ_d from Eq. (7–198).
2. Calculate T from Eq. (7–192).
3. Calculate ϵ_t and ϵ_l from Eqs. (7–195) and (7–196).
4. Check t_d and α by Eqs. (7–194) and (7–193a).
5. Check λ by Eq. (7–199).
6a. If the set of calculated values for t_d, α and λ is not sufficiently close to the set assumed, repeat step (1) to (5).
6b. If the set of calculated values for t_d, α and λ is sufficiently close to the set assumed, proceed to calculate θ from Eq. (7–197). This will provide a pair of $T - \theta$ values, which give a point on the theoretical torque-twist curve.
7. Select other values of ϵ_{ds} and repeat steps (1) to (6) for each ϵ_{ds} value. This will provide a number of pairs of $T - \theta$ values to plot a whole torque-twist curve.

To illustrate this trial-and-error procedure the following calculation is made from Beam $M2$.

1. Select $\epsilon_{ds} = 0.002$ and assume $t_d = 1.90$ in., $\alpha = 42.5°$ and $1/\lambda = 0.45$. From Table 7.4 $k_1 = 0.8267$.

$\sigma_d = (0.8267)(0.45)(4430) = 1,648$ psi
$A_0 = (10 - 1.90)(15 - 1.90) = 106.1$ in.^{2}
$p_0 = 2(10 - 1.90 + 15 - 1.90) = 42.4$ in.

2. $T = 2A_0\, t_d\, \sigma_d \sin \alpha \cos \alpha$
   $= 2(106.1)(1.90)(1,648) \sin 42.5 \cos 42.5$
   $= 331.0$ in.-k

3. $\epsilon_l = \left(\dfrac{A_0^2 \sigma_d}{p_0 T \cot \alpha} - \dfrac{1}{2}\right) \epsilon_{ds}$

   $= \left[\dfrac{(106.1)^2(1.648)}{(42.4)(331.0) \cot 42.5°} - \dfrac{1}{2}\right] 0.002 = 0.00142 < \epsilon_{ly}$

   $\epsilon_t = \left(\dfrac{A_0^2 \sigma_d}{p_0 T \tan \alpha} - \dfrac{1}{2}\right) \epsilon_{ds}$

   $= \left[\dfrac{(106.1)^2(1.648)}{(42.4)(331) \tan 42.5} - \dfrac{1}{2}\right] 0.002 = 0.00189 > \epsilon_{ty}$

4. $t_d = \dfrac{A_l f_l}{p_0 \sigma_d} + \dfrac{A_t f_t}{s \sigma_d} = \dfrac{(1.76)(0.00142)(29.000)}{(42.4)(1.648)} + \dfrac{(0.11)(51.8)}{(4.125)(1.648)}$

   $= 1.88$ in. ≈ 1.90 in.   O.K.

   $\cos \alpha = \sqrt{\dfrac{A_l f_l}{p_0 \sigma_d t_d}} = \sqrt{\dfrac{(1.76)(0.00142)(29.000)}{(42.4)(1.648)(1.90)}} = 0.739$

   $\alpha = 42.3° \approx 42.5°$ O.K.

5. $\lambda = \sqrt{\dfrac{\epsilon_l + \epsilon_t + \epsilon_{ds}}{(\epsilon_{ds}/2)} - 0.3}$

   $= \sqrt{\dfrac{0.00142 + 0.00189 + 0.002}{0.001} - 0.3} = 2.24$

   $\dfrac{1}{\lambda} = 0.447 \approx 0.45$ O.K.

6b. Since the calculated values of t_d, α and λ are close to the assumed values, calculate θ:

$$\theta = \dfrac{\epsilon_{ds}}{2 t_d \sin \alpha \cos \alpha} = \dfrac{0.002}{2(1.90) \sin 42.5 \cos 42.5}$$
$$= 0.001056 \text{ rad./in.} = 0.0605 \text{ deg./in.}$$

7. The T and θ values for other ϵ_{ds} values are summarized in the following Table 7.6

The T and θ values in Table 7.6 are used to plot the theoretical torque-twist curve in Fig. 7.38. It can be seen that the theoretical values compared very well with the experimental values near the ultimate load stage. The theoretical maximum torque is only 4.6% less than the experimental value,

Table 7.6 Summary of calculation for Beam M2

ϵ_{ds} ($\times 10^{-3}$)	T (in.-k)	ϵ_l ($\times 10^{-3}$)	ϵ_t ($\times 10^{-3}$)	t_d (in.)	α (deg.)	$1/\lambda$	θ ($\times 10^{-3}$ deg./in.)
0.5	144.0	0.595	0.756	1.43	42.5	0.38	20.1
1.0	235.1	0.963	1.24	1.62	42.5	0.40	35.5
1.5	293.7	1.23	1.60	1.77	42.6	0.43	48.7
2.0	331.0	1.42	1.89	1.90	42.5	0.45	60.5
2.5	342.6	1.56	2.34	1.97	41.5	0.46	73.3
3.0	335.6	1.57	2.86	2.05	40.0	0.46	85.1
3.5	327.9	1.58	3.15	2.15	39.5	0.47	95.0

while the theoretical angle of twist at maximum torque is almost identical to the experimental value.

The truss model theory which takes into account the softening of concrete, not only can predict the strength and the angle of twist at failure, it can also predict the strains in the reinforcement and concrete at high loads. The calculated strains ϵ_t and ϵ_l in Table 7.6 show that the yielding of stirrups occurred at a torque of 318 in.-kips, whereas the longitudinal steel barely yields at the termination of the test. These yield strains are substantiated by the actual measurements of steel strains reported in Ref. 15 and are indicated on the torque-twist curve in Fig. 7.38. The concrete strain at the surface, ϵ_{ds}, is about 0.0025 at failure. This value is also close to the average measured strains reported in Ref. 15.

In addition, the effective thickness, t_d, of about 2 in. at maximum torque (compared to a 10 in. by 15 in. cross section) is consistent with the observation of crack pattern on the saw-cut sections of test beams, one of which is shown in Fig. 3.14 (c). The softening coefficient, $1/\lambda$, of about 0.46 at failure also appears to be reasonable.

The success of the truss model in predicting the various strains and parameters at high loads is logical, as concrete is extensively cracked, so that a real specimen approaches the theoretical model. The prediction of the torque-twist curve at lower load stages, however, are significantly below the experimental curve as shown in Fig. 7.38. This is so because the concrete has not yet been extensively cracked (or not cracked at all) and the concrete core obviously affects the torsional behavior of a member. A transition from an uncracked member, which follows St.-Venant's theory, to a fully cracked member, which can be predicted by the truss model, is very much expected.

REFERENCES

1. Ritter, W., "Die Bauweise Hennebique," Schweizerishe Bauzeitung, Zurich, February 1899.
2. Morsch, E., "Der Eisenbetonbau, seine Anwendung und Theorie," 1st ed., Wayss and Frey-

tag, A. G., Im Selbstverlag der Firma, Neustadt a. d. Haardt, May 1902, 118 pp.; "Der Eisenbetonbau, seine Theorie und Anwendung," 2nd ed., Verlag von Konrad Wittmer, Stuttgart, 1906, 252 pp.; 3rd ed. translated into English by E. P. Goodrich, McGraw-Hill Book Co., New York, 1909, 368 pp.

3. Rausch, E., "Design of Reinforced Concrete in Torsion" (Berechnung des Eisenbetons gegen Verdrehung), Technische Hochschule, Berlin, 1929, 53 pp. (in German). A second edition was published in 1938. The third edition was titled "Drillung (Torsion), Schub and Scheren in Stahlbetonbau," Deutcher Ingenieur-Verlag GmbH, Dusseldorf, 1953, 168 pp.

4. Hsu, T. T. C., "Post-Cracking Torsional Rigidity of Reinforced Concrete Sections," *Journal of the American Concrete Institute,* Proc., Vol. 70, No. 5, May 1973, pp. 352–360.

5. Lampert, P. and B. Thurlimann, "Torsionsversuche an Stahlbetonbalken" (Torsion Tests of Reinforced Concrete Beams), Bericht Nr. 6506–2, June 1968; "Torsion-Biege-Versuche an Stahlbetonbalken" (Torsion-Bending Tests on Reinforced Concrete Beams), Bericht Nr. 6506–3, January 1969, Institut fur Baustatik, ETH, Zurich. (in German)

6. Elfgren, L., "Reinforced Concrete Beams Loaded in Combined Torsion, Bending and Shear," Publication 71:3, Division of Concrete Structures, Chalmers University of Technology, Goteborg, Sweden, 1972.

7. Wagner, H., "Ebene Blechwandtrager mit Sehr Dunnem Stegblech" (Flat Sheet Metal Girders with Very Thin Metal Web), *Zeitschrift fur Flugtechnik und Motorluftschiffahr,* Vol. 20, No. 8 to 12, Berlin, 1929 (in German). Translated into English as Technical Memorandum of the National Advisory Committee for Aeronautics, TM604–606, Washington, D.C., 1931.

8. Collins, M., "Torque-Twist Characteristics of Reinforced Concrete Beams," *Inelasticity and Non-Linearity in Structural Concrete,* University of Waterloo Press, Waterloo, Ontario, 1973, pp. 211–232.

9. CEB-FIP, "Model Code for Concrete Structures," CEB-FIP International Recommendations, third edition, Comite Euro-International du Beton (CEB), 1978.

10. Thurlimann, B., "Shear Strength of Reinforced and Prestressed Concrete—CEB Approach," ACI-CEB-PCI-FIP Symposium, *Concrete Design: U.S. and European Practices,* ACI Publication SP-59, American Concrete Institute, Detroit, 1979, pp. 93–115.

11. Thurlimann, B., "Torsional Strength of Reinforced and Prestressed Concrete Beams—CEB Approach," ACI-CEB-PCI-FIP Symposium, *Concrete Design: U.S. and European Practices,* ACI Publication SP-59, American Concrete Institute, Detroit, 1979. pp. 117–143.

12. Collins, M. P. and D. Mitchell, "Shear and Torsion Design of Prestressed and Non-Prestressed Concrete Beams," *Journal of the Prestressed Concrete Institute,* Vol. 25, No. 5, September–October 1980, pp. 32–100.

13. Vecchio, F. and M. P. Collins, "Stress–Strain Characteristics of Reinforced Concrete in Pure Shear," IABSE Colloquium, *Advanced Mechanics of Reinforced Concrete,* Delft, 1981, Final Report, pp. 211–225.

14. Hsu, T. T. C., "Is the 'Staggering Concept' of Shear Design Safe?" *Journal of the American Concrete Institute,* Proc., Vol. 79, No. 6, November–December 1982.

15. Hsu, T. T. C., "Torsion of Structural Concrete–Behavior of Reinforced Concrete Rectangular Members," *Torsion of Structural Concrete,* SP-18, American Concrete Institute, Detroit, 1968, pp. 261–306; also Development Dept. Bulletin D135, Portland Cement Association, Skokie, Illinois (with added appendixes).

16. Hognestad, E., N. W. Hanson, and D. McHenry, "Concrete Stress Distribution in Ultimate Strength Design," *Journal of the American Concrete Institute,* Proc., Vol. 52, No. 4, December 1955, pp. 455–479.

17. Hsu, T. T. C. and Y. L. Mo, "Softening of Concrete in Torsional Members," Research Report No. ST-TH-001-83, Dept. of Civil Engineering, University of Houston, Houston, Texas; (submitted to the *Journal of the American Concrete Institute*).

Plate 8: Reinforced concrete wall beams supporting cantilever flower bins. Severe torsional cracking was observed in the beams and slabs, 1976.

8
Space Frames

Space frames are the most common type of structures used in buildings. In the past, space frames in a building were broken down into a series of plane frames for analysis. In doing this, the torsional effect of the members is neglected. This treatment of space frames is justifiable in those situations where torsional stiffness of a member is much less than the flexural stiffness. However, in many other cases the torsional moment will be significant in a space structure, and the torsional stiffness cannot be neglected. In this chapter, many example problems will be given, so that an engineer can develop an ability to discern those cases where the torsion effect should be considered.

Torsional effect is particularly significant in reinforced concrete space frames, for the following two reasons. First, let us compare a reinforced concrete space frame to a steel space frame. Reinforced concrete members usually have bulky sections, such as a rectangular solid section, whereas steel members are usually made up of thin walls, such as an I-section. The former will have a ratio of torsional stiffness to flexural stiffness many times higher than that of the latter. Consequently, relative to the flexural moment, torsional moment will be much greater in reinforced concrete space frames than in steel space frames. Second, in reinforced concrete members the efficiency of torsional reinforcement is several times less than that of flexural reinforcement (see Section 9.3.1). That is, a torsional moment will require several times (by volume) more reinforcement than a flexural moment of equal magnitude. Therefore, in a reinforced concrete space frame the requirement for torsional steel may be comparable to flexural steel even though, in magnitude, the torsional moment is many times less than the flexural moment.

To analyze torsional moments in space frames, three methods will be introduced in this chapter: the flexibility method, the stiffness method, and the moment distribution method. Each will be convenient for a particular type of space frame. In presenting these methods, it will be assumed that the reader is familiar with their application to plane frames subjected to in-plane loadings. In such frames only the flexural stiffnesses of the members are involved. In this chapter, however, we shall generalize these methods to include torsional stiffnesses of the members so that they will be applicable to space frames.

Space frames are often statically indeterminate to a high order even in relatively simple problems. The analysis will involve many unknowns and will be tedious. A powerful tool to reduce the number of unknowns is the theory of symmetry; thus an in-depth study of this theory will be presented first, and its application will extend throughout the chapter.

8.1 SYMMETRY OF SPACE STRUCTURES

8.1.1 Application of Symmetry in Structures

The concept of symmetry in a structure subjected to loadings is often considered to be a matter of common sense. Take the simple frame and loadings in Fig. 8.1 (a), for example; anyone can see that both the structure and the loadings are symmetric. Common sense, however, often fails to help us identify more sophisticated types of symmetry. The frame in Fig. 8.2 (a) will throw many structural engineers into confusion. Is the structure in Fig. 8.2 (a) symmetric? Are the loadings on the structure symmetric? The answer to the first question is yes and to the second is no. These two correct answers frequently surprise many engineers.

If the symmetric structure in Fig. 8.2 (a) is cut at the origin, three pairs of internal forces, F_x, F_y, and M_z, as shown in Fig. 8.2 (c), are possible under arbitrary coplanar forces. However, under the pair of antisymmetric loads P, it is possible to determine from the theory of symmetry that $F_x = F_y = 0$. Only one unknown, M_z, remains to be determined. In this way, utilization of symmetry helps us to reduce a third-degree indeterminate structural problem to a first-degree indeterminate structural problem. The complexity of the solution is greatly reduced.

The theory of symmetry also tells us that the flexural rotation about the z-axis at the origin is zero. This condition of compatibility can be used to determine the unknown M_z. Once M_z is solved for, all the force, shear, and moment diagrams can be constructed through statics alone.

Throughout history humankind has exhibited a liking for symmetric objects. It has been suggested that such aesthetic preference may have stemmed from the symmetric appearance of the humans themselves. From a practical point of view, symmetric structures are often economical because of the intrinsic repetition of forms. In some cultures, the requirement for symmetric forms has even been included in building specifications. For whatever reasons,

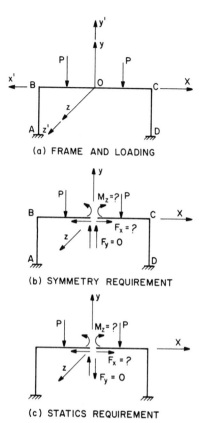

(a) FRAME AND LOADING

(b) SYMMETRY REQUIREMENT

(c) STATICS REQUIREMENT

Fig. 8.1. Planar symmetric structure.

symmetric structures are very common. For the analysis of such structures, the theory of symmetry can be a powerful tool. The theory of symmetry described below originated from the lecture notes of A. C. Scordelis.

8.1.2 Definition of Symmetry of Structures

The geometric and stiffness properties of any structure may be defined with respect to a rectangular Cartesian coordinate system (x, y, z). This same structure can also be described with respect to another rectangular coordinate system (x', y', z'). If a structure can be *identically described* in these two coordinate systems and the two coordinate system have one or more of the following three types of relationships between them, then the structure is termed symmetric. The three types of relationships are:

1. $x' = -x,\ y' = y,\ z' = z$ planar symmetry
2. $x' = -x,\ y' = -y,\ z' = z$ axial symmetry
3. $x' = -x,\ y' = -y,\ z' = -z$ point symmetry

To illustrate this definition of symmetry, three space frames are given in Figs. 8.3 (a), 8.4 (a), and 8.5 (a). First, let us look at the structure in Fig. 8.3 (a). Upon the structure we first impose the rectangular coordinate system (x, y, z). Then the second rectangular coordinate system (x', y', z') is placed according to the relationship $x' = -x$, $y' = y$, and $z' = z$. It can be seen that the structure can be identically described in those two coordinate systems. By "identically described" we mean that for the point a described as (3, 0, 0) in the (x, y, z) coordinate system, there is a corresponding point a' described also as (3, 0, 0) in the (x', y', z') coordinate system. Similarly, points b and b' are both described identically as (5, 3, 0) in their respective coordi-

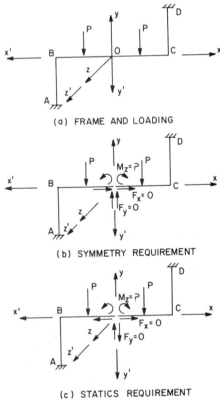

(a) FRAME AND LOADING

(b) SYMMETRY REQUIREMENT

(c) STATICS REQUIREMENT

Fig. 8.2. Axial symmetric structure.

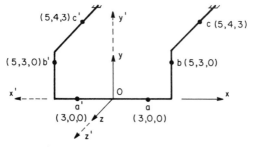

(a) PLANAR SYMMETRIC STRUCTURE

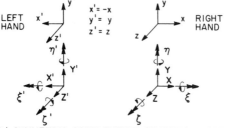

(b) SYMMETRIC COMPLEMENTS (EXTERNAL LOADS, REACTIONS, ROTATIONS, DISPLACEMENTS)

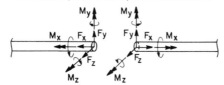

(c) SYMMETRIC COMPLEMENTS (INTERNAL FORCES AND MOMENTS)

Fig. 8.3. Planar symmetry.

nate systems. Those two pairs of points, a–a' and b–b', are called symmetric points. The pair of symmetric points c–c' has identical coordinates (5, 4, 3).

The relationship $x' = -x$, $y' = y$, and $z' = z$ between the two coordinate systems distinguishes the structure as planar symmetric. This term is derived from the observation that the structure is symmetric about the y–z plane. A line joining any pair of symmetric points a–a', b–b', and c–c' will be bisected by the y–z plane, which is the plane of symmetry. The two halves of the structure bisected by the y–z plane are therefore mirror images of each other. Such visual aid is called the *principle of reflection* and is very useful in the rapid identification of planar symmetry.

The structure in Fig. 8.4 (a) illustrates the case of axial symmetry. Two rectangular coordinate systems (x, y, z) and (x', y', z') are imposed on the

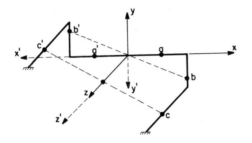

(a) AXIAL SYMMETRIC STRUCTURE

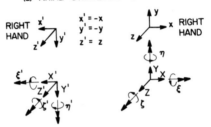

(b) SYMMETRIC COMPLEMENTS (EXTERNAL LOADS, REACTIONS, ROTATIONS, DISPLACEMENTS)

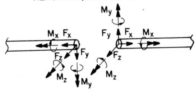

(c) SYMMETRIC COMPLEMENTS (INTERNAL FORCES, AND MOMENTS)

Fig. 8.4. Axial symmetry.

structure. They have the relationship $x' = -x$, $y' = -y$, and $z' = z$. It can be seen that the structure is identically described in these two coordinate systems. Such a structure is termed axial symmetric because a line connecting any pair of symmetric points, $a–a'$, $b–b'$, or $c–c'$, is always bisected by the axis z. This z-axis is called the axis of symmetry. If we rotate one-half of the structure about the axis of symmetry by 180°, the two halves of the structure will coincide exactly. Such visual aid is called the *principle of rotation* and is convenient in identifying axial symmetry.

The structure in Fig. 8.5 (a) illustrates the case of point symmetry. Again the two rectangular coordinate systems (x, y, z) and (x', y', z') are imposed, having the relationship $x' = -x$, $y' = -y$, and $z' = -z$. It can be seen that the structure is identically described in these two coordinate systems.

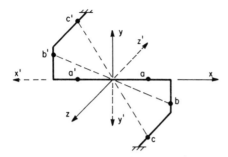

(a) POINT SYMMETRIC STRUCTURE

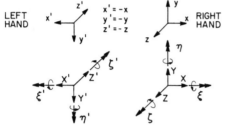

(b) SYMMETRIC COMPLEMENTS (EXTERNAL LOADS, REACTIONS, ROTATIONS, DISPLACEMENTS)

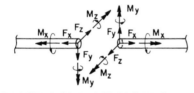

(c) SYMMETRIC COMPLEMENTS (INTERNAL FORCES AND MOMENTS)

Fig. 8.5. Point symmetry.

Such a structure is called point symmetry because the lines connecting the pairs of symmetric points, a–a', b–b', and c–c', always pass through the point of origin. Unfortunately, no simple visual aid is available for identification of point symmetry.

Based on the definition of symmetry given above, we have concluded that there are three types of symmetry in structures, namely, planar symmetry, axial symmetry, and point symmetry. The reader may have encountered the term "antisymmetric structures" in literature. This term has always been vague and imprecise, and has been abandoned in our terminology.

We shall now introduce the important concept of symmetric complements. Symmetric complements are a pair of quantities (forces, displacements, etc.) that are identical with respect to their respective coordinate systems. Symmetric complements are shown in Fig. 8.3 (b) for a structure with planar symmetry. Members of each pair of symmetric complements, X–X', Y–Y', Z–Z', ξ–ξ', η–η', and ζ–ζ', are identical in the (x, y, z) and (x', y', z') coordinate systems. The quantities X, Y, Z, X', Y', and Z' may be forces or displacements. The quantities ξ, η, ζ, ξ', η' and ζ' may be moments or rotations. For the symmetric complements X–X', Y–Y', and Z–Z' each member of a pair of quantities simply follows the positive direction of its respective coordinate axis in the two systems. In the case of ξ–ξ', η–η', and ζ–ζ', however, it is the vectors of the symmetric complements that must follow the positive direction of the coordinate axes. These vectors must be obtained by either the right-hand convention or the left-hand convention consistent with the respective coordinate systems. For this structure with planar symmetry the (x, y, z) coordinate gives the right-hand convention, while the (x', y', z') coordinate gives the left-hand convention.

Symmetric complements in axial symmetry are given in Fig. 8.4 (b). The principle for determining the directions of symmetric complements is the same as discussed above. However, it should be noted that the (x', y', z') coordinate has a right-hand convention, and the directions of ξ', η', and ζ' have been determined accordingly.

Symmetric complements in point symmetry are given in Fig. 8.5 (b). The (x', y', z') coordinate has a left-hand convention, resulting in the directions of ξ', η', and ζ' indicated.

Symmetric complements for internal forces must be treated with care. The pair of internal forces must be acting on symmetrical cut surfaces. These surfaces are given in Figs. 8.3 (c), 8.4 (c), and 8.5 (c) together with the symmetric complements.

Now that the symmetric complements have been defined, we can also define an antisymmetric complement. Antisymmetric complements are obtained by reversing the direction of one of the two quantities in a pair of symmetric complements.

8.1.3 Theorem of Symmetry

Using the concept of symmetric complements we can easily deduce the symmetry or the antisymmetry of loadings. A symmetric loading is one in which one-half of the load is the symmetric complement of the other half. An antisymmetric loading is one in which one-half of the load is the antisymmetric complement of the other half. Similar deduction applies to the symmetry or antisymmetry of reactions, displacements, and internal forces.

The terminology for symmetry having been clarified, we can state the following two fundamental theorems of symmetry:

1. *Symmetric* external loadings applied to a *symmetric* structure produce *symmetric* reactions, displacements, and internal forces.
2. *Antisymmetric* external loadings applied to a symmetric structure produce *antisymmetric* reactions, displacements, and internal forces.

The power of these two theorems can be observed by the fact that any unsymmetric loading may be replaced by the superposition of symmetric and antisymmetric loadings. Take the symmetric horizontal frame in Fig. 8.6, for example. The frame is subjected to two arbitrary loads, W and P, in Fig. 8.6 (a). This unsymmetric loading can be replaced by the superposition of the symmetric loadings in Fig. 8.6 (b) and the antisymmetric loadings in Fig. 8.6 (c). The *symmetric* loadings are obtained by replacing each given force, W and P, by one-half its original value plus the *symmetric* complement of this one-half force. The *antisymmetric* loading is obtained by replacing each given force, W and P, by one-half its original value plus the *antisymmetric* complement of this one-half force.

8.1.4 Internal Forces and External Displacements at Origin

For a symmetric structure subjected to symmetric or antisymmetric loadings, certain internal forces and certain displacements at the origin must be zero. This knowledge is often very useful in reducing the number of unknowns in the solution of an indeterminate structure.

If a cut is made at the origin to expose a pair of internal forces, this pair of internal forces must satisfy two requirements. (1) *Symmetry* requires that the pair of internal forces must be consistent with the pair of symmetric complements. (2) *Statics* requires that the pair of internal forces must be equal and opposite. If these two requirements contradict each other, they can be satisfied only if the internal forces assume the particular value of zero.

Any displacement at the origin should also satisfy two requirements simulta-

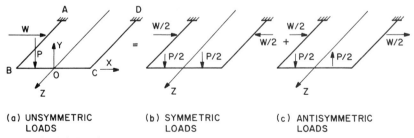

(a) UNSYMMETRIC LOADS   (b) SYMMETRIC LOADS   (c) ANTISYMMETRIC LOADS

Fig. 8.6. Resolve unsymmetric loads into symmetric and antisymmetric loads.

neously. (1) *Symmetry* requires that the pair of symmetric points adjacent to the origin must be displaced according to the direction of the symmetric complement. (2) *Geometry* requires that this pair of symmetric points must be equal and in the same direction. If these two requirements contradict each other, they can be satisfied simultaneously only if the displacement at the orgin is zero.

If the structure with planar symmetry in Fig. 8.3 (a) is subjected to a symmetric loading, then the first theorem of symmetry requires that the three pairs of internal forces (F_x, F_y, and F_z) on the cut faces at the origin (see Fig. 8.3a) should be in the same direction as the three pairs of symmetric complements X–X', Y–Y', and Z–Z' (see Fig. 8.3b). At the same time statics requires that each pair of internal forces must have opposite direction. Since the two pairs of internal forces F_y and F_z cannot satisfy both the requirement of symmetry and the requirement of statics, their values must be zero (i.e., $F_y = F_z = 0$). On the other hand, since the pair of internal forces, F_x, can satisfy both requirements simultaneously, the value of the pair will be unknown (i.e., $F_x = ?$). Similar reasoning can be applied to the three pairs of internal moments (M_x, M_y, and M_z) at the origin. The first theorem of symmetry requires that these three pairs of internal moments in Fig. 8.3 (c) should have the direction shown by the three symmetric complements ξ–ξ', η–η', and ζ–ζ' in Fig. 8.3 (b). Again, statics requires that each pair of internal moment must have opposite direction. Since the pair of M_x cannot satisfy both requirements, $M_x = 0$; and since the two pairs of M_y and M_z can satisfy both requirements, $M_y = ?$ and $M_z = ?$.

Certain displacements and rotations at the origin may also become zero under symmetric loading. The displacements Δ_x and the rotations, ϕ_y and ϕ_z, at both sides of the cut surfaces at the origin must displace or rotate in opposite direction according to the symmetric complements X–X', η–η', and ζ–ζ', respectively. However, continuity or geometry requires that Δ_x, ϕ_y, and ϕ_z on both sides of the cut surface must move in the same direction. This contradiction can be resolved only if $\Delta_x = \phi_y = \phi_z = 0$. Similar reasoning will result in $\Delta_y = ?$, $\Delta_z = ?$, and $\phi_x = ?$.

In summary, the following conclusions can be made at the origin of a planar symmetric structure subjected to symmetric loading:

$$F_y = F_z = M_x = \Delta_x = \phi_y = \phi_z = 0 \qquad (8\text{–}1)$$

$$F_x, M_y, M_z, \Delta_y, \Delta_z, \text{ and } \phi_x \text{ all} = ? \qquad (8\text{–}2)$$

Comparison of Eqs. (8–1) and (8–2) shows that for a particular symmetry and loading, if an internal force is zero, the corresponding displacement is unknown, or vice versa. This observation is logical because the statics requirement for the internal forces is just opposite to the continuity requirement for the displacements. Equations (8-1) and (8-2) are listed in the second and ninth columns of Table 8.1.

If the structure with planar symmetry in Fig. 8.3 (a) is subjected to an *antisymmetric* loading, then the second theorem of symmetry requires that the six pairs of internal forces (F_x, F_y, F_z, M_x, M_y, and M_z) in Fig. 8.3 (c) on the cut faces at the origin should each have one of its two quantities reversed. At the same time statics requires that each pair of internal forces must have opposite direction. Under this circumstance, F_x, M_y, and M_z cannot satisfy both the symmetry and statics requirements. Therefore, $F_x = M_y = M_z = 0$. In contrast, since F_y, F_z, and M_x can satisfy both requirements, F_y, F_z, and M_x all = ?. These conclusions are listed in the third column of Table 8.1.

Comparison of the second and third columns in Table 8.1 leads to an interesting observation: Quantities that are zero (0) under symmetric loading are unknown (?) under antisymmetric loading, and vice versa. This is logical

Table 8.1 Internal forces and external displacements at origin for three types of symmetric structures subjected to symmetric and antisymmetric loadings (Figs. 8.3, 8.4 and 8.5)

INT. FORCES	PLANAR SYMM. ABOUT y–z AXIS		AXIAL SYMM. ABOUT z-AXIS		POINT SYMM. ABOUT ORIGIN		EXT. DISPL.	PLANAR SYMM. ABOUT y–z PLANE		AXIAL SYMM. ABOUT z-AXIS		POINT SYMM. ABOUT ORIGIN	
	SL	AL	SL	AL	SL	AL		SL	AL	SL	AL	SL	AL
F_x	?	0	?	0	?	0	Δ_x	0	?	0	?	0	?
F_y	0	?	?	0	?	0	Δ_y	?	0	0	?	0	?
F_z	0	?	0	?	?	0	Δ_z	?	0	?	0	0	?
M_x	0	?	?	0	0	?	ϕ_x	?	0	0	?	?	0
M_y	?	0	?	0	0	?	ϕ_y	0	?	0	?	?	0
M_z	?	0	0	?	0	?	ϕ_z	0	?	?	0	?	0
1	2		3	4	5	6	7	8	9	10	11	12	13	14

Note: SL = Symmetric loading; AL = Antisymmetric loading; ? = unknown quantities; 0 = zero.

Two Observations:
1. Quantities that are zero (0) under symmetric loading are unknowns (?) under antisymmetric loading, and vice versa.
2. For a particular symmetry and loading, if an internal force is zero, the corresponding displacement is unknown, or vice versa.

because a pair of symmetric complements produced by symmetric loading have one of its quantities just opposite to the corresponding quantity in a pair of antisymmetric complements produced by antisymmetric loading.

We shall now look at the axial symmetric structure in Fig. 8.4 (a). If this structure is subjected to symmetric loading, the theorem of symmetry requires that the six pairs of internal forces shown in Fig. 8.4 (c) at the origin be identical to the six symmetric complements shown in Fig. 8.4 (b). Recalling the requirement of statics that each pair of internal forces must be in opposite directions, it can be seen that the two pairs of internal forces, F_z and M_z, cannot satisfy both requirements. Hence $F_z = M_z = 0$. In contrast, since F_x, F_y, M_x, and M_y can satisfy both requirements, we conclude that F_x, F_y, M_x, and M_y all $= ?$. This information is listed in the fourth column of Table 8.1.

Once the six pairs of internal forces at the origin of an axial symmetric structure are known under symmetric loading, it is quite simple to write down in the fifth column of Table 8.1 the six pairs of internal forces produced by antisymmetric loading. This is done by interchanging the 0 and ? values in the fourth and fifth columns.

For the displacements at the origin of an axial symmetric structure, it can be seen in Table 8.1 that column 11 is a reversal of column 4, and column 12 is a reversal of column 5. This relationship has already been explained.

For the point symmetric structure given in Fig. 8.5 (a) the information about internal forces and displacements at the origin is given in the sixth, seventh, thirteenth and fourteenth columns of Table 8.1.

8.1.5   Examples 8.1, 8.2, 8.3, and 8.4

Theory of symmetry has been expounded in Sections 8.1.2, 8.1.3, and 8.1.4. We can now summerize the procedures for determining the "zero" or "unknown" values of the internal forces at the origin of a symmetric structure:

Step 1. Impose two rectangular coordinate systems (x, y, z) and (x', y', z') on the structure such that the structure can be identically described in the two coordinate systems. Based on the relationship between these two coordinate systems, determine the type of symmetry for the structure. It may be planer, axial, or point symmetry. If the visual aids are used, planar symmetry can be determined by the principle of reflection and axial symmetry by the principle of rotation.

Step 2. Determine the type of loading (symmetric or antisymmetric) according to the symmetric complements. The symmetric complements are determined according to Figs. 8.3 (b), 8.4 (b), and 8.5 (b). For forces and displacements, the pair of symmetric complements should be in the direction of the pair of corresponding axes x–x', y–y', or z–z'. For moments or rotations, the pair of quantities should also be determined by the intrinsic left-hand or right-hand nature of the two coordinate systems. If the visual aids are used, the symmetric complements can be determined by the principle of reflection in planar symmetric structures or by the principle of rotation in axial symmetric structures.

Step 3. Following the two theorems of symmetry, determine the direction of the pairs of internal forces at the origin that belong to the same type of symmetry as the loading. If the pair of internal forces are in the same direction, the pair cannot satisfy the requirement of statics, and, therefore, the value of the pair must be zero. If the pair of internal forces are in opposite directions, the pair can satisfy the requirement of statics and should be an unknown. The displacements at the origin can be obtained by reversing the answer for the corresponding internal forces. If the internal force is zero, the corresponding displacement must be unknown, and vice versa.

EXAMPLE 8.1

Using these procedures, we shall now determine the zero values of the internal forces at the origin of the structure in Fig. 8.2.

Step 1. The structure in Fig. 8.2 (a) can be identically described in two coordinate systems (x, y, z) and (x', y', z') as shown in Fig. 8.2 (a). The coordinate systems have the relationship $x' = -x$, $y' = -y$, and $z' = z$. Therefore, the structure is axial symmetrical. This conclusion can also be made by applying the principle of rotation. If one-half of the structure, *OBA*, is rotated 180° about the z-axis, it will coincide with the other half of the structure, *OCD*.

Step 2. The pair of external loads P is antisymmetric because the two loads are in the same direction, while the pair of symmetric complements should be in opposite direction like the pair of y–y' axes. The antisymmetry of the loading can also be determined by the principle of rotation. If one-half of the structure, *OBA*, is rotated 180° about the z-axis, the load P on *OBA* would be opposite to the load P on the other half of the structure, *OCD*.

Step 3. Under antisymmetric loading the three pairs of internal forces at the origin, F_x, F_y, and M_z, are shown in Fig. 8.2 (b). They are antisymmetric because when compared to the symmetric complements in Fig. 8.4 (b), the direction of one force in each pair has been reversed. In Fig. 8.2 (b) the pair of internal forces are in the same direction for F_x and F_y. This cannot satisfy the requirement of statics as shown in Fig. 8.2 (c). Therefore $F_x =$

$F_y = 0$. The pair of internal moments M_z, in Fig. 8.2 (b), however, are in opposite direction. This can satisfy the requirement of statics, and M_z should be an unknown.

The displacements at the origin are obtained by reversing the answer for the corresponding internal forces: Δ_x and $\Delta_y = ?$; and $\phi_z = 0$.

The structure in Fig. 8.2 can also be considered as point symmetric if the z'-axis is reversed. However, a moment of reflection will show that the loading remains antisymmetric, and the antisymmetric internal forces are identically shown in Fig. 8.2 (b). Therefore the conclusion would be exactly the same.

EXAMPLE 8.2

The zero values of internal force at the origin of the frame in Fig. 8.1 (a) will be determined by visual aids as follows:

Step 1. The structure is planar symmetric because one-half of the structure is the mirror image of the other half with respect to the symmetric y–z plane.

Step 2. The two loads P are also symmetric according to the principle of reflection. They are mirror images with respect to the y–z plane.

Step 3. The internal forces at the origin, which satisfy the requirement of symmetry, are given in Fig. 8.1 (b). These internal forces are symmetric, since members of each pair of forces are mirror images with respect to the symmetric y–z plane. The internal force $F_y = 0$ because it does not satisfy the requirement of statics shown in Fig. 8.1 (c). In contrast, F_x and $M_z = ?$, as they are consistent with the requirement of statics. By reversing the solution for internal forces the corresponding displacements are: $\Delta_x = \phi_z = 0$ and $\Delta_y = ?$.

It is interesting to note that the frame *ABCD* in Fig. 8.1 can also be considered as axial symmetric about the y-axis. Using the principle of rotation, the prediction for the internal forces at the origin will remain the same.

EXAMPLE 8.3

An interesting space structure is shown in Fig. 8.7. This structure can be considered point symmetric or axial symmetric depending on the coordinate systems imposed upon it. Point symmetry can be obtained by placing the two coordinate systems (x, y, z) and (x', y', z') as shown in Fig. 8.7 (a). These two coordinate systems have the relationship $x' = -x$, $y' = -y$, and $z' = -z$. The pair of points H and H' are symmetric points because a line drawn between H and H' would pass through the origin O. The

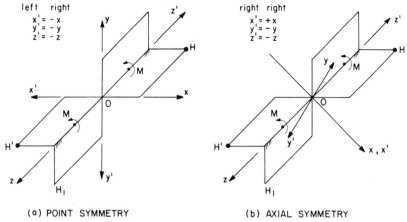

(a) POINT SYMMETRY   (b) AXIAL SYMMETRY

Fig. 8.7. Example for point and axial symmetry.

two moments M are also symmetric. Therefore, the internal forces and displacements at origin should satisfy the requirement of symmetry, given in Table 8.1, resulting in:

$$
\begin{aligned}
F_x &= ? & \Delta_x &= 0 \\
F_y &= ? & \Delta_y &= 0 \\
F_z &= ? & \Delta_z &= 0 \\
M_x &= 0 & \phi_x &= ? \\
M_y &= 0 & \phi_y &= ? \\
M_z &= 0 & \phi_z &= ?
\end{aligned}
$$

The structure can also be axial symmetric if the two coordinate system (x, y, z) and (x', y', z') are placed as shown in Fig. 8.7 (b). The relationship between the two coordinate systems are $x' = x$, $y' = -y$, $z' = -z$. The axis of rotation is the x-axis, which is $45°$ from the x-axis in Fig. 8.7 (a). The pair of points H and H' are no longer symmetric. The symmetric point for H should be H_1. The two moments M have become antisymmetric, and the internal forces and displacements at origin should be:

$$
\begin{aligned}
F_x &= ? & \Delta_x &= 0 \\
F_y &= 0 & \Delta_y &= ? \\
F_z &= 0 & \Delta_z &= ? \\
M_x &= ? & \phi_x &= 0 \\
M_y &= 0 & \phi_y &= ? \\
M_z &= 0 & \phi_z &= ?
\end{aligned}
$$

It can be seen that analysis for the same structure and the same loading can be very different when the coordinate systems are imposed in a different manner.

EXAMPLE 8.4

The spiral staircase shown in Fig. 8.8 will be analyzed using the theory of symmetry to determine the zero values of the internal forces at the origin.

Step 1. The coordinate system (x, y, z) is placed as shown in Fig. 8.8. Using the principle of rotation, one-half of the structure will coincide with the other half when rotated 180° about the z-axis. Therefore, the structure is axial symmetric about the z-axis.

Step 2. The vertical uniformly distributed loading is antisymmetric. This is so because when one-half of the structure is rotated 180° about the z-

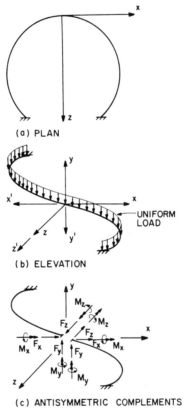

(a) PLAN

(b) ELEVATION

(c) ANTISYMMETRIC COMPLEMENTS

Fig. 8.8. Spiral staircase.

axis, the downward vertical load on the rotating half of the structure will become an upward vertical load.

Step 3. The antisymmetric internal forces at the origin determined by the principle of rotation are given in Fig. 8.8 (c). In each pair of internal forces, the forces will be zero if they are in the same direction and will be unknown if they are in opposite directions. This will give:

$$F_x = 0 \qquad \Delta_x = ?$$
$$F_y = 0 \qquad \Delta_y = ?$$
$$F_z = ? \qquad \Delta_z = 0$$
$$M_x = 0 \qquad \phi_x = ?$$
$$M_y = 0 \qquad \phi_y = ?$$
$$M_z = ? \qquad \phi_z = 0$$

Note that the answer for the displacements is simply the reversal of the answer in the corresponding internal forces.

8.2 PLANE FRAMES

8.2.1 Characteristics of Plane Frames

An interesting characteristic of plane frames that can considerably simplify their analysis is introduced here. Fig. 8.9 (a) shows a fixed-ended horizontal plane frame *ABCD* subjected to the most general types of arbitrary external loadings. These loadings can always be resolved into six types of basic external loads, $\bar{F}_x$, $\bar{F}_y$, $\bar{F}_z$, $\bar{M}_x$, $\bar{M}_y$, and $\bar{M}_z$. If the fixed-end condition is removed at point *D*, there will, in general, be six unknown reaction components, F_x, F_y, F_z, M_x, M_y, and M_z, as shown in Fig. 8.9 (b). This general case of external loadings and unknown reactions can be separated into two types of forces: (1) forces in the plane of the frame consisting of three external loadings $\bar{F}_x$, $\bar{F}_z$, and $\bar{M}_y$, plus three unknown reactions F_x, F_z, and M_y, as shown in Fig. 8.9 (c); and (2) forces perpendicular to the plane of the frame, consisting of three external loadings $\bar{F}_y$, $\bar{M}_x$, and $\bar{M}_z$, plus three unknown reactions F_y, M_x, and M_z, as shown in Fig. 8.9 (d). A separate analysis of these two types of forces, each having three unknowns, is considerably simpler than the analysis of one general force system having six unknowns.

The characteristics of a plane frame can be stated as follows: If a plane frame is subjected to an in-plane external loading, the unknown reactions must be in the plane, and the three perpendicular reactions must be zero. If a plane frame is subjected to perpendicular external loadings, the unknown reactions must be perpendicular, and the three in-plane reactions must be zero.

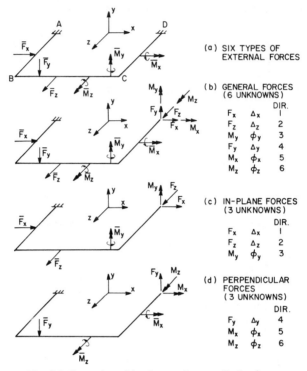

Fig. 8.9. Separation of in-plane and perpendicular forces.

To prove these characteristics of plane frames we designate the directions of the three in-plane reaction forces F_x, F_z, and M_y (and the three displacements Δ_x, Δ_z, and ϕ_y) as 1, 2, and 3, respectively, as shown in Fig. 8.9 (b) and (c). Similarly, the directions of the three perpendicular reaction forces F_y, M_x, and M_z (and the three displacements Δ_y, ϕ_x, and ϕ_z) are taken as 4, 5, and 6, respectively, as shown in Fig. 8.9 (b) and (d). Using the flexibility method discussed below in Section 8.3, we can write six equations for the six zero displacement conditions at support D by means of matrix notation:

$$
\begin{bmatrix}
a_{11} & a_{12} & a_{13} & a_{14} & a_{15} & a_{16} \\
a_{21} & a_{22} & a_{23} & a_{24} & a_{25} & a_{26} \\
a_{31} & a_{32} & a_{33} & a_{34} & a_{35} & a_{36} \\
a_{41} & a_{42} & a_{43} & a_{44} & a_{45} & a_{46} \\
a_{51} & a_{52} & a_{53} & a_{54} & a_{55} & a_{56} \\
a_{61} & a_{62} & a_{63} & a_{64} & a_{65} & a_{66}
\end{bmatrix}
\begin{bmatrix}
F_x \\
F_z \\
M_y \\
F_y \\
M_x \\
M_z
\end{bmatrix}
=
\begin{bmatrix}
-\Delta_{1c} - \Delta_{1p} \\
-\Delta_{2c} - \Delta_{2p} \\
-\Delta_{3c} - \Delta_{3p} \\
-\Delta_{4c} - \Delta_{4p} \\
-\Delta_{5c} - \Delta_{5p} \\
-\Delta_{6c} - \Delta_{6p}
\end{bmatrix}
\quad (8\text{-}3)
$$

| Flexibility matrix | Reaction vector | Displacement vector |

The matrix a is the well-known flexibility matrix. A typical element a_{ij}, where i and j vary from 1 to 6, is the displacement of support D in the i direction caused by a unit force in the j direction. In the displacement vector, a typical displacement, Δ_{ic}, where $i = 1, 2 \ldots 6$, represents the displacement in the i direction caused by the coplanar external loadings in the plane of the frame; and Δ_{ip} is the displacement in the i direction caused by external loadings perpendicular to the frame.

The matrix Eq. (8–3) can be separated into two independent matrix equations:

$$
\begin{bmatrix} a_{11} & a_{12} & a_{13} \\ a_{21} & a_{22} & a_{23} \\ a_{31} & a_{32} & a_{33} \end{bmatrix}
\begin{bmatrix} F_x \\ F_z \\ M_y \end{bmatrix} =
\begin{bmatrix} -\Delta_{1c} \\ -\Delta_{2c} \\ -\Delta_{3c} \end{bmatrix}
\tag{8–4}
$$

and:

$$
\begin{bmatrix} a_{44} & a_{45} & a_{46} \\ a_{54} & a_{55} & a_{56} \\ a_{64} & a_{65} & a_{66} \end{bmatrix}
\begin{bmatrix} F_y \\ M_x \\ M_z \end{bmatrix} =
\begin{bmatrix} -\Delta_{4p} \\ -\Delta_{5p} \\ -\Delta_{6p} \end{bmatrix}
\tag{8–5}
$$

Equation (8–4) consists of only in-plane quantities, whereas Eq. (8–5) is for perpendicular quantities. Four conditions must be satisfied to make this separation. They are:

$$
\begin{bmatrix} a_{41} & a_{42} & a_{43} \\ a_{51} & a_{52} & a_{53} \\ a_{61} & a_{62} & a_{63} \end{bmatrix} = 0
\tag{8–6}
$$

$$
\begin{bmatrix} -\Delta_{4c} \\ -\Delta_{5c} \\ -\Delta_{6c} \end{bmatrix} = 0
\tag{8–7}
$$

$$
\begin{bmatrix} a_{14} & a_{15} & a_{16} \\ a_{24} & a_{25} & a_{26} \\ a_{34} & a_{35} & a_{36} \end{bmatrix} = 0
\tag{8–8}
$$

$$
\begin{bmatrix} -\Delta_{1p} \\ -\Delta_{2p} \\ -\Delta_{3p} \end{bmatrix} = 0
\tag{8–9}
$$

Equations (8–6) and (8–7) are indeed satisfied for a plane frame because out-of-plane displacements in the 4, 5, or 6 directions cannot be produced by in-plane forces in the 1, 2, or 3 directions. This is so because both the plane frame and the in-plane forces are symmetric about the frame's own plane. According to the theory of symmetry a symmetric in-plane force acting on a symmetric structure cannot produce nonsymmetric out-of-plane displacement. Equations (8–8) and (8–9), on the other hand, are satisfied from Maxwell's reciprocal theorem.

8.2.2 Simplification of Loadings on Plane Frames

Combining the characteristics of the plane frame discussed in Section 8.2.1 and the theory of symmetry in Section 8.1, it is often possible to resolve arbitrary loadings, which cause six unknowns in a frame, into a set of four simple loadings, each involving only one or two unknowns. The following example illustrates such a simplification.

Figure 8.10 (a) shows a fixed-ended rectangular horizontal frame $ABCD$ that is symmetric about the y–z plane. It is subjected to unsymmetric loadings, consisting of a load W and a load P. If we cut the frame at the origin O, there will be six unknown forces, F_x, F_y, F_z, M_x, M_y, and M_z.

According to the characteristics of a plane frame, these arbitrary loadings can be resolved into a set of perpendicular loadings and a set of in-plane loadings as shown in Fig. 8.10 (b) and (c), respectively. The former will

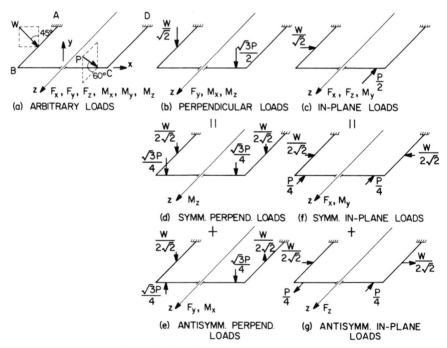

Fig. 8.10. Simplification of arbitrary loadings.

create only three unknowns, F_y, M_x, M_z, at the origin, while the latter gives the other three unknowns, F_x, F_z, and M_y.

The perpendicular loadings in Fig. 8.10 (b) can be further resolved into two sets of loadings (illustrated in Fig. 8.6). One is symmetrical as shown in Fig. 8.10 (d), and the other is antisymmetrical as shown in Fig. 8.10 (e). The symmetrical perpendicular loadings will produce only one unknown, M_z, at the origin, while the antisymmetrical perpendicular loadings will induce two unknowns, F_y and M_x, at the origin.

Similarly, the in-plane loadings in Fig. 8.10 (c) can be resolved into a set of symmetrical loadings shown in Fig. 8.10 (f) and another set of antisymmetrical loadings shown in Fig. 8.10 (g). The former produces two unknowns, F_x and M_y at the origin, while the latter gives only one unknown, F_z, at the origin.

In conclusion, arbitrary loadings on a symmetric plane frame, which cause six unknowns at the origin as shown in Fig. 8.10 (a), can be resolved into four sets of loadings:

1. A set of symmetrical perpendicular loadings giving one unknown, M_z, at the origin, Fig. 8.10 (d).
2. A set of antisymmetrical perpendicular loadings giving two unknowns, F_y and M_x, at the origin, Fig. 8.10 (e).
3. A set of symmetrical in-plane loadings giving two unknowns, F_x and M_y, at the origin, Fig. 8.10 (f).
4. A set of antisymmetrical in-plane loadings giving one unknown, F_z, at the origin, Fig. 8.10 (g).

The analysis of these four sets of loadings, each producing one or two unknowns, could be clearer in concept than the analysis of one set of arbitrary loadings in connection with six unknowns.

8.3 FLEXIBILITY METHOD

8.3.1 General Description

The flexibility method is also known as the consistent deformation method. In the well-known analysis of plane frames subjected to in-plane loadings, the flexibility method involves only the bending flexibilities of structural members. In the frames subjected to out-of-plane loadings, however, this method should be generalized to include also the torsional flexibilities of structural members. In this section it is assumed that the reader is familiar with the flexibility method used in plane frames under in-plane loading. We shall demonstrate here how this method can be extended to frames subjected to out-of-plane loadings. The general flexibility method for analyzing statically indeterminate structures includes the following steps:

1. Recognize the n degrees of static indeterminacy in a structure. This should include displacement restraints and flexural rotational restraints as well as torsional rotational restraints.

2. Remove n-number of restraints (either displacements or rotations) so that the structure becomes statically determinate. This will expose n-number of unknown forces or moments (including both torsional and bending moments).

3. Draw the bending and torsional moment diagrams on the statically determinate structure caused by each of the unknowns as well as the external loadings. These moments will produce displacements and rotations at the location where the restraints are removed. The method used to calculate displacements and rotations may be the moment-area concept, the conjugate beam analogy, the virtual work method, Castigliano's theorem, or any other convenient means. The displacement caused by a unit force (or a rotation caused by a unit moment) is known as a flexibility element of the structure. The evaluation of all the flexibility elements of the structure is the crux of this method, whereby the name "flexibility method" is derived.

4. Write n-number of equations to restore the n-number of displacements and rotations where the restraints have been removed. From these simultaneous equations, the n-number of unknown forces and moments can be determined. Since the equations are established on the basis of restoring the displacements, this method is also known as the consistent deformation method. Any convenient method can be used to solve simultaneous equations. The quadratic formula is easy for two equations. For more than two equations iteration procedures can be used in hand calculation. However, some programmable hand calculators, such as the T1 59 and HP 97, have built-in programs to solve up to ten simultaneous equations. The large electronic computers, of course, have capacities to solve hundreds of simultaneous equations.

5. Once the unknowns are solved, the actual bending and torsional moments caused by each unknown can be determined. The final bending and torsional moments are simply the sum of the moments caused by all the unknowns plus those induced by the external loads.

8.3.2 Important Details

Three important points should be emphasized in using the flexibility method:

1. Symmetry and Plane Frame. One should try one's best to utilize the theory of symmetry (Section 8.1) and the characteristics of plane frames (Section 8.2) to reduce the number of unknowns. For a symmetric structure, it is often wise to cut the structure at the origin and to expose the unknowns there. Some of the unknowns may be zero due to symmetry. For a plane structure, it is often wise to separate the external loading into a set of in-plane loadings plus a set of perpendicular loadings. Each set of loadings will produce a maximum of three unknowns at each section instead of six.

2. *Flexibility*. In evaluating the flexibility of a structure, both the flexural and torsional flexibilities of the members must be included. For a prismatic member of length l, the relative flexural rotation of the ends is computed from:

$$\phi = \int_0^l M \frac{dx}{EI} \tag{8-10}$$

where M is the bending moment along an infinitesimal length dx, and EI is the flexural rigidity of the section. The ratio (dx/EI) is the bending flexibility of the length dx. The relative torsional rotation of the ends, on the other hand, is computed by St. Venant's theory:

$$\phi = \int_0^l T \frac{dx}{GC} \tag{8-11}$$

where T is the torque along an infinitesimal length dx, and GC is the torsional rigidity of the section. The ratio (dx/GC) is the "torsional flexibility" of the length dx. If a uniform T exists along the whole length, then:

$$\phi = T \frac{l}{GC} \tag{8-12}$$

and (l/GC) is the "torsional flexibility" of the member.

3. *Sign Convention*. To be consistent in a space frame, the sign of forces and moments will have to be related to an imposed coordinate system. An external force or load is positive if it points toward the positive direction of a coordinate axis. An external moment is positive if its moment vector based on a right-hand screw points toward the positive direction of a coordinate axis. The sign of an internal moment or force, however, should be determined not only from its *direction* with respect to the coordinate axis, but also from the *face* on which it acts. This face is also related to the coordinate axis, and should first be defined. A positive face is one that has an outward normal in the positive direction of a coordinate axis. Conversely, a negative face will have an outward normal opposite to the positive direction of a coordinate axis.

The sign convention for internal shears and moments can now be stated as follows: An internal shear is defined as positive when the shear acting on a positive face points toward the positive direction of a coordinate axis, or if the shear acting on a negative face points toward the negative direction. An internal shear is negative if the shear acting on the positive face points toward the negative direction of a coordinate axis, or if the shear acting on a negative face points toward the positive direction. Similarly, a positive moment is one that acts on a positive face with the moment vector based on a right-hand screw pointing toward the positive direction of an axis. For a bending moment, the axis for the moment vector is different from the axis for the outward normal of the surface. For a torsional moment, however, these two axes are the same.

A bending moment diagram will be plotted on the tension side of a flexural member. This is the side where tensile reinforcement is required. A positive torque or shear diagram will be plotted on either the upper or the front side of the member, and a negative torque or shear diagram on the lower or back side.

The flexibility method will be illustrated by the following four examples.

8.3.3 Examples 8.5, 8.6, 8.7, and 8.8

EXAMPLE 8.5

The rectangular horizontal plane frame shown in Figs. 8.9 and 8.10 will now be used to demonstrate the flexibility method of analysis. The frame $ABCD$ is symmetrical and is subjected to a 30 kips vertical load at the origin O, as shown in Fig. 8.11 (a). The $(x, y, z,)$ coordinate system and the dimensions of the frame are given. A cross section of 10 in. by 20 in. is assumed throughout the frame. It is also assumed that the concrete cross section is uncracked, and that the Poisson's ratio $\nu = 0.2$. The aim of this analysis is to find the bending and torsional moment diagrams.

The frame $ABCD$ in Fig. 8.11 (a) is, in general, statically indeterminate to the sixth degree. If a cut is made at any section, there may be six unknowns F_x, F_y, F_z, M_x, M_y, and M_z. However, three of the six unknowns must be zero, $F_x = F_z = M_y = 0$, because this plane frame is subjected only to perpendicular loading (see Section 8.2 and Fig. 8.10b). Furthermore, if the cut is made at the origin O, two more unknowns are equal to zero, $F_y = M_x = 0$, because the external load is also symmetrical about the y–z plane (see Section 8.1 and Fig. 8.10d). As a result, only one unknown, M_z, remains to be solved.

In view of the symmetry about the y–z plane, it is only necessary to analyze half of the structure, ABO, which is shown in Fig. 8.11 (c) through (h).

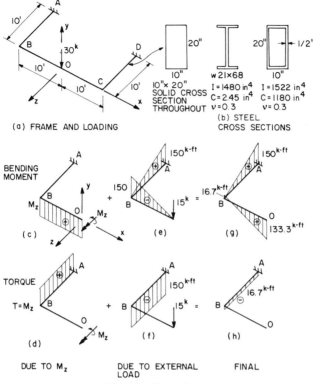

(a) FRAME AND LOADING

(b) STEEL CROSS SECTIONS

Fig. 8.11. Example 8.5.

The unknown M_z produces bending moments and torques, as shown in Fig. 8.11 (c) and (d), respectively. The external load on *ABO* should be 15 kips, which is one-half of the 30-kips external load, owing to symmetry. This 15-kips load also induces bending moments and torques, as given in Fig. 8.11 (e) and (f), respectively.

To find the unknown M_z we can write one equation to restore the rotation about the z-axis to zero at the origin. $\phi_z = 0$ is a condition demanded by symmetry. The rotations at the origin due to the unknown M_z and the 15-kip external load will be calculated by the moment-area concept, since this method is most convenient for cantilevers:

$$\phi_z = \underbrace{\frac{M_z(10)}{EI}}_{\text{Fig. 8.11 (c)}} + \underbrace{\frac{M_z(10)}{GC}}_{\text{(d)}} - \underbrace{\frac{0.5(150)(10)}{EI}}_{\text{(e)}} - \underbrace{\frac{150(10)}{GC}}_{\text{(f)}} = 0 \qquad (a)$$

$$M_z = \frac{75 + \dfrac{EI}{GC} 150}{1 + \dfrac{EI}{GC}} \tag{b}$$

It can be seen from the above equation that the unknown M_z is a function of the ratio of the flexural rigidity EI to the torsional rigidity GC. For the 10 in. by 20 in. concrete cross section this rigidity ratio is:

$$I = \frac{10(20)^3}{12} = 6{,}667 \text{ in.}^4 \tag{c}$$

$$C = \beta x^3 y = 0.229(10)^3(20) = 4{,}580 \text{ in.}^4 \tag{d}$$

$$\frac{E}{G} = 2(1 + \nu) = 2(1 + 0.2) = 2.4 \tag{e}$$

$$\frac{EI}{GC} = \frac{2.4(6{,}667)}{4{,}580} = 3.49 \tag{f}$$

Substituting this rigidity ratio into Eq. (b):

$$M_z = \frac{75 + (3.49)(150)}{1 + 3.49} = 133.3 \text{ k-ft} \tag{g}$$

Substituting this value of M_z into Fig. 8.11 (c) and adding this bending moment diagram to that in Fig. 8.11 (e), we arrive at the final bending moment diagram in Fig. 8.11 (g). Similarly, Eq. (g) is substituted into Fig. 8.11 (d). The addition of torque diagrams in Figs. 8.11 (d) and (f) gives the final torque diagram in Fig. 8.11 (h).

The final moment diagrams in Fig. 8.11 (g) and (h) show that the torque at support A is only $\frac{1}{9}$ of the bending moment at the same section. Such an order-of-magnitude difference is quite common in reinforced concrete frames. However, this torque is very significant because the cost to provide reinforcement for this torsional steel will be of the same magnitude as the cost for the flexural steel. The relative efficiency of torsional steel vs. flexural steel given in Section 9.3.1 shows that for the same weight of reinforcement the torsional steel is four to eight times less efficient than the flexural steel. In other words, for the same magnitude of moment, the torsional steel will cost an order of magnitude more than the flexural steel.

In contrast to a reinforced concrete frame, it is interesting to study two cases in which the frame in Fig. 8.11 (a) is made up of structural steel. As shown in Fig. 8.11 (b) the first steel cross section is a W21 × 68 wide-flange I-section, and the second is a 10 in. by 20 in. steel box ½ in. thick. For W21 × 68:

$$\frac{EI}{GC} = \frac{2.6(1,480)}{2.45} = 1,571 \tag{h}$$

From Eq. (b):

$$M_z = \frac{75 + 1,571(150)}{1 + 1,571} = 149.95 \text{ k-ft} \tag{i}$$

$$T_{AB} = 150 - 149.95 = 0.05 \text{ k-ft} \tag{j}$$

Such a small torque on member AB can indeed be neglected. This explains why torsion is generally not considered in the analysis of a steel frame made up of I-sections.

For the steel box:

$$\frac{EI}{GC} = \frac{2.6(1,522)}{1,180} = 3.35 \tag{k}$$

$$M_z = \frac{75 + 3.35(150)}{1 + 3.35} = 132.8 \text{ k-ft} \tag{l}$$

$$T_{AB} = 150 - 132.8 \text{ k-ft} = 17.2 \text{ k-ft} \tag{m}$$

This torque of 17.2 k-ft is also small when compared to the bending moment of 150 k-ft at support A. However, for accurate design of the steel box, this torque should be included.

EXAMPLE 8.6

A horizontal plane frame with fixed ends is made up of two 10 in. by 20 in. concrete members AB and BC connected at a 90° angle, shown in Fig. 8.12 (a). A vertical 20-kip external load is acting at joint B. Draw the bending and torsional moment diagrams.

Since this plane frame is subjected to a perpendicular load, the in-plane internal forces F_x, F_z, and M_y must be zero at any section. Also, in view of the symmetry of the frame and the load about the y–z plane, the symmetric complements of the remaining internal forces F_y, M_x, and M_z should be mirror images as shown at the origin in Fig. 8.12 (b). Since the pair of internal forces F_y and the pair of internal moments M_x cannot satisfy the

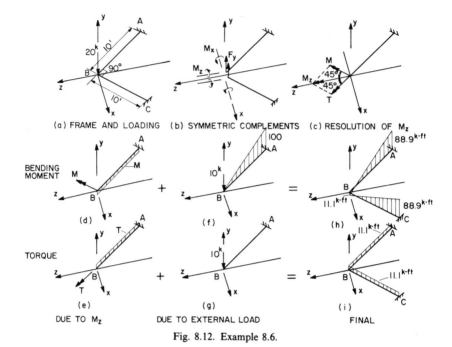

(a) FRAME AND LOADING (b) SYMMETRIC COMPLEMENTS (c) RESOLUTION OF M_z

BENDING MOMENT

(d) (f) (h)

TORQUE

(e) (g) (i)

DUE TO M_z DUE TO EXTERNAL LOAD   FINAL

Fig. 8.12. Example 8.6.

static requirements, they must also be zero. The only unknown left at the origin is the moment M_z.

To draw the flexural and torsional moment diagrams due to M_z, the term M_z was resolved into two 45° components M and T, where $M = T = M_z/\sqrt{2}$, as shown in Fig. 8.12 (c). The bending moment component M and the torque component T produce bending and torsional moment diagrams in Fig. 8.12 (d) and (e), respectively. Notice that only one-half of the frame, AB, is drawn. The other half is identical to it because of symmetry.

The external load acting on the member AB should be 10 kips, i.e., one-half of the 20-kips external load. The bending moment diagram caused by the 10-kips load is shown in Fig. 8.12 (f). The torque diagram caused by the 10-kips load happens to be zero, as shown in Fig. 8.12 (g).

Symmetry also requires that the rotation about the z-axis $\phi_z = 0$. From this condition the unknown M_z (or $M = T = M_z/\sqrt{2}$) can be solved:

$$\phi_z = \frac{1}{\sqrt{2}}\frac{M(10)}{EI} + \frac{1}{\sqrt{2}}\frac{T(10)}{GC} - \frac{1}{\sqrt{2}}\frac{0.5(100)(10)}{EI} = 0 \qquad \text{(a)}$$

$$M = T = \frac{50}{1 + \dfrac{EI}{GC}} = \frac{50}{1 + 3.49} = 11.1 \text{ k-ft} \qquad \text{(b)}$$

The calculation of $EI/GC = 3.49$ was given by Eqs. (c) through (f) in Example 8.5 for a 10 in. by 20 in. solid uncracked concrete member.

Now that the moment diagrams in Fig. 8.12 (d) and (e) are determined, they are added to the moment diagrams induced by external loading, shown in Fig. 8.12 (f) and (g), respectively, to arrive at the final bending moment and torque diagrams in Fig. 8.12 (h) and (i).

EXAMPLE 8.7

A reinforced concrete plane frame ABC with fixed ends is shown in Fig. 8.13 (a). Both members AB and BC have identical square cross sections. The lengths of members and the (x, y, z) coordinate system are given. A torque $T = 100$ k-ft is acting at a section 8 ft from support A. Determine the bending and torsional moment diagrams.

This frame is not symmetric. However, we can still utilize the characteristics of plane frames subjected to perpendicular loadings to know that only three unknowns M_x, M_y, and F_z exist at any section. These three unknowns are shown in Fig. 8.13 (b) when the support C is removed. The bending moment and torque diagrams produced by these three unknowns are given in Fig.

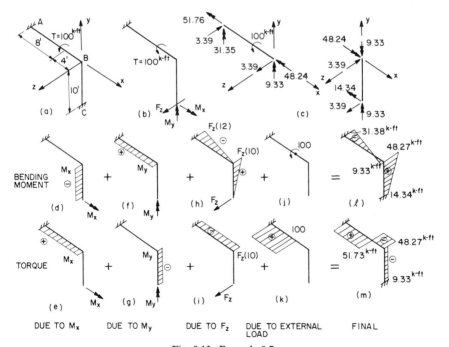

Fig. 8.13. Example 8.7.

8.13 (d), (e), (f), (g), (h), and (i). The bending and torsional moment diagrams induced by the external load are also given, in Fig. 8.13 (j) and (k).

To find the three unknowns, M_x, M_y, and F_z, we can write three equations to restore the original fixed-end condition of support C, i.e., $\phi_x = 0$, $\phi_y = 0$, and $\Delta_z = 0$. However, it is wise first to evaluate the rotations and displacement, ϕ_x, ϕ_y, and Δ_z, due to each of the three unknowns, M_x, M_y, and F_z, individually, and due to the external load.

Rotations and displacement due to M_x [see Fig. 8.13 (d) and (e)]:

$$\phi_x = \frac{M_x(10)}{EI} + \frac{M_x(12)}{GC} = \left(\frac{10}{EI} + \frac{12}{GC}\right) M_x \tag{a}$$

$$\phi_y = 0 \tag{b}$$

$$\Delta_z = -\frac{M_x(10)}{EI} \frac{(10)}{2} - \frac{M_x(12)}{GC} (10) = -\left(\frac{50}{EI} + \frac{120}{GC}\right) M_x \tag{c}$$

Note that the factors $(10/EI + 12/GC)$ and $-(50/EI + 120/GC)$ in Eq. (a) and (c), respectively, are flexibility elements of the structure.

Rotations and displacements due to M_y [see Fig. 8.13 (f) and (g)]:

$$\phi_x = 0 \tag{d}$$

$$\phi_y = \frac{M_y(12)}{EI} + \frac{M_y(10)}{GC} = \left(\frac{12}{EI} + \frac{10}{GC}\right) M_y \tag{e}$$

$$\Delta_z = -\frac{M_y(12)}{EI} \frac{(12)}{2} = -\left(\frac{72}{EI}\right) M_y \tag{f}$$

Rotations and displacement due to F_z [see Fig. 8.13 (h) and (i)]:

$$\phi_x = -\frac{F_z(10)}{EI} \frac{(10)}{2} - \frac{F_z(10)(12)}{GC} = -\left(\frac{50}{EI} + \frac{120}{GC}\right) F_z \tag{g}$$

$$\phi_y = -\frac{F_z(12)}{EI} \frac{(12)}{2} = -\left(\frac{72}{EI}\right) F_z \tag{h}$$

$$\Delta_z = \frac{F_z(10)}{EI} \frac{(10)}{2} \left(\frac{2}{3} 10\right) + \frac{F_z(12)}{EI} \left(\frac{12}{2}\right) \left(\frac{2}{3} 12\right) + \frac{F_z(10)(12)}{GC} (10)$$

$$= \left(\frac{909.33}{EI} + \frac{1,200}{GC}\right) F_z \tag{i}$$

Rotations and displacement due to external load [see Fig. 8.13 (j) and (k)]:

$$\phi_x = \frac{100(8)}{GC} = \frac{800}{GC} \tag{j}$$

$$\phi_y = 0 \tag{k}$$

$$\Delta_z = -\frac{100(8)}{GC}(10) = -\frac{8,000}{GC} \tag{l}$$

In Eqs. (a) through (l), the positive or negative value of a rotation and displacement can be obtained as follows: First, a rotation produced by its corresponding moment (or a displacement produced by its corresponding force) must be positive. In this case, ϕ_x, ϕ_y, and Δ_z produced by M_x, M_y, and F_z, respectively, must be positive. Second, to determine the sign of a displacement (or rotation) at location 1 caused by a force at another location 2, we can compare the sign of the moment diagram produced by this force at location 2 to the sign of the moment diagram of the same member produced by the corresponding force of the displacement at location 1. If the moment diagrams have the same signs, the displacement is positive. If the moment diagrams have opposite signs, the displacement is negative.

The three equations $\phi_x = 0$, $\phi_y = 0$, and $\Delta_z = 0$ at support C can now be established. Adding Eqs. (a), (d), (g), and (j) we obtain:

$$\left(\frac{10}{EI} + \frac{12}{GC}\right) M_x - \left(\frac{50}{EI} + \frac{120}{GC}\right) F_z + \frac{800}{GC} = 0 \tag{m}$$

Similarly:

$$\left(\frac{12}{EI} + \frac{10}{GC}\right) M_y - \left(\frac{72}{EI}\right) F_z = 0 \tag{n}$$

$$-\left(\frac{50}{EI} + \frac{120}{GC}\right) M_x - \left(\frac{72}{EI}\right) M_y + \left(\frac{909.33}{EI} + \frac{1,200}{GC}\right) F_z - \frac{8,000}{GC} = 0 \tag{o}$$

These three equations can be expressed more concisely in terms of matrix notation:

$$\begin{bmatrix} \left(\dfrac{10}{EI} + \dfrac{12}{GC}\right) & 0 & -\left(\dfrac{50}{EI} + \dfrac{120}{GC}\right) \\[2mm] 0 & \left(\dfrac{12}{EI} + \dfrac{10}{GC}\right) & -\left(\dfrac{72}{EI}\right) \\[2mm] -\left(\dfrac{50}{EI} + \dfrac{120}{GC}\right) & -\left(\dfrac{72}{EI}\right) & \left(\dfrac{909.33}{EI} + \dfrac{1,200}{GC}\right) \end{bmatrix} \begin{bmatrix} M_x \\[2mm] M_y \\[2mm] F_z \end{bmatrix} = \begin{bmatrix} -\dfrac{800}{GC} \\[2mm] 0 \\[2mm] \dfrac{8,000}{GC} \end{bmatrix} \tag{p}$$

| Flexibility matrix | Unknown force vector | Displacement vector |

Notice that the flexibility matrix is symmetrical and satisfies Maxwell's reciprocal theorem.

The rigidity ratio EI/GC for a square section is calculated below, using x to represent the dimension of the section:

$$I = \frac{x^4}{12} \tag{q}$$

$$C = 0.141 x^4 \tag{r}$$

$$\frac{E}{G} = 2(1 + \nu) = 2.4 \tag{s}$$

$$\frac{EI}{GC} = 2.4 \left(\frac{1}{12}\right)\left(\frac{1}{0.141}\right) = 1.418 \tag{t}$$

Multiplying Eq. (p) by EI and substituting $EI/GC = 1.418$ into Eq. (p) gives:

$$
\begin{bmatrix}
27.016 & 0 & -220.15 \\
0 & 26.18 & -72 \\
-220.15 & -72 & 2,610.93
\end{bmatrix}
\begin{bmatrix}
M_x \\
M_y \\
F_z
\end{bmatrix}
=
\begin{bmatrix}
-1,134.4 \\
0 \\
11,344
\end{bmatrix}
\tag{u}
$$

The solution of Eq. (u) can be most conveniently obtained using a programmable hand calculator:

$$M_x = -14.343 \text{ k-ft} \tag{x}$$

$$M_y = 9.3307 \text{ k-ft} \tag{y}$$

$$F_z = 3.3937 \text{ k} \tag{z}$$

Substituting these values of M_x, M_y, and F_z into Fig. 8.13 (d), (f), and (h), respectively, and adding these bending moment diagrams to that caused by external load in Fig. 8.13 (j), we arrive at the final bending moment diagram in Fig. 8.13 (l). Similarly, the values of M_x, M_y, and F_z are used in Fig. 8.13 (e), (g), and (i), respectively. The summation of torque diagrams in Figs. 8.13 (e), (g), (i), and (k) gives the final torque diagram in Fig. 8.13 (m).

It is also interesting to find from simple statics the internal forces at joint B and the reactions at support A. The results are given in Fig. 8.13 (c).

EXAMPLE 8.8

A reinforced concrete staircase is supported by a space frame $ABCD$ as shown in Fig. 8.14 (a). The frame has a square cross section throughout and is fixed at the ends. The dimensions of the frame and the (x, y, z) coordinate system are given. The frame is subjected to a vertical uniform load of 2k/ft. Determine the bending and torsional moment diagrams.

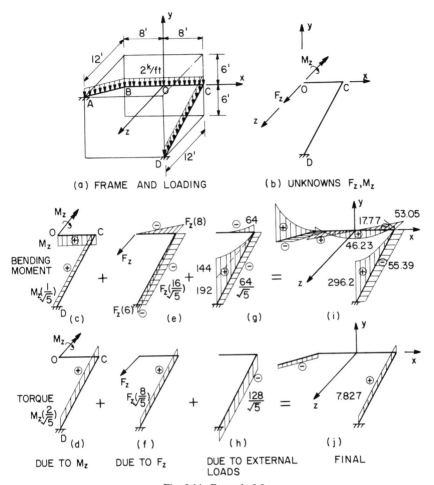

Fig. 8.14. Example 8.8.

This space frame is axial symmetric about the z-axis, and the loading is antisymmetric. From Table 8.1, column 5, it can be seen that only two unknowns, F_z and M_z, exist at the origin. They are shown in Fig. 8.14 (b) acting at point O on one-half of the frame OCD. The bending and the torque diagrams created by M_z and F_z are shown in Fig. 8.14 (c), (d), (e), and (f). The bending and torsional moment produced by the uniform external loading are given in Fig. 8.14 (g) and (h).

Because of symmetry we can write two equations, $\phi_z = 0$ and $\Delta_z = 0$, at the origin, according to Table 8.1, column 12. To evaluate the flexibility elements, we shall calculate ϕ_z and Δ_z caused by M_z, F_z, and the external loads.

Rotation and displacement due to M_z:

$$\phi_z = \frac{M_z}{EI}(8) + \frac{M_z}{\sqrt{5}EI}(6\sqrt{5})\left(\frac{1}{\sqrt{5}}\right) + \frac{2M_z}{\sqrt{5}GC}(6\sqrt{5})\left(\frac{2}{\sqrt{5}}\right)$$

$$= \left(\frac{23.89}{\sqrt{5}EI} + \frac{24}{\sqrt{5}GC}\right)M_z \tag{a}$$

$$\Delta_z = -\frac{M_z}{\sqrt{5}EI}(6\sqrt{5})\left(\frac{2}{\sqrt{5}}\right)(8) + \frac{2M_z}{\sqrt{5}GC}(6\sqrt{5})\left(\frac{1}{\sqrt{5}}\right)(8)$$

$$= \left(-\frac{96}{\sqrt{5}EI} + \frac{96}{\sqrt{5}GC}\right)M_z \tag{b}$$

Rotation and displacement due to F_z:

$$\phi_z = -\frac{F_z(16)}{\sqrt{5}EI}(6\sqrt{5})\left(\frac{1}{\sqrt{5}}\right) + \frac{F_z(8)}{\sqrt{5}GC}(6\sqrt{5})\left(\frac{2}{\sqrt{5}}\right)$$

$$= \left(-\frac{96}{\sqrt{5}EI} + \frac{96}{\sqrt{5}GC}\right)F_z \tag{c}$$

$$\Delta_z = \frac{F_z(8)}{EI}\left(\frac{8}{2}\right)\left(\frac{2}{3}8\right) + \frac{F_z(6)}{EI}\left(\frac{6\sqrt{5}}{2}\right)\left(\frac{2}{3}6\sqrt{5}\right)\left(\frac{1}{\sqrt{5}}\right)$$

$$+ \frac{F_z(16)}{\sqrt{5}EI}(6\sqrt{5})\left(\frac{2}{\sqrt{5}}\right)(8) + \frac{F_z(8)}{\sqrt{5}GC}(6\sqrt{5})\left(\frac{1}{\sqrt{5}}\right)(8)$$

$$= \left(\frac{2{,}277.62}{\sqrt{5}EI} + \frac{384}{\sqrt{5}GC}\right)F_z \tag{d}$$

Rotation and displacement due to external loading:

$$\phi_z = -\frac{64}{EI}\left(\frac{8}{3}\right) - \frac{64}{\sqrt{5}EI}(6\sqrt{5})\left(\frac{1}{\sqrt{5}}\right) - \frac{128}{\sqrt{5}GC}(6\sqrt{5})\left(\frac{2}{\sqrt{5}}\right)$$

$$= -\left(\frac{765.62}{\sqrt{5}EI} + \frac{1{,}536}{\sqrt{5}GC}\right) \tag{e}$$

$$\Delta_z = -\frac{192}{EI}\left(\frac{6\sqrt{5}}{2}\right)\left(\frac{2}{3}6\sqrt{5}\right)\left(\frac{1}{\sqrt{5}}\right) - \frac{144}{EI}\left(\frac{6\sqrt{5}}{3}\right)\left(\frac{3}{4}6\sqrt{5}\right)\left(\frac{1}{\sqrt{5}}\right)$$

$$+ \frac{64}{\sqrt{5}EI}(6\sqrt{5})\left(\frac{2}{\sqrt{5}}\right)(8) - \frac{128}{\sqrt{5}GC}(6\sqrt{5})\left(\frac{1}{\sqrt{5}}\right)(8)$$

$$= -\left(\frac{11{,}856}{\sqrt{5}EI} + \frac{6{,}144}{\sqrt{5}GC}\right) \tag{f}$$

Now the equation $\phi_z = 0$ can be established by adding the three rotations in Eqs. (a), (c), and (e) and equating the sum to zero. Similarly, the equation $\Delta_z = 0$ can be obtained by adding the three displacements in Eqs. (b), (d), and (f) and equating the sum to zero. These two equations can be expressed in matrix form as follows:

$$
\begin{bmatrix}
\left(\dfrac{23.89}{\sqrt{5}EI} + \dfrac{24}{\sqrt{5}GC}\right) & \left(-\dfrac{96}{\sqrt{5}EI} + \dfrac{96}{\sqrt{5}GC}\right) \\[3mm]
\left(-\dfrac{96}{\sqrt{5}EI} + \dfrac{96}{\sqrt{5}GC}\right) & \left(\dfrac{2{,}277.62}{\sqrt{5}EI} + \dfrac{384}{\sqrt{5}GC}\right)
\end{bmatrix}
\begin{bmatrix} M_z \\[3mm] F_z \end{bmatrix}
=
\begin{bmatrix}
\dfrac{765.62}{\sqrt{5}EI} + \dfrac{1{,}536}{\sqrt{5}GC} \\[3mm]
\dfrac{11{,}856}{\sqrt{5}EI} + \dfrac{6{,}144}{\sqrt{5}GC}
\end{bmatrix}
$$

(g)

For square concrete sections, the rigidity ratio $EI/GC = 1.418$ has been calculated from Eqs. (q) through (t) in Example 8.7. Multiplying Eq. (g) by $\sqrt{5}EI$ and substituting this EI/GC into Eq. (g) give:

$$
\begin{bmatrix} 57.922 & 40.128 \\ 40.128 & 2{,}822.13 \end{bmatrix}
\begin{bmatrix} M_z \\ F_z \end{bmatrix}
=
\begin{bmatrix} 2{,}943.67 \\ 20{,}568.19 \end{bmatrix}
$$

(h)

The solution of Eq. (h) can easily be obtained:

$$M_z = 46.227 \text{ k-ft} \tag{i}$$
$$F_z = 6.631 \text{ k} \tag{j}$$

Substituting these values of M_z and F_z into Fig. 8.14 (c), (d), (e), and (f), the bending and torsional moment diagrams in these figures can be determined. Summing the three bending moment diagrams in Figs. 8.14 (c), (e), and (g) results in the final bending moment diagram shown in Fig. 8.14 (i). Similarly, adding the three torsional moment diagrams in Figs. 8.14 (d), (f), and (h) gives the final torsional moment diagram in Fig. 8.14 (j).

8.4 STIFFNESS METHOD

8.4.1 General Description

The stiffness method will now be introduced. It is assumed that the reader is familiar with this method in the analysis of plane frames subjected to in-plane loadings. In this special case of analysis, only the bending stiffness of the structural members are involved. In this section, the stiffness method will be generalized to include the torsional stiffness of structural members for the analysis of frames subjected to out-of-plane loadings.

The general procedures of the stiffness method are as follows:

1. Recognize the n-number of rotations and displacements at all the joints in a structure that are not zero under the external loadings. Both bending and torsional rotations as well as both in-plane and out-of-plane displacements should be included.

2. Select these n-numbers of rotations and displacements at the joints as unknowns and fix them so that they become zero. Since the rotations and displacements are selected as unknowns, the stiffness method is sometimes referred to as the slope-deflection method.

3. Allow the rotations and displacements to move one at a time. For each movement draw the bending and torsional moment diagrams and identify the bending or torsional moments at the end of each member adjacent to the moving joint. The end moment created by a unit joint movement (rotation or displacement) is defined as the stiffness of the member. The sum of the stiffnesses of all the members adjacent to a joint is called the joint stiffness. The name "stiffness method" stems from the fact that the evaluation of stiffnesses of the members and of the joints lies at the heart of this method.

4. Draw all the bending and torsional moment diagrams due to the external loadings for the fixed-end members. The bending and torsional moments at the end of the members are called the fixed end moment (*FEM*).

5. Write n-number of equations for the equilibrium of the joints. These equations may represent the equilibrium of moments or the equilibrium of forces, and should include the n-number of unknown rotations and displacements. From these simultaneous equations, the n-number of unknowns can be solved. Any convenient methods for solving simultaneous equations can be used here.

6. Once the unknown rotations and displacements are solved, the actual bending and torsional moment diagrams caused by each unknown rotation or displacement can be determined. The final bending and torsional moments are simply the sum of the moments caused by all the unknowns plus those induced by the external loads.

8.4.2 Important Details

In using the stiffness method, the following four concepts should be studied.

1. Symmetry and Plane Frame. The theory of symmetry (Section 8.1) and the characteristics of plane frames (Section 8.2) should be used as much as possible to reduce the number of unknowns. For a symmetric structure subjected to symmetric or antisymmetric loads, some of the joint rotations or displacements are often zero due to planar, axial, or point symmetry.

Also, if two joints are located symmetrically about a plane, axis, or point, the rotations and displacements should be symmetric, and the magnitudes of a pair of symmetric rotations (or displacements) should be the same. For a plane structure, it is often wise to separate the external loading into a set of in-plane loadings plus a set of perpendicular loadings. Each set will cause a maximum of three unknown rotations or displacements at each joint, instead of six.

2. *Stiffness.* In evaluating the stiffness of a member or a joint, both the bending and torsional stiffnesses of the members must be included. The defini-

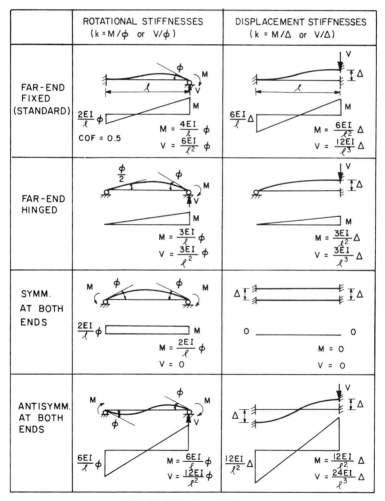

Fig. 8.15. Bending stiffnesses.

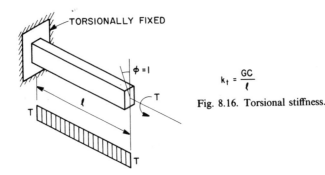

$$k_t = \frac{GC}{l}$$

Fig. 8.16. Torsional stiffness.

tions of various bending stiffnesses (k) are reviewed in Fig. 8.15. On the other hand, torsional stiffness k_t is defined simply as the torque T required to produce a unit twisting rotation ($\phi = 1$) of a member as illustrated in Fig. 8.16. For a prismatic member subjected to uniform St. Venant torsion:

$$T = \frac{GC}{l} \phi \qquad (8\text{--}13)$$

where l is the length of the member and GC is the torsional rigidity of the section. Taking $\phi = 1$, the torsional stiffness is:

$$k_t = \frac{GC}{l} \qquad (8\text{--}14)$$

3. Fixed End Moments. The bending moment diagrams and the fixed end bending moments for a member subjected to three common types of lateral forces are given in Fig. 8.17. The fixed end torsional moments for a member subjected to a torque, however, will be derived below. A straight nonuniform member *AB*, shown in Fig. 8.18 (a), is fixed at both ends and subjected to a torque at point *C*. The length *a* for *AC* has a torsional stiffness k_{ta}, and the length *b* for *CB* has a torsional stiffness k_{tb}. Let the torsional rotation at *C* be ϕ_c. The torque to the left of point *C* is:

$$T_a = k_{ta} \phi_c \qquad (8\text{--}15)$$

and the torque to the right of point *C* is:

$$T_b = k_{tb} \phi_c \qquad (8\text{--}16)$$

Consider point *C* as a joint, and isolate a small length of beam from the

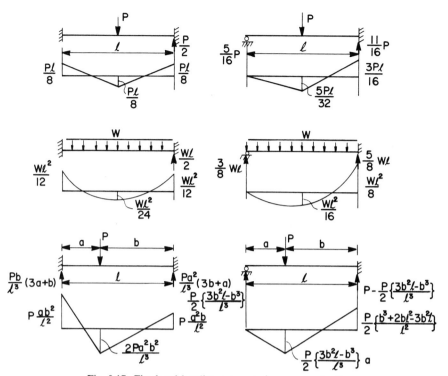

Fig. 8.17. Fixed end bending moments for common loadings.

left of C to the right of C. Taking the moment equilibrium of this element about the beam axis, we obtain:

$$T_a + T_b = T \tag{8-17}$$

Substituting Eqs. (8–15) and (8–16) into Eq. (8–17) gives:

$$\phi_c = \frac{T}{(k_{ta} + k_{tb})} \tag{8-18}$$

Note that the denominator $(k_{ta} + k_{tb})$ is the stiffness of the joint C, and Eq. (8–18) simply states that the factor relating the torque T and the rotation ϕ_c at the joint is the stiffness of the joint $(k_{ta} + k_{tb})$. Substituting Eq. (8–18) into Eq. (8–15) and Eq. (8–16), we have:

$$T_a = \left(\frac{k_{ta}}{k_{ta} + k_{tb}} \right) T \tag{8-19}$$

$$T_b = \left(\frac{k_{tb}}{k_{ta} + k_{tb}} \right) T \tag{8-20}$$

419

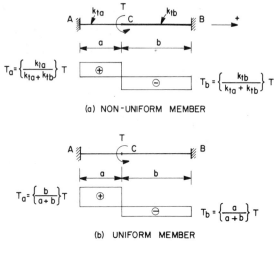

(a) NON-UNIFORM MEMBER

(b) UNIFORM MEMBER

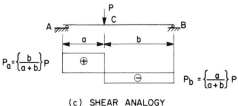

(c) SHEAR ANALOGY

Fig. 8.18. Torsional fixed end moments.

The ratios $k_{ta}/(k_{ta} + k_{tb})$ and $k_{tb}/(k_{ta} + k_{tb})$ in Eqs. (8–19) and (8–20) are known as the distribution factors for members a and b, respectively. It can be seen from Fig. 8.18 (a) that the torsional fixed end moment is simply the torque T times the distribution factor of the adjacent member.

If the member is uniform, Eq. (8–19) and Eq. (8–20) become:

$$T_a = \left(\frac{b}{a+b}\right) T \qquad (8\text{–}21)$$

$$T_b = \left(\frac{a}{a+b}\right) T \qquad (8\text{–}22)$$

For this case the torque diagram and the fixed end moments are shown in Fig. 8.18 (b). It is interesting to observe an analogy between this torque diagram and the shear diagram shown in Fig. 8.18 (c). This shear diagram is obtained from a load P acting at the same location C on a simply supported beam AB.

4. Sign Convention. In the flexibility method, a sign convention was introduced that is convenient for expressing the internal forces and moments of a member and for plotting the shear, bending moment, and torque diagrams.

This sign convention will continue to be used in the stiffness method for the same purpose. It will be referred to as the "member sign convention." However, in writing the force or moment equilibrium equations for the joints, a special sign convention has to be established. Figure 8.19 (a) shows a typical joint with all the forces acting on it. A force is considered positive if it acts on the *joint* and points toward the positive direction of a coordinate axis, regardless of the sign of the face on which the force acts. Figure 8.19 (b) shows the same joint with all the moments acting on it. A positive moment is one that acts on the *joint* with the moment vector based on a right-hand screw pointing toward the positive direction of a coordinate axis. Therefore, all the forces and moments shown in Fig. 8.19 are positive. In this special sign convention, it is important to emphasize that the forces or moments

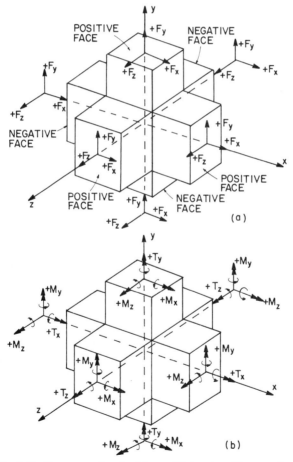

Fig. 8.19. Sign convention for joint equilibrium in stiffness method.

must be acting on the *joint* and not on the member. For this reason, it will be referred to as the "joint sign convention."

Comparing the joint sign convention to the member sign convention, it can be seen that the two are identical on the positive faces of a joint. However, these two sign conventions are opposite on the negative faces of a joint. For a pair of shears or moments, the sign of the shear or moment acting on the negative face of a joint (based on the joint sign convention) must be opposite to the sign of the other shear or moment acting on the adjacent member (based on the member sign convention). In other words, to determine a shear or moment on the negative face of a joint from the shear or moment diagram of an adjacent member, its sign must be changed.

The following three examples will illustrate the stiffness method. In general, the stiffness method is more convenient in highly statically indeterminate frames where more than two members are connected at each joint.

8.4.3 Examples 8.9, 8.10, and 8.11

EXAMPLE 8.9

The example problem 8.7, which was solved by the flexibility method, will now be solved by the stiffness method. Use of the same example can best

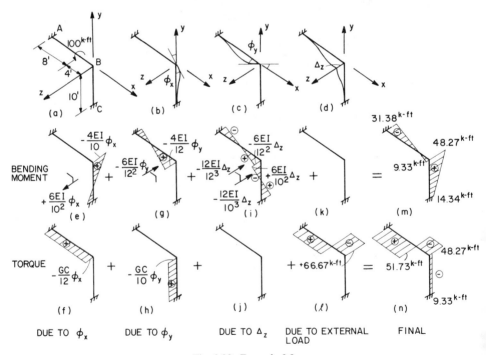

Fig. 8.20. Example 8.9.

illustrate the differences between these two methods. Figure 8.20 (a) shows the reinforced concrete frame and the loading. Both members AB and BC have identical square cross sections. Utilizing the characteristics of plane frames subjected to perpendicular loadings, only two types of rotation, ϕ_x and ϕ_y, and one type of displacement, Δ_z, are possible at any section. Select ϕ_x, ϕ_y, and Δ_z at the joint B as the three unknowns. The flexural deformations produced by these three unknowns are shown individually in Fig. 8.20 (b), (c), and (d). The bending moment diagrams produced by ϕ_x, ϕ_y, Δ_z and the external load are given in Fig. 8.20 (e), (g), (i) and (k), respectively. The magnitudes of the bending moments and the shear forces are obtained from the list of bending stiffnesses in Fig. 8.15 (far-end fixed in this example). The torsional moment diagrams produced by ϕ_x, ϕ_y, Δ_z and the external load are given in Fig. 8.20 (f), (h), (j), and (l), respectively. The magnitudes of the torques are obtained from the torsional stiffness in Eq. (8–14) and from the torsional fixed end moment equation in Fig. 8.18. All the bending and torsional moment diagrams are plotted according to the member sign convention. The shears and moments acting on joint B are given in Fig. 8.20. Notice that the moments on the joint have signs opposite to those of the moment diagrams of the adjacent members because they all act on the negative faces of the joint, and their signs have been based on the joint sign convention.

To find the three unknowns, ϕ_x, ϕ_y, and Δ_z, we can write three equations for the equilibrium of the joint as follows:

$$\Sigma M_x = 0 \qquad -\frac{4EI}{10}\phi_x - \frac{GC}{12}\phi_x + \frac{6EI}{10^2}\Delta_z + 66.67 = 0 \qquad \text{(a)}$$

$$\Sigma M_y = 0 \qquad -\frac{4EI}{12}\phi_y - \frac{GC}{10}\phi_y - \frac{6EI}{12^2}\Delta_z = 0 \qquad \text{(b)}$$

$$\Sigma F_z = 0 \qquad +\frac{6EI}{10^2}\phi_x - \frac{6EI}{12^2}\phi_y - \frac{12EI}{12^3}\Delta_z - \frac{12EI}{10^3}\Delta_z = 0 \qquad \text{(c)}$$

Eqs. (a), (b), and (c) can be expressed in one matrix equation:

$$
\begin{bmatrix}
-\left(\dfrac{4EI}{10}+\dfrac{GC}{12}\right) & 0 & +\dfrac{6EI}{100} \\[2ex]
0 & -\left(\dfrac{4EI}{12}+\dfrac{GC}{10}\right) & -\dfrac{6EI}{144} \\[2ex]
+\dfrac{6EI}{100} & -\dfrac{6EI}{144} & -\left(\dfrac{EI}{144}+\dfrac{12EI}{1,000}\right)
\end{bmatrix}
\begin{bmatrix}
\phi_x \\[2ex] \phi_y \\[2ex] \Delta_z
\end{bmatrix}
=
\begin{bmatrix}
-66.67 \\[2ex] 0 \\[2ex] 0
\end{bmatrix}
\quad \text{(d)}
$$

Stiffness matrix   Displ. vector   Force vector

Taking $EI/GC = 1.418$ as in Example 8.7, Eq. (d) is simplified to:

$$\begin{bmatrix} -0.45877 & 0 & 0.06 \\ 0 & -0.40385 & -0.04167 \\ 0.06 & -0.04167 & -0.01894 \end{bmatrix} \begin{bmatrix} EI\phi_x \\ EI\phi_y \\ EI\Delta_z \end{bmatrix} = \begin{bmatrix} -66.67 \\ 0 \\ 0 \end{bmatrix} \quad (e)$$

The solution of Eq. (e) by a programmable hand calculator is:

$$EI\phi_x = 313.1 \quad (f)$$
$$EI\phi_y = -132.3 \quad (g)$$
$$EI\Delta_z = 1,281.6 \quad (h)$$

Substitute Eqs. (f), (g), and (h) into Fig. 8.20 (e) through (j) and remember $EI/GC = 1.418$. Adding Fig. 8.20 (e), (g), (i), and (k) results in the final bending moment diagram in Fig. 8.20 (m). Similarly, the summation of Fig. 8.20 (f), (h), (j), and (l) gives the final torsional moment diagram in Fig. 8.20 (n). It can be seen that these moment diagrams are exactly the same as those in Fig. 8.13 (l) and (m), which were solved by the flexibility method in Example 8.7.

EXAMPLE 8.10

A reinforced concrete T-shape frame, tested and reported in Section 9.3.2 in Chapter 9, is shown in Fig. 8.21 (a). Beam AB represents a 6 in. by 9 in. floor beam 9 ft long simply supported at point A and framing into a spandrel beam CD at joint B. The 6 in. by 12 in. spandrel beam CD is 9 ft long and is simply supported at both ends in all directions, except torsionally fixed about the x-axis. The floor beam AB is subjected to a uniform load w. Assuming that the frame is uncracked, determine the bending and torsional moment diagrams in the two beams.

The T-shape frame is a plane frame subjected to perpendicular loading. Therefore, only three types of rotation and displacement, ϕ_x, ϕ_z, and Δ_y, are possible at each section. The frame and loading are also symmetric about the y–z plane, and ϕ_z must be zero along floor beam AB, including joint B. As a result, only an unknown rotation ϕ_x and an unknown displacement Δ_y exist at the joint. The flexural deformation produced by ϕ_x and Δ_y are shown in Fig. 8.21 (b) and (c), respectively. The bending moment diagrams created by ϕ_x, Δ_y, and the external load w are shown in Fig. 8.21 (d), (f), and (h), respectively. The magnitudes of the bending moments and shear forces are obtained from Fig. 8.15 for ϕ_x and Δ_y, and from Fig. 8.17 for the external load w. The rotation ϕ_x will induce a torque diagram according

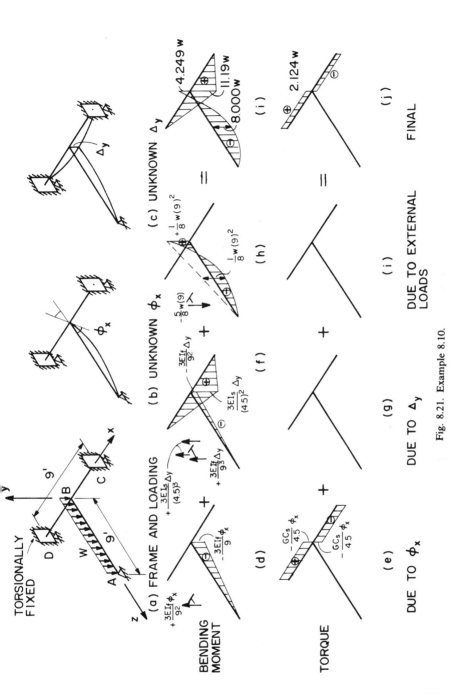

Fig. 8.21. Example 8.10.

425

to Eq. (8–14) as shown in Fig. 8.21 (e). No torques are produced by Δ_y and w as shown in Fig. 8.21 (g) and (i). The values of shears and moments acting on joint B are given using the joint sign convention. Notice that the signs for moments are identical to those of the moment diagrams of the adjacent members for the two positive faces of the joint and are opposite for the one negative face.

To find the two unknowns, ϕ_x and Δ_y, we can write two equations for the equilibrium of joint B as follows:

$$\Sigma M_x = 0 \qquad -\left(\frac{3EI_f}{9} + 2\frac{GC_s}{4.5}\right)\phi_x - \frac{3EI_f}{9^2}\Delta_y + \frac{1}{8}w(9)^2 = 0 \tag{a}$$

$$\Sigma F_y = 0 \qquad +\frac{3EI_f}{9^2}\phi_x + \left(\frac{3EI_f}{9^3} + 2\frac{3EI_s}{(4.5)^3}\right)\Delta_y - \frac{5}{8}w(9) = 0 \tag{b}$$

where I_f is the moment of inertia of the floor beam AB, and I_s and C_s are the moment of inertia and torsional constant, respectively, of the spandrel beam CD.

To convert GC_s and EI_s into EI_f we shall calculate the rigidity ratios (EI_f/GC_s) and (EI_f/EI_s):

$$I_f = \frac{6(9)^3}{12} = 364.5 \text{ in.}^4 \tag{c}$$

$$I_s = \frac{6(12)^3}{12} = 864 \text{ in.}^4 \tag{d}$$

$$C_s = 0.229(6)^3(12) = 593.6 \text{ in.}^4 \tag{e}$$

$$\frac{E}{G} = 2(1 + 0.2) = 2.4 \tag{f}$$

$$\frac{EI_f}{GC_s} = \frac{2.4(364.5)}{593.6} = 1.4737 \tag{g}$$

$$\frac{EI_f}{EI_s} = \frac{364.5}{864} = 0.4219 \tag{h}$$

Using the two rigidity ratios from Eqs. (g) and (h), Eqs. (a) and (b) are expressed in matrix form:

$$\begin{bmatrix} 0.63491 & 0.03704 \\ 0.03704 & 0.16018 \end{bmatrix} \begin{bmatrix} EI_f\phi_x \\ EI_f\Delta_y \end{bmatrix} = \begin{bmatrix} 10.125w \\ 5.625w \end{bmatrix} \tag{i}$$

The solution of Eq. (i) is:

$$EI_f \phi_x = 14.09w \qquad \text{(j)}$$

$$EI_f \Delta_y = 31.86w \qquad \text{(k)}$$

Equations (j) and (k) are used to calculate the moment diagrams in Fig. 8.21 (d), (e), (f), and (g). The summation of Fig. 8.21 (d), (f), and (h) gives the final bending moment diagram in Fig. 8.21 (i), while the final torsional moment diagram is given in Fig. 8.21 (j). It can be seen that the final torsional moment in the spandrel beam is produced by the rotation ϕ_x alone.

EXAMPLE 8.11

A typical reinforced concrete plane frame is isolated from a multi-story building, and is shown in Fig. 8.22 (a). It consists of a typical two-span continuous floor beam supported by columns above and below the beam. The far ends of the columns are assumed to be fixed. All the columns are identical, each having a 14 in. by 14 in. cross section and a 10-ft length. The continuous beam has a 12 in. by 24 in. cross section for both the 20-ft and 24-ft spans. A uniform torsional moment of 2.4 k-ft/ft is acting on the 20-ft span, and a concentrated torque of 30 k-ft on the 24-ft span. Determine the bending and torsional moment diagrams for the frame.

Since this plane frame is subjected to perpendicular loadings, only three types of movement are possible at each joint, namely, ϕ_y, ϕ_z, and Δ_x. Furthermore, since the frame is symmetric about the z-axis and is subjected to symmetric loading, ϕ_y and Δ_x should also vanish along the beam (see Table 8.1). As a result, only ϕ_z is possible along the beam. We shall call the three unknown rotations at the joints A, B, and C, ϕ_A, ϕ_B, and ϕ_C, respectively. The subscript z has been omitted for convenience. It is understood that all the rotations are about the z-axis. Since ϕ_z is the only rotation possible along the beam, the beam will be subjected only to torsional moments about the z-axis, T_z, and the columns can develop only bending moment about the z-axis, M_z.

Using the stiffness method, each of the three joint rotations will be allowed to rotate individually, while the other joints are assumed to be fixed. The moment diagrams caused by the rotation of joint A, ϕ_A, are shown in Fig. 8.22 (b). Both the bending and torsional moment diagrams are plotted in the same figure. It is understood that the moment diagrams on the beam are for torsional moment, T_z, and the moment diagrams on the columns are for bending moment, M_z. The magnitude of the beam moments on joint A is simply the product of the rotation ϕ_A and the torsional stiffness of beam AB, k_{tAB}. The sign of $k_{tAB}\phi_A$ at joint A is opposite to the sign of the torsional moment diagram because this torsional moment is acting on a negative face of joint A. The torsional moment acting on joint B has the

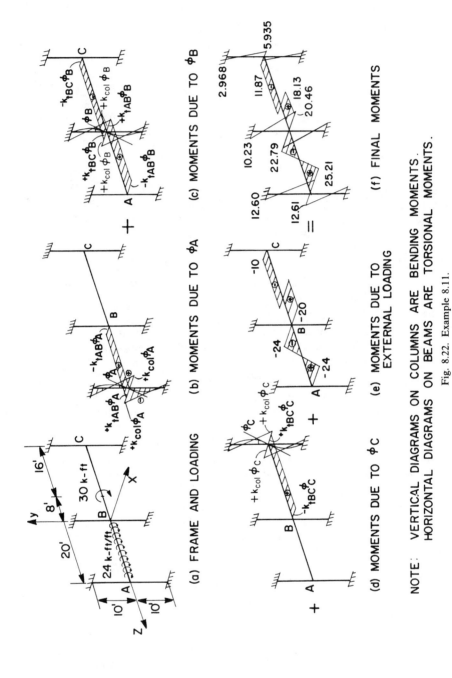

(a) FRAME AND LOADING

(b) MOMENTS DUE TO ϕ_A

(c) MOMENTS DUE TO ϕ_B

(d) MOMENTS DUE TO ϕ_C

(e) MOMENTS DUE TO
EXTERNAL LOADING

(f) FINAL MOMENTS

NOTE: VERTICAL DIAGRAMS ON COLUMNS ARE BENDING MOMENTS.
HORIZONTAL DIAGRAMS ON BEAMS ARE TORSIONAL MOMENTS.

Fig. 8.22. Example 8.11.

same sign as the torsional moment diagram because it is acting on a positive face. Similarly, the magnitude of the column moments on joint A is $k_{col}\phi_A$, where k_{col} is the bending stiffness of one column. The sign of the moment for the top column is *positive* because this moment is acting on the *positive* face of the joint adjacent to a *positive* moment diagram. The sign of the moment for the bottom column is also *positive* because this moment is acting on the *negative* face of the joint adjacent to a *negative* moment diagram.

The moment diagrams caused by ϕ_B and ϕ_C are shown in Fig. 8.22 (c) and (d), respectively. The torsional moment diagrams created by the external loading are given in Fig. 8.22 (e). The fixed end moments in this figure are calculated as follows:

$$FEM_{AB} = FEM_{BA} = \frac{2.4(20)}{2} = 24 \text{ k-ft} \tag{a}$$

$$FEM_{BC} = 30\left(\frac{16}{8+16}\right) = 20 \text{ k-ft} \qquad \text{(see Fig. 8.18)} \tag{b}$$

$$FEM_{CB} = 30\left(\frac{8}{8+16}\right) = 10 \text{ k-ft} \qquad \text{(see Fig. 8.18)} \tag{c}$$

To solve the three unknowns, ϕ_A, ϕ_B, and ϕ_C, we can write three moment equilibrium equations about the z-axis, one for each of the three joints:

$$(\Sigma M_z)_A = 0 \qquad (2k_{col} + k_{tAB})\,\phi_A - k_{tAB}\phi_B - 24 = 0 \tag{d}$$

$$(\Sigma M_z)_B = 0$$
$$-k_{tAB}\phi_A + (2k_{col} + k_{tAB} + k_{tBC})\phi_B - k_{tBC}\phi_C - (24+20) = 0 \tag{e}$$

$$(\Sigma M_z)_C = 0 \qquad -k_{tBC}\phi_B + (2k_{col} + k_{tBC})\phi_C - 10 = 0 \tag{f}$$

Equations (d), (e), and (f) can be expressed more conveniently in matrix form:

$$\begin{bmatrix} (2k_{col} + k_{tAB}) & -k_{tAB} & 0 \\ -k_{tAB} & (2k_{col} + k_{tAB} + k_{tBC}) & -k_{tBC} \\ 0 & -k_{tBC} & (2k_{col} + k_{tBC}) \end{bmatrix} \begin{bmatrix} \phi_A \\ \phi_B \\ \phi_C \end{bmatrix} = \begin{bmatrix} 24 \\ 44 \\ 10 \end{bmatrix} \tag{g}$$

All the stiffnesses will now be evaluated in terms of the flexural rigidity of the column, EI:

$$\text{Column} \quad I = \frac{(14)^4}{12} = 3{,}201 \text{ in.}^4 \tag{h}$$

$$\text{Beam} \quad C = 0.229 \, (12)^3(24) = 9{,}497 \text{ in.}^4 \tag{i}$$

$$\frac{E}{G} = 2(1 + 0.2) = 2.4 \tag{j}$$

$$\frac{EI}{GC} = \frac{2.4(3{,}201)}{9{,}497} = 0.809 \tag{k}$$

$$k_{col} = \frac{4EI}{10} = 0.4EI \tag{l}$$

$$k_{tAB} = \frac{GC}{20} = \frac{EI}{20(0.809)} = 0.0618EI \tag{m}$$

$$k_{tBC} = \frac{GC}{24} = \frac{EI}{24(0.809)} = 0.0515EI \tag{n}$$

Substituting Eqs. (l), (m), and (n) into Eq. (g) results in:

$$\begin{bmatrix} 0.8618 & -0.0618 & 0 \\ -0.0618 & 0.9133 & -0.0515 \\ 0 & -0.0515 & 0.8515 \end{bmatrix} \begin{bmatrix} EI\phi_A \\ EI\phi_B \\ EI\phi_C \end{bmatrix} = \begin{bmatrix} 24 \\ 44 \\ 10 \end{bmatrix} \tag{o}$$

The solution of Eq. (o) is:

$$EI\phi_A = 31.516 \tag{p}$$

$$EI\phi_B = 51.146 \tag{q}$$

$$EI\phi_C = 14.837 \tag{r}$$

Substituting Eqs. (l), (m), (n), (p), (q), and (r) into Fig. 8.22 (b), (c), and (d), the moment diagrams in these figures can be determined. Summing the moment diagrams in Fig. 8.22 (b), (c), (d), and (e), we obtain the final moment diagram in Fig. 8.22 (f). Comparing Fig. 8.22 (f) to Fig. 8.22 (e), it can be seen that the torsional moment diagrams for the beam in these two figures do not differ greatly. This observation implies that the small 14 in. by 14 in. columns have provided very large torsional restraint at the supports for the beam.

It should be emphasized that this stiffness method is extremely simple and straight forward for the solution of this highly statically indeterminate frame. The simplicity originated from the symmetry of the columns above and below the beam, enabling the number of unknowns at each joint to be reduced from three to one. If the columns are roughly the same above and below the beam but not identical, the assumption of symmetry can still be made to utilize this very simple method and to arrive at a good approximate answer.

8.5 MOMENT DISTRIBUTION METHODS

8.5.1 General Descriptions

The moment distribution method is a variation of the stiffness method discussed in Section 8.4. The reader is assumed to be familiar with the conventional moment distribution method of plane frames subjected to in-plane loadings. For such structures only the flexural stiffnesses of the members are involved. In this section we shall generalize the moment distribution method so that it will be applicable to space frames and out-of-plane loadings. For these types of structures and loadings, torsional stiffnesses of the members must also be included.

The moment distribution method and the stiffness method are the same in the following basic concepts. First, rotations and displacements at the joints are selected as unknowns. Second, the shears and moments on the joint are related to the unknowns by the stiffnesses (see Figs. 8.15 and 8.16). Third, the shears and moments on the joints are related to the external load by the predetermined formulas for fixed end moment (see Figs. 8.17 and 8.18). Fourth, the rotation and displacement unknowns are determined from the force and moment equilibrium conditions of the joints. The difference between the two methods lies in how the equilibrium conditions of the joints are met. In the stiffness method, these equilibrium conditions are met by writing equations of equilibrium and solving these simultaneous equations to find the unknowns explicitly. In the moment distribution method, however, the conditions of equilibrium are met by allowing the joints to rotate in successive sequence until the moments at each joint are in equilibrium. This procedure is called the relaxation process. In this method it is not necessary to know the magnitude of the unknowns at any stage of the relaxation process.

In the distribution of the unbalanced end moments among the members of a rotating joint, the distributed end moments are proportional to the stiffness of the members. These moments are "carried over" to the far ends of the members if they are assumed to be temporarily fixed. Therefore, it is necessary to generalize two concepts in the conventional moment distribution method:

1. *Distribution factor* (*D.F.*): This factor is defined as:

$$D.F. = \frac{\text{Stiffness of a member adjacent to a joint}}{\text{Stiffness of joint}} = \frac{k}{\Sigma k}$$

where the stiffness of a joint is the sum of all member stiffnesses at a joint. At each joint both the bending and torsional stiffnesses must be included.

2. *Carry-over factor* (*C.O.F.*): The carry-over factor is 0.5 for a prismatic flexural member with the far-end fixed. This can be deduced from the bending moment diagram in Fig. 8.15. The carry-over factor should be −1 for a prismatic torsional member with the far-end fixed. This can easily be seen from Fig. 8.23. The signs for both carry-over factors are derived from the joint sign convention discussed in Section 8.4.2(4).

The procedures for moment distribution can be summarized as follows:

1. Recognize the *n*-number of rotations and displacements at all the joints in a structure that are not zero under the external loadings.

2. *Unbalanced fixed end moments:* Fix temporarily all *n*-number of rotations and displacements at the joints, and draw all the bending and torsional

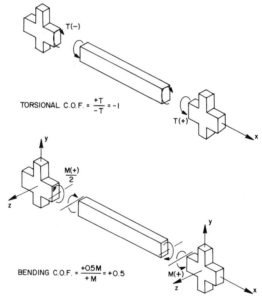

Fig. 8.23. Carry-over factors (joint sign convention).

moment diagrams due to the external loadings, resulting in unbalanced fixed end moments on the joints.

3. *Distributed moments:* Allow each joint rotation to occur individually such that moment equilibrium is restored one joint at a time. In this process, the unbalanced end moments at a joint are distributed among the members adjacent to the joint according to the distribution factors.

4. *Carry-over moments:* The distributed end moments at a joint are carried over to all its adjacent joints according to the carry-over factors. This process creates a new generation of smaller unbalanced end moments at the joints.

5. Repeat steps (3) and (4) until the unbalanced end moments at the joints are small enough to be neglected.

6. Sum all the end moments on a member including the fixed end moments, the distributed moments, and the carry-over moments. This should be done for all the members in the frame.

7. Draw bending and torsional moment diagrams for all the members from the final end moments of each member. In converting an end moment on the joint to an end moment on the member, the sign must be changed if the end moment is acting on a negative face of the joint [see Section 8.4.2 (4)].

As compared to the stiffness method, the moment distribution method has the following advantage. The moment diagrams can be obtained directly from the final end moments without our knowing the magnitude of the unknown rotations and displacements, thus avoiding the solution of simultaneous equations to find the rotations and displacements and the conversion of these rotations and displacements back into the moment diagrams. However, the moment distribution method also has a very serious weakness. The complexity of the moment distribution process increases very rapidly as the number of unknown rotations and displacements increases. Therefore, this method is convenient only for frames with small numbers of unknowns. It should be considered as a special short-cut method for some particular cases of frames and loadings. The following three examples illustrate the moment distribution method.

8.5.2 Examples 8.12, 8.13, and 8.14

EXAMPLE 8.12

The frame and loadings given in Example 8.11 are shown again in Fig. 8.24 (a). This same problem will now be solved by the moment distribution method. The discussions in Example 8.11 are still valid regarding the symme-

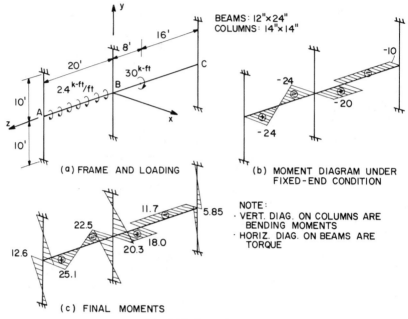

Fig. 8.24. Example 8.12.

try of frames and loadings and the recognition of three unknown rotations, ϕ_A, ϕ_B and ϕ_C, about the z-axis.

Before the process of moment distribution is made, we shall first evaluate the four fixed end moments, the required stiffnesses of the members, and the required distribution factors.

Fixed end moments:

$$FEM_{AB} = FEM_{BA} = 2.4\left(\frac{20}{2}\right) = 24\text{ k-ft}$$

$$FEM_{BC} = 30\left(\frac{16}{24}\right) = 20\text{ k-ft}$$

$$FEM_{CB} = 30\left(\frac{8}{24}\right) = 10\text{ k-ft}$$

The moment diagrams under fixed end conditions are shown in Fig. 8.24 (b).

Stiffness (about z-axis):

All columns: $\quad I = \dfrac{(14)^4}{12} = 3{,}201 \text{ in.}^4$

$$k_{\text{col}} = \frac{4EI}{10} = \frac{4E(3{,}201)}{10} = 1{,}280\ E$$

Beam AB: $\quad C_{AB} = C_{BA} = \beta x^3 y = 0.229(12)^3(24) = 9{,}497 \text{ in.}^4$

$$k_{tAB} = k_{tBA} = \frac{GC_{AB}}{20} = \frac{E(9{,}497)}{2.4(20)} = 197.9\ E$$

Beam BC: $\quad C_{BC} = C_{CB} = 9{,}497 \text{ in.}^4$

$$k_{tBC} = k_{tCB} = \frac{GC_{BC}}{24} = \frac{E(9{,}497)}{2.4(24)} = 164.9\ E$$

Distribution factors (about z-axis):

Joint A: $\quad \Sigma k_A = 2k_{\text{col}} + k_{tAB} = 2(1{,}280\ E) + 197.9\ E = 2{,}758\ E$

$$(D.F.)_{\text{col}} = \frac{2(1{,}280)}{2{,}758} = 0.928$$

$$(D.F.)_{AB} = \frac{197.9}{2{,}758} = 0.072$$

Joint B: $\quad \Sigma k_B = 2k_{\text{col}} + k_{tBA} + k_{tBC}$
$$= 2(1{,}280\ E) + (197.9\ E) + 164.9\ E = 2{,}924\ E$$

$$(D.F.)_{\text{col}} = \frac{2(1{,}280)}{2{,}924} = 0.876$$

$$(D.F.)_{BA} = \frac{197.9}{2{,}924} = 0.068$$

$$(D.F.)_{BC} = \frac{164.9}{2{,}924} = 0.056$$

Joint C: $\quad \Sigma k_C = 2k_{\text{col}} + k_{tCB} = 2(1{,}280\ E) + 164.9\ E = 2{,}726\ E$

$$(D.F.)_{\text{col}} = \frac{2(1{,}280)}{2{,}726} = 0.940$$

$$(D.F.)_{CB} = \frac{164.9}{2{,}726} = 0.060$$

The process of moment distribution can be tabulated in Table 8.2. The first line indicates the joints, and the second line gives the members adjacent to the joints. The third line records the distribution factors for the end moments of the members, while the fourth line registers the unbalanced fixed end moments. With all these preliminaries completed, the actual process of moment distribution can begin.

Table 8.2   Moment distribution of example 8.12

1	Joints	A			B			C
2	Members	2 Col.	AB	BA	2 Col.	BC	CB	2 Col.
3	D.F.	0.928	0.072	0.068	0.876	0.056	0.060	0.940
4	FEM		−24	−24		−20	−10	
5	D.M.	+22.3	+1.7	+3.0	+38.5	+2.5	+0.6	+9.4
6	C.O.M.		−3.0	−1.7		−0.6	−2.5	
7	D.M.	+2.8	+0.2	+0.2	+2.0	+0.1	+0.2	+2.3
8	Sum	+25.1	−25.1	−22.5	+40.5	−18.0	−11.7	+11.7
9	One col.	+12.6			+20.3			+5.85

Line 5 shows the distribution of the unbalanced fixed end moment (or the sum of all fixed end moments) among the adjoining members when a joint is allowed to rotate. The distributed moments are calculated simply by multiplying the unbalanced fixed end moment by the distribution factors. Line 6 shows the carry-over of end moments from one joint to its neighbors. One must ascertain whether the moment to be carried over is the bending type or the torsion type, so that the appropriate carry-over factor is used. In this particular case, the torsional carry-over factor of −1 must be used, since both beams are transferring torsional moments. These carry-over moments become the second generation of unbalanced fixed end moments to be distributed again. Line 7 shows the second cycle of moment distribution. It can be seen that the distributed moments are quite small, and no further carry-over of moments is necessary. In line 8, all the moments at each particular end of a member are summed. It should be noticed that the distribution factor for the columns at a joint has been taken as twice the value for one column (to include the identical top and bottom columns). Therefore, the final end moment for each column should be half of that given in line 8. These values for one column are given in line 9. From the final end moments in lines 8 and 9, the torsional moment diagram for the beams and the bending moment diagram for the columns are plotted in Fig. 8.24 (c).

EXAMPLE 8.13

The same example problem used in Example 8.5 will now be solved by the moment distribution method. The frame and loading are shown again, in Fig. 8.25 (a). It can be seen that this plane frame is subjected to a perpendicular loading. Only three types of movement, ϕ_x, ϕ_z, and Δ_y, are possible at any section. Furthermore, both the frame and loading are symmetric about the y–z plane. The rotations and displacements at joint B will be mirror images of those at joint C. Consequently, only three unknowns, ϕ_x, ϕ_z, and Δ_y, at one joint need to be considered, and only half a frame will be plotted. The symmetry of ϕ_x and Δ_y has a special significance here. When

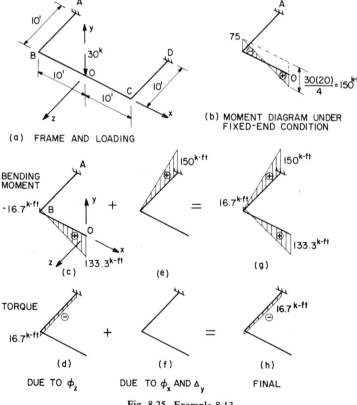

(a) FRAME AND LOADING

(b) MOMENT DIAGRAM UNDER FIXED-END CONDITION

BENDING MOMENT

(c)

(e)

(g)

TORQUE

(d)

(f)

(h)

DUE TO ϕ_z DUE TO ϕ_x AND Δ_y FINAL

Fig. 8.25. Example 8.13.

the rotations ϕ_x and the displacements Δ_y occur symmetrically at joints B and C, the frame acts like two statically determinate cantilevers (AB and DC) subjected to vertical loading, inducing no stresses in beam BC. Consequently, no moment distribution is necessary when ϕ_x and Δ_y are allowed to move. The only moment distribution required is the one associated with ϕ_z, the joint rotation about the z-axis.

Since moment distribution at the joints is required only about the z-axis, calculations will be made only for those fixed end moments, stiffnesses, and distribution factors that are related to the joint rotation ϕ_z.

$$FEM_{BC} = FEM_{CB} = \frac{30(20)}{8} = 75 \text{ k-ft}$$

The moment diagram corresponding to this fixed end moment is shown in Fig. 8.25 (b).

$$I_{BC} = \frac{10(20)^3}{12} = 6,667 \text{ in.}^4$$

$$k_{BC} = \frac{2EI_{BC}}{l_{BC}} = \frac{2E(6,667)}{20} = 667 \, E$$

$$C_{BA} = 0.229(10)^3(20) = 4,580 \text{ in.}^4$$

$$\frac{E}{G} = 2(1 + 0.2) = 2.4$$

$$k_{tBA} = \frac{GC_{BA}}{l_{BA}} = \frac{E(4,580)}{2.4(10)} = 191 \, E$$

$$(D.F.)_{BC} = \frac{k_{BC}}{k_{BC} + k_{tBA}} = \frac{667E}{667E + 191E} = 0.777$$

$$(D.F.)_{BA} = \frac{k_{tBA}}{k_{BC} + k_{tBA}} = \frac{191E}{667E + 191E} = 0.223$$

Notice that $k_{BC} = 2EI_{BC}/l_{BC}$ is the flexural stiffness for symmetric end moments given in Fig. 8.15. Using this stiffness, no carry-over of moments is necessary between joints B and C. The moment distribution process becomes extremely simple, as shown in Table 8.3.

The bending and torsional moment diagrams calculated from the end moments in Table 8.3 are plotted in Fig. 8.25 (c) and (d), respectively. The moment at midspan of beam BC is:

$$M_{\pounds BC} = \frac{30(20)}{4} - 16.7 = 133.3 \text{ k-ft}$$

The moment diagram produced by ϕ_x and Δ_y at the joint is shown in Fig. 8.25 (e). It is obtained by placing a 15-kip load (which is one-half of

Table 8.3 Moment distribution in Example 8.13

Joints	A	B	
Members	AB	BA	BC
D.F.		0.223	0.777
FEM D.M. C.O.M.	−16.7	+16.7	−75.0 +58.3
Sum	−16.7	+16.7	−16.7

the 30-kip external load due to symmetry) at joint B. No torsional moments
can be induced by ϕ_x and Δ_y.

Summing the bending moments in Fig. 8.25 (c) and (e) gives the final
bending moment in Fig. 8.25 (g). The final torsional moment diagram is
shown in Fig. 8.25 (h), which is the same as the torsional moment diagram
in Fig. 8.25 (d) due to ϕ_z alone.

EXAMPLE 8.14

The example problem shown in Fig. 8.26 (a) was solved by the flexibility
method in Example 8.7 and by the stiffness method in Example 8.9. It will
now be solved by the moment distribution method. As discussed in Example
8.9, this plane frame ABC is subjected to perpendicular loading so that only
two rotations, ϕ_x and ϕ_y, and one displacement, Δ_z, are possible at joint
B. The purpose of this example problem is primarily to illustrate the treatment
of the unknown displacement, Δ_z.

The procedure of the moment distribution method with lateral displacement
Δ_z can be divided into three steps.

Step 1. Assume that the displacement of the joint is prevented (i.e.,
$\Delta_z = 0$.). Find the fixed end moments due to external loads and make moment
distribution for ϕ_x and ϕ_y. The resulting moment diagrams determine the
shears in the frame and the forces at the joint required to prevent the displacement, Δ_z.

The fixed end torsional moments due to the 100 k-ft torque are:

$$FEM_{BA} = 100\left(\frac{8}{8+4}\right) = 66.7 \text{ k-ft}$$

$$FEM_{AB} = 100\left(\frac{4}{8+4}\right) = 33.3 \text{ k-ft}$$

These *FEM*s are plotted in Fig. 8.26 (b). Since these *FEM*s are about the x-axis, moment distribution is necessary only for joint rotation ϕ_x. The stiffnesses and distribution factors for this moment distribution about the x-axis are:

For square section: $\quad \dfrac{EI}{GC} = 1.418$

$$K_{tBA} = \frac{GC}{12} = \frac{EI}{1.418(12)} = 0.0587EI$$

$$K_{BC} = \frac{4EI}{10} = 0.4EI$$

$$(D.F.)_{BA} = \frac{0.0587}{0.0587 + 0.4} = 0.128$$

$$(D.F.)_{BC} = \frac{0.4}{0.0587 + 0.4} = 0.872$$

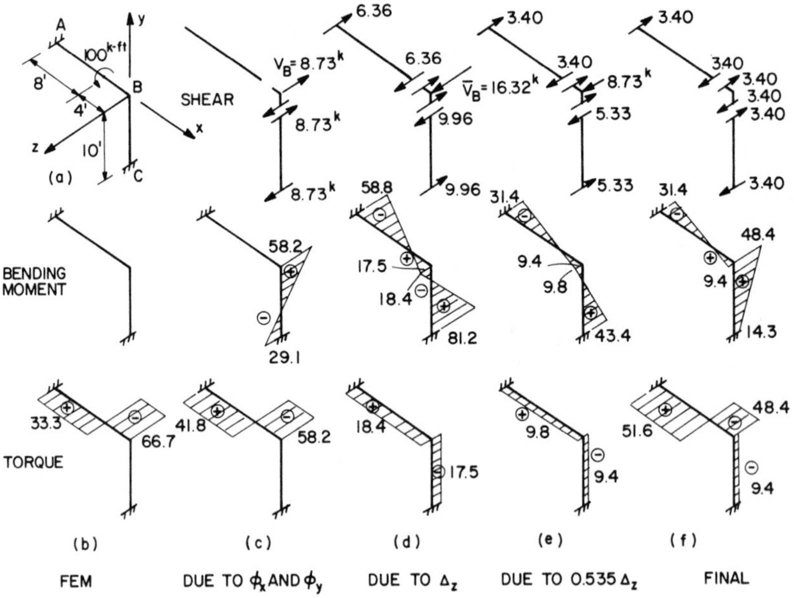

Fig. 8.26. Example 8.14.

The moment distribution for ϕ_x is performed simply in Table 8.4.

Table 8.4 Moment distribution for ϕ_x

Joints	A	B		C
Members	$AB(T_x)$	$BA(T_x)$	$BC(M_x)$	$CB(M_x)$
D.F.		0.128	0.872	
FEM D.M. C.O.M.	+33.3 +8.5	+66.7 −8.5	−58.2	 −29.1
Sum	+41.8	+58.2	−58.2	−29.1

The bending and torsional moment diagrams based on the end moments in Table 8.4 are plotted in Fig. 8.26 (c). Notice that both the bending and torsional moment diagrams adjacent to joint B have signs opposite to those shown in Table 8.4 because both faces of the joint are negative according to the given coordinate system. Figure 8.26 (c) includes the end shears of the members and the forces V_B at joint B required to prevent the displacement Δ_z. V_B is easily calculated from statics:

$$V_B = \frac{58.2 + 29.1}{10} = 8.73 \text{ k}$$

Step 2. Allow the displacement Δ_z to move. Some convenient fixed end moments can be assumed to be caused by the unknown Δ_z. Moment distribution about ϕ_x and ϕ_y will then be made to distribute these new fixed end moments. The resulting moment diagrams determine the end shears of the members and the force at the joint required to produce the moment diagrams.

The fixed end bending moments due to Δ_z are selected to be 100 k-ft for member BA about the y-axis and 144 k-ft for member BC about the x-axis. The ratio of these two *FEM*s should be the same as the ratio of the two member stiffnesses.

$$\frac{k_{BC}}{k_{BA}} = \frac{\left(\dfrac{EI}{10^2}\right)}{\left(\dfrac{EI}{12^2}\right)} = \frac{144}{100}$$

The *FEM*s should be distributed for the rotation about the x-axis, ϕ_x, as well as the rotation about the y-axis, ϕ_y. The moment distribution for ϕ_x is performed in Table 8.5.

Table 8.5 Moment distribution for ϕ_x due to Δ_z

Joints	A	B		C
Members	$AB(T_x)$	$BA(T_x)$	$BC(M_x)$	$CB(M_x)$
D.F.		0.128	0.872	
FEM			+144	+144
D.M.		−18.4	−125.6	
C.O.M.	+18.4			−62.8
Sum	+18.4	−18.4	+18.4	+81.2

The moment distribution for ϕ_y is performed next in Table 8.6, but this is preceded by the calculation of the stiffnesses and distribution factors for moment distribution about the y-axis:

$$k_{BA} = \frac{4EI}{12} = 0.333\,EI$$

$$k_{tBC} = \frac{GC}{10} = \frac{EI}{1.418(10)} = 0.0705\,EI$$

$$(D.F.)_{BA} = \frac{0.333}{0.333 + 0.0705} = 0.825$$

$$(D.F.)_{BC} = \frac{0.0705}{0.333 + 0.0705} = 0.175$$

Table 8.6 Moment distribution for ϕ_y due to Δ_z

Joints	A	B		C
Members	$AB(M_y)$	$BA(M_y)$	$BC(T_y)$	$CB(T_y)$
D.F.		0.825	0.175	
FEM D.M. C.O.M.	−100 +41.2	−100 +82.5	+17.5 −17.5	
Sum	−58.8	−17.5	+17.5	−17.5

Based on the end moments in Tables 8.5 and 8.6, the bending and torsional moment diagrams are plotted in Fig. 8.26 (d). Also shown are the end shears of the members and the force $\overline{V}_B$ required to produce these moment diagrams:

$$\overline{V}_B = \frac{58.8 + 17.5}{12} + \frac{18.4 + 81.2}{10} = 6.36 + 9.96$$
$$= 16.32 \text{ k}$$

Step 3. Proportion the moment diagrams and the end shears due to Δ_z such that $\overline{V}_B = V_B$. To accomplish this we calculate the scale ratio:

$$\frac{V_B}{\overline{V}_B} = \frac{8.73}{16.32} = 0.535$$

Multiplying all the moment diagrams and end shears in Fig. 8.26 (d) by this scale ratio (0.535), we obtain the moment diagrams and end shears in Fig. 8.26 (e). Summing the moment diagrams and end shears in Fig. 8.26 (e) with those in Fig. 8.26 (c) we arrive at the final moment diagrams and end shears in Fig. 8.26 (f).

This example problem illustrates the complexity of the moment distribution method when an unknown displacement is present and is closely interacting with the rotations at the joint. Therefore, the moment distribution method is seldom used in problems with more than one unknown displacement.

Plate 9: Torsional cracking near the support of a spandrel beam in a reinforced concrete parking garage in South Florida, 1964.

9
Spandrel Beams

9.1 INTRODUCTION

A spandrel beam in a reinforced concrete space frame is an exterior beam that carries slab, joists, and/or beams on one side only and is supported by the exterior columns. Typical spandrel beams are shown in Fig. 9.1 (a). Such beams are subjected to large torsional moment in addition to bending moments and shear forces. In modern terminology, a spandrel beam is also loosely used to mean any edge beam that is subjected to large torsional moments as a result of carrying slab, joists, and/or beams from one side. It is not necessary that a spandrel beam be located in the exterior surface of a structure. This latter, more general, terminology is implied in this book.

Before cracking, a reinforced concrete structure behaves reasonably elastically. The forces and moments obtained from an elastic analysis based on uncracked sections are permitted by most building codes for the design of the member components. For the design of spandrel beams such an elastic method has also been implied in the ACI Codes prior to 1977.

Design of spandrel beams based on elastic analysis, however, has often been found to require large torsional moments and, consequently, a large amount of torsional reinforcement. Such excessive and uneconomical design of torsional reinforcement is not supported by experience and observation. Recent tests[4-9] have shown that the large torsional moment in the spandrel beam may be dissipated by redistribution of moments after torsional cracking of the spandrel beam. Utilizing this inelastic behavior of a reinforced concrete structure, a new torsional limit design method has been incorporated into the 1977 and 1983 ACI Codes.

Torsional limit design of a member is possible only if the structure is statically indeterminate such that the torsional moment in the member can be redistributed to the adjacent members. To identify such structures the 1977 or 1983 ACI Code Commentary states:

11.6.2 and 11.6.3—In designing for torsion in reinforced concrete structures, two conditions may be distinguished:

(a) The torsional moment cannot be reduced by redistribution of internal forces (Section 11.6.2). This is referred to as "equilibrium torsion," since the torsional moment is required for the structure to be in equilibrium.

(b) The torsional moment can be reduced by redistribution of internal forces after cracking (Section 11.6.3). This will occur, for example, in a spandrel beam if the torsion arises from the member twisting in order to maintain compatibility of deformations. This type of torsion is referred to as "compatibility torsion."

Example drawings to illustrate these two catagories of structures are also given in Figs. 11–6 (a) and (b) of the Commentary. For condition (a), torsional

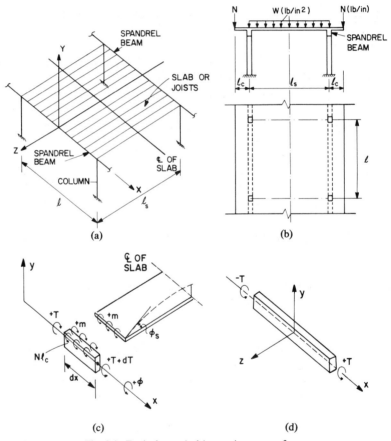

Fig. 9.1. Typical spandrel beams in a space frame.

reinforcement must be provided to resist the total design torsional moments, normally based on elastic analysis. For condition (b), the maximum factored torsional moments could be reduced to $\phi(4\sqrt{f_c'}\ \Sigma x^2 y/3)$ based on the torsional limit design method.

It should be pointed out that the use of the two terms "equilibrium torsion" and "compatibility torsion" is not precise because both equilibrium and compatibility conditions are involved in the analysis of either condition (a) or condition (b). One should not be misled to believe that only equilibrium conditions are used in condition (a) or that only compatibility conditions are utilized in condition (b). Another common misunderstanding is to assume that "equilibrium torsion" occurs only in statically determinate structures, while "compatibility torsion" occurs only in statically indeterminate structures. Although the latter statement is true, the former is not, because "equilibrium torsion" could occur in a statically indeterminate structure. In view of all these ambiguities, the author prefers to avoid these two terms.

In this chapter, we will first introduce the elastic analysis of spandrel beams, in Section 9.2. Then the torsional limit design method will be studied in Section 9.3. Finally a design example is given in Section 9.4.

9.2 Elastic Analysis

The purpose of studying elastic analysis of spandrel beams is threefold. First, elastic analysis describes correctly the torsional behavior of a spandrel beam before cracking. Second, the torsional moment may be used for design of torsional reinforcement if the design torque is less than the cracking torque. Third, if the design torsional moment is greater than the cracking torque, then the elastic analysis shows the extent of redistribution of moments required between the spandrel beam and the adjacent flexural members in the torsional limit design method.

The method for the elastic analysis of spandrel beams has been studied by several researchers.[1-3] The method of elastic analysis introduced herein for spandrel beams was developed by Saether and Prachand,[1] and is based on the following four assumptions:

1. The frame is symmetrical about the vertical plane through the center line of the slab and about the y–z plane as shown in Fig. 9.1 (a).
2. The slab is treated as a series of narrow individual beams spanning between spandrel beams. In other words, the twisting resistance of the slab is neglected.
3. The vertical deflection of the spandrel beam is small and therefore can be assumed to be zero. When the beam twists, the rotation will also

produce a vertical displacement at the beam–slab interface. This vertical displacement will be restrained by the vertical stiffness of the slab. The effect of this vertical slab restraint on the spandrel beam will be neglected.

4. In the usual design, the center of twist of the spandrel beam will be below the horizontal centroidal axis of the slab. When the beam twists, the horizontal displacement of the top portion of the beam will be restrained by the large horizontal stiffness of the slab. The effect of this horizontal slab restraint on the spandrel beam will be neglected.

These four assumptions are established to simplify the mathematical derivation so as to highlight the basic concept of the transfer of torsional moment from the slab to the spandral beam. All the last three assumptions, 2, 3, and 4, will cause the analysis to veer toward the conservative side. The slab restraints on spandrel beams, which are neglected in Assumptions 3 and 4, have been studied by Salvadori[2] and Shoolbred and Holland.[3] It was found that the effect of the vertical slab restraint (Assumption 3) is small, but the effect of the horizontal slab restraint (Assumption 4) could be significant, especially for deep spandrel beams. Some economy could be obtained if the horizontal slab restraint were taken into account according to Refs. 2 and 3.

9.2.1 Interior Span

Let us first analyze an interior span of a continuous spandrel beam for a one-bay symmetrical frame, shown in Fig. 9.1 (a) and (b). The spandrel beam supports a slab that has a length l_s and is subjected to a load w. To be general, the spandrel beam also supports a cantilever slab with a line load N at the end. This line load N represents the weight of the exterior wall. The length of the cantilever is designated l_c.

A general description of the behavior of the spandrel beam subjected to torsional moment would be helpful before the derivation of equations. If the spandrel beam is infinitely rigid, the slab will deform in the same manner along the length of the spandrel beam. The moment per unit length, m, transferred from the slab to the spandrel beam should be uniform as shown in Fig. 9.2 (a), and the distribution of the torque, T, for the spandrel beam should be a straight line. However, since the spandrel beam does twist and the maximum torsional rotation does occur at the midspan, the transferred moment, m, will be distributed in a nonuniform fashion with a minimum at midspan, as shown in Fig. 9.2 (b). For such an m-diagram, the T-diagram will not be a straight line, but should increase rapidly at support as indicated in the figure. The following derivation will determine mathematically the shape of these T and m diagrams.

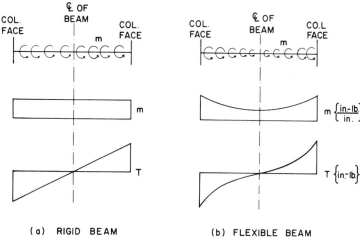

(a) RIGID BEAM   (b) FLEXIBLE BEAM
Fig. 9.2. T-diagram and m-diagram for spandrel beam.

A differential length dx of the spandrel beam is isolated in Fig. 9.1 (c), including m and T acting on it. The positive directions of m and T as well as the positive direction of the torsional rotation ϕ are also indicated. The sign convention for T and ϕ is based on the member sign convention explained in Section 8.3.2.(3). Taking the equilibrium of moment about the x-axis:

$$T + dT - T - mdx + N\ell_c dx = 0$$
$$\frac{dT}{dx} = m - N\ell_c \tag{9-1}$$

According to St. Venant's Eq. (1–60) the relationship between T and ϕ of the spandrel beam is:

$$T = GC\frac{d\phi}{dx} \tag{9-2}$$

where

G = shear modulus of elasticity
C = torsional constant of spandrel beam = $\beta x^3 y$ (β is given in Table 1.1)

Now consider a unit width of slab and designate the end rotation of the slab ϕ_s, with the positive direction as indicated in Fig. 9.1 (c). This rotation, ϕ_s, should be equal to the spandrel beam rotation ϕ owing to the compatability of rotations between the slab and the spandrel beam. The moment m induced by the load w and the rotation $\phi_s = \phi$ can be expressed by the following relationship:

$$m = \frac{K_s}{2}\phi + \frac{w l_s^2}{12} \qquad (9\text{--}3)$$

where $K_s = 4EI_s/l_s$ = flexural stiffness of slab per unit width (in.-lb/in.). I_s is the moment of inertia of the slab per unit width (in.^{3}), and l_s is the clear span of the slab (in.).

Equations (9–1), (9–2), and (9–3) are three equations for solving the three unknowns T, m, and ϕ. Substituting first the moment m from Eqs. (9–3) into (9–1), we have:

$$\frac{dT}{dx} = \frac{K_s}{2}\phi + \underbrace{\frac{w l_s^2}{12} - N l_c}_{M_F} \qquad (9\text{--}4)$$

where $w l_s^2/12 - N l_c$ is defined as the fixed end moment per unit length M_F. Then we can solve the two unknowns T and ϕ from the two simultaneous Eqs. (9–2) and (9–4). Differentiating Eq. (9–4) with respect to x, and eliminating $d\phi/dx$ from these two equations, we obtain:

$$\frac{d^2T}{dx^2} = \frac{K_s}{2GC}T \qquad (9\text{--}5)$$

Let $n = \sqrt{\dfrac{K_s}{2GC}}$; then:

$$\frac{d^2T}{dx^2} - n^2 T = 0 \qquad (9\text{--}5a)$$

The general solution of Eq. (9–5a) is:

$$T = A \sinh nx + B \cosh nx \qquad (9\text{--}6)$$

where A and B are constants to be determined by boundary conditions.

Before using the boundary conditions we recall the first assumption, that the spandrel beam is symmetrical about the y–z plane. In view of the anti-symmetrical nature of the torsional moment T with respect to the y–z plane (shown in Fig. 9.1d), the symmetrical function cosh nx must be eliminated in Eq. (9–6) by taking $B = 0$. Hence Eq. (9–6) becomes:

$$T = A \sinh nx \qquad (9\text{–}7)$$

Differentiating Eq. (9–7) with respect to x and substituting dT/dx into Eq. (9–1):

$$m = An \cosh nx + N\ell_c \qquad (9\text{–}8)$$

Equations (9–7) and (9–8) show that the T-diagram in Fig. 9.2 (b) is an antisymmetrical sinhnx function, and the m-diagram is a symmetrical coshnx function.

The unknown constants A in Eqs. (9–7) and (9–8) can be derived from the following boundary condition. Referring to Fig. 9.3 (a), the torsional rotation of the beam ϕ at $x = \ell/2$ must be equal to the rotation at the upper end of the column, ϕ_c. To find ϕ we differentiate Eq. (9–7) with respect to x and substitute dT/dx into Eq. (9–4). Hence:

$$\phi = \frac{2}{K_s} (An \cosh nx - M_F) \qquad (9\text{–}9)$$

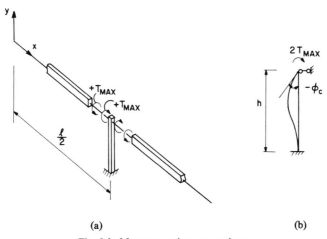

(a)   (b)

Fig. 9.3. Moments acting on a column.

At $x = \frac{l}{2}$:

$$\phi = \frac{2}{K_s}\left(An \cosh \frac{nl}{2} - M_F\right) \qquad (9\text{--}10)$$

To find ϕ_c, we notice from Fig. 9.3 (a) and (b) that an isolated column is subjected to two maximum torques, T_{max}, from both sides of the spandrel beam. These two maximum torques are taken to be equal if the frame is assumed to have an infinite number of identical spans. Therefore:

$$\phi_c = -\frac{2T_{max}}{K_c} \qquad (9\text{--}11)$$

where $K_c = 4EI_c/h$ = flexural stiffness of a column (in.-lb). I_c is the moment of inertia of a column (in.^{4}), and h is the height of the column (in.). The negative sign in Eq. (9–11) is necessary because a positive T_{max} on the column produces a ϕ_c that is opposite to the positive direction of ϕ defined in Fig. 9.1 (c). The maximum torque, T_{max}, occurs at $x = l/2$ and can be expressed by Eq. (9–7) as:

$$T_{max} = A \sinh \frac{nl}{2} \qquad (9\text{--}12)$$

Substituting T_{max} into Eq. (9–11) gives:

$$\phi_c = -\frac{2}{K_c}\left(A \sinh \frac{nl}{2}\right) \qquad (9\text{--}13)$$

Now the unknown constant A can be obtained by equating ϕ from Eq. (9–10) to ϕ_c from Eq. (9–13):

$$A = \frac{M_F}{n \cosh \dfrac{nl}{2} + \dfrac{K_s}{K_c} \sinh \dfrac{nl}{2}} \qquad (9\text{--}14)$$

Substituting A into Eq. (9–7):

$$T = \frac{M_F}{n \cosh \dfrac{n\mathscr{l}}{2} + \dfrac{K_s}{K_c} \sinh \dfrac{n\mathscr{l}}{2}} \sinh nx \qquad (9\text{-}15)$$

In Eq. (9–15), both n and K_s/K_c have units of 1/in., and M_F has units of in.-lb/in.

The maximum torque, $T_{\max}$, occurs at $x = \mathscr{l}/2$:

$$T_{\max} = \frac{\sinh \dfrac{n\mathscr{l}}{2}}{n \cosh \dfrac{n\mathscr{l}}{2} + \dfrac{K_s}{K_c} \sinh \dfrac{n\mathscr{l}}{2}} M_F \qquad (9\text{-}16)$$

To simplify the calculation of Eq. (9–16), a diagram is provided using the following two nondimensional parameters:

$$P = n\mathscr{l} = \mathscr{l} \sqrt{\frac{K_s}{2GC}} = \sqrt{\frac{2EI_s}{\mathscr{l}_s GC}}$$

$$Q = \frac{K_s}{K_c} \mathscr{l} = \frac{\mathscr{l} I_s h}{\mathscr{l}_s I_c}$$

Then:

$$T_{\max} = C_1 \frac{\mathscr{l}}{2} M_F \qquad (9\text{-}17)$$

where:

$$C_1 = \frac{\sinh \dfrac{P}{2}}{\dfrac{P}{2} \cosh \dfrac{P}{2} + \dfrac{Q}{2} \sinh \dfrac{P}{2}} \qquad (9\text{-}18)$$

In Eq. (9–18) C_1 is shown to be a function of P and Q. P is the square root of the ratio of the flexural stiffness of the slab to the torsional stiffness of a spandrel beam, while Q is the ratio of the flexural stiffness of the slab to the flexural stiffness of a column. C_1 is plotted in Fig. A1 in Appendix A. It can be seen that C_1 increases when P or Q decreases. This means

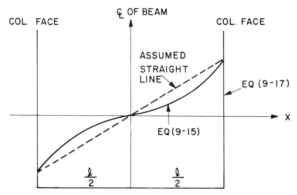

Fig. 9.4. Torque diagram for interior span.

that, for a constant slab, the torque in the spandrel beam increases when the torsional rigidity of the spandrel beam or the flexural stiffness of the column increases. This tendency explains why large torsional moments frequently occur in the spandrel beams at the lower levels of high-rise buildings. At these locations, the columns are very stiff, and the spans of the slabs or joists are large.

In practical design, it is only necessary to calculate T_{max} at the end of a spandrel beam using Eq. (9–17) and Fig. A1. The torque diagram may be assumed conservatively to be a straight-line distribution from a maximum at support to zero at the midspan (see Fig. 9.4).

9.2.2 Exterior Span

The study of an exterior span of a continuous spandrel beam will be focused on a three-span and one-bay symmetrical frame as shown in Fig. 9.5. It is believed that the behavior of an exterior span in a three-span spandrel beam is typical and representative. For this frame a coordinate system (x_1, y_1, z_1) is imposed on the left exterior span, and another coordinate system (x_2, y_2, z_2) is imposed on the interior span. All the forces and moments in connection with the exterior span will have a subscript of 1, and those related to the interior span will have a subscript of 2. The exterior column will be labeled column a, while the interior column will be column b. The torsional rigidity of the beam GC and the flexural stiffness of the slab K_s will be assumed constant throughout the structure.

For the exterior span we can rewrite Eq. (9–5) as:

$$\frac{d^2 T_1}{dx_1^2} = \frac{K_s}{2GC} T_1 \qquad (9\text{–}19)$$

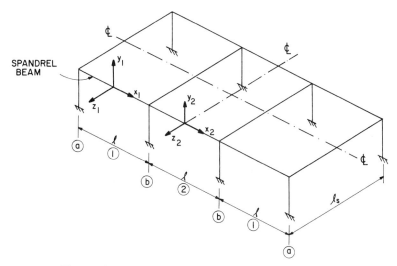

Fig. 9.5. Spandrel beam in a three-span, one-bay space frame.

where T_1 and x_1 refer to the exterior span of the spandrel beam. The general solution of Eq. (9–19) can be written in the form of an exponential function:

$$T_1 = A_1 e^{nx_1} + B_1 e^{-nx_1} \qquad (9\text{-}20)$$

where $n = \sqrt{k_s/2GC}$. The constants A_1 and B_1 will be determined from the following two boundary conditions:

$$\text{At } x_1 = -\frac{\ell}{2}: \qquad \phi_1 = \phi_{ca} \qquad (9\text{-}21)$$

$$\text{At } x_1 = +\frac{\ell}{2}: \qquad \phi_1 = \phi_{cb} \qquad (9\text{-}22)$$

In Eqs. (9–21) and (9–22), ϕ_1 is the torsional rotation of the spandrel beam in the exterior span. ϕ_{ca} and ϕ_{cb} are the flexural rotations at the top ends of columns a and b, respectively.

To find ϕ_1 we differentiate Eq. (9–20):

$$\frac{dT_1}{dx_1} = A_1 n e^{nx_1} - B_1 n e^{-nx_1} \qquad (9\text{-}23)$$

Substituting dT_1/dx_1 for dT/dx in Eq. (9–4), we have:

$$\phi_1 = \frac{2}{K_s} [A_1 n e^{nx_1} - B_1 n e^{-nx_1} - M_F] \tag{9-24}$$

For the interior span of the spandrel beam, Eq. (9–5) can be rewritten as:

$$\frac{d^2 T_2}{dx_2^2} = \frac{K_s}{2GC} T_2 \tag{9-25}$$

where T_2 and x_2 refer to the interior span of the spandrel beam. The general solution of Eq. (9–25) is:

$$T_2 = A_2 e^{nx_2} + B_2 e^{-nx_2} \tag{9-26}$$

where $n = \sqrt{k_s/2GC}$. Since T_2 must be antisymmetrical, $B_2 = -A_2$ and:

$$T_2 = A_2(e^{nx_2} - e^{-nx_2}) \tag{9-27}$$

The constant A_2 can be determined from the boundary condition:

$$\text{At } x_2 = -\frac{l}{2}: \qquad \phi_2 = \phi_{cb} \tag{9-28}$$

In Eq. (9–28) ϕ_2 is the torsional rotation of the spandrel beam in the interior span. To find ϕ_2 we differentiate Eq. (9–27):

$$\frac{dT_2}{dx_2} = A_2 n(e^{nx_2} + e^{-nx_2}) \tag{9-29}$$

Substituting dT_2/dx_2 for dT/dx in Eq. (9–4) results in:

$$\phi_2 = \frac{2}{K_s} [A_2 n(e^{nx_2} + e^{-nx_2}) - M_F] \tag{9-30}$$

The rotations ϕ_{ca} and ϕ_{cb} at the top of columns a and b, respectively, can also be expressed in terms of the unknown constants A_1, B_1, and A_2. Introducing T_1 from Eq. (9–20) and noticing $x_1 = -l/2$:

$$\phi_{ca} = \frac{T_1}{K_{ca}} = \frac{1}{K_{ca}} (A_1 e^{-nl/2} + B_1 e^{nl/2}) \tag{9-31}$$

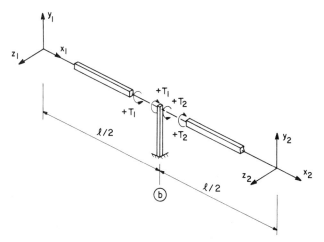

Fig. 9.6. Moments acting on an interior column.

where $K_{ca} = 4EI_{ca}/h$ = flexural stiffness of exterior column a. The term I_{ca} is the moment of inertia of column a, and h is the height of the column. Similarly, introducing T_1 and T_2 from Eqs. (9–20) and (9–27) and noticing $x_1 = l/2$ and $x_2 = -l/2$:

$$\phi_{cb} = \frac{(-T_1 + T_2)}{K_{cb}} = \frac{1}{K_{cb}} [-A_1 e^{nl/2} - B_1 e^{-nl/2}$$

$$+ A_2(e^{-nl/2} - e^{nl/2})] \quad (9\text{–}32)$$

where $K_{cb} = 4EI_{cb}/h$ = flexural stiffness of interior column b. The term I_{cb} is the moment of inertia of column b. The negative sign before T_1 in Eq. (9–32) is required because a positive T_1 on column b will produce a negative rotation ϕ at the top of the column. This member sign convention is illustrated in Fig. 9.6.

The three boundary conditions in Eqs. (9–21), (9–22), and (9–28) can now be expressed in terms of the unknown constants A_1, B_1, and A_2:

Substitute Eqs. (9–24) and (9–31) into Eq. (9–21):

$$A_1 e^{-nl/2} \left(n - \frac{K_s}{2K_{ca}} \right) - B_1 e^{nl/2} \left(n + \frac{K_s}{2K_{ca}} \right) = M_F \quad (9\text{–}33)$$

Substitute Eqs. (9–24) and (9–32) into Eq. (9–22):

$$A_1 e^{n l/2} \left(n + \frac{K_s}{2K_{cb}} \right) - B_1 e^{-n l/2} \left(n - \frac{K_s}{2K_{cb}} \right)$$

$$+ A_2 (e^{n l/2} - e^{-n l/2}) \frac{K_s}{2K_{cb}} = M_F \qquad (9\text{–}34)$$

Substitute Eqs. (9–30) and (9–32) in Eq. (9–28):

$$A_1 e^{n l/2} \frac{K_s}{2K_{cb}} + B_1 e^{-n l/2} \frac{K_s}{2K_{cb}}$$

$$+ A_2 \left[n(e^{n l/2} + e^{-n l/2}) + \frac{K_s}{2K_{cb}} (e^{n l/2} - e^{-n l/2}) \right] = M_F \qquad (9\text{–}35)$$

The three unknown constants A_1, B_1, and A_2 can be solved by the three simultaneous Eqs. (9–33), (9–34), and (9–35). Substituting the two constants A_1 and B_1 into Eq. (9–20) will give the torsional moments T_1 for any section in the exterior span.

To simplify the above three equations, let:

$$P = n l = l \sqrt{\frac{K_s}{2GC}} = l \sqrt{\frac{2EI_s}{l_s GC}}$$

$$Q_a = l \frac{K_s}{K_{ca}} = l \frac{I_s h}{l_s I_{ca}}$$

$$Q_b = l \frac{K_s}{K_{cb}} = l \frac{I_s h}{l_s I_{cb}}$$

Then Eqs. (9–33), (9–34), and (9–35) become:

$$A_1 e^{-p/2} \left(\frac{P}{2} - \frac{Q_a}{4} \right) - B_1 e^{p/2} \left(\frac{P}{2} + \frac{Q_a}{4} \right) = \frac{l}{2} M_F \qquad (9\text{–}36)$$

$$A_1 e^{p/2} \left(\frac{P}{2} + \frac{Q_b}{4} \right) - B_1 e^{-p/2} \left(\frac{P}{2} - \frac{Q_b}{4} \right)$$

$$+ A_2 (e^{p/2} - e^{-p/2}) \frac{Q_b}{4} = \frac{l}{2} M_F \qquad (9\text{–}37)$$

$$A_1 e^{p/2} \frac{Q_b}{4} + B_1 e^{-p/2} \frac{Q_b}{4}$$

$$+ A_2 \left[(e^{p/2} - e^{-p/2}) \frac{P}{2} + (e^{p/2} - e^{-p/2}) \frac{Q_b}{4} \right] = \frac{l}{2} M_F \qquad (9\text{–}38)$$

Eqs. (9–36), (9–37), and (9–38) have been solved for three special cases:

Case 1: The interior columns b are twice as stiff as the exterior columns a, i.e., $Q = \frac{1}{2} Q_a = Q_b$. In this case the torque diagrams will be identical for exterior and interior spans.

Case 2: All the interior and exterior columns have equal stiffnesses, i.e., $Q = Q_a = Q_b$.

Case 3: The interior columns b are one-half as stiff as the exterior columns a, i.e., $Q = Q_a = \frac{1}{2} Q_b$.

For each of the three cases the torsional moments are given at the following three sections:

1. Exterior end of exterior span
2. Interior end of exterior span
3. Exterior end of first interior span

The torsional moments are calculated by the following formula:

$$T = C_i \frac{l}{2} M_F$$

where $C_i = C_1, C_2, \ldots C_7$, are functions of P and Q and are given by graphs listed below. These graphs are available in Appendix A.

	EXTERIOR END, EXTERIOR SPAN	INTERIOR END, EXTERIOR SPAN	EXTERIOR END, FIRST INTERIOR SPAN
Case 1	C_1 (Fig. A1)	C_1 (Fig. A1)	C_1 (Fig. A1)
Case 2	C_2 (Fig. A2)	C_3 (Fig. A3)	C_4 (Fig. A4)
Case 3	C_5 (Fig. A5)	C_6 (Fig. A6)	C_7 (Fig. A7)

When the stiffness ratio of exterior to interior columns falls in between Case 1 and Case 2, or between Case 2 and Case 3, an interpolation of the two cases can be used to yield an approximate solution. Observation of Figs. A1 through A7 shows that the absolute values of C_1 through C_7 always increase with decreasing Q. That is, the torque in the spandrel beam always increases when the column stiffness increases. The absolute values of C_1

through C_7 also increases with the decrease of P, except for C_3 and C_6 in the lower range of P (see Figs. A3 and A6).

9.3 LIMIT DESIGN METHOD

9.3.1 Advantages of the Limit Design Method

In Section 9.2 the elastic analysis for a spandrel beam supporting a floor slab in an indeterminate space frame was studied. Although an elastic analysis of spandrel beams based on an uncracked section can reasonably describe the behavior before cracking, this method of analysis is open to questions after torsional cracking. As shown in Fig. 3.13 of Chapter 3, the torsional stiffness of a reinforced concrete member drops significantly after cracking to a small fraction of its value before cracking. Such a reduction of torsional stiffness will cause a large redistribution of moments from the spandrel beams to the floor slab or joists. The results of this redistribution are the significant decrease of *torsional* moment in the spandrel beam and the corresponding increase of *flexural* moments in the floor slab or joists.

The redistribution of moments after torsional cracking can be utilized in design for great economic benefit because reinforcement is much more efficient in carrying *bending* moment than in carrying *torsional* moment. The relative efficiency of flexural reinforcement and torsional reinforcement can be obtained by equating the torsional strength of a reinforced concrete member, T_n, to the flexural strength, M_n.

The torsional strength of a rectangular reinforced concrete member T_n can be estimated by Rausch's theory according to Eq. (3–18):

$$T_n = 2\frac{x_1 y_1 A_t f_y}{s} \qquad (9\text{–}39)$$

Let ρ be the volume percentage of torsional stirrups, expressed by:

$$\rho = \frac{2A_t(x_1 + y_1)}{xys}$$

Then assuming equal volumes of longitudinal steel and stirrups (i.e., $m = 1$), the total volume percentage of torsional steel, ρ_t, including both longitudinal bars and stirrups, is two times ρ:

$$\rho_t = \frac{4A_t(x_1 + y_1)}{xys} \qquad (9\text{–}40)$$

Substituting A_t from Eq. (9–40) into Eq. (9–39), the torsional strength, T_n, can be expressed in terms of total percentage of steel, ρ_t:

$$T_n = \frac{x_1 y_1 xy}{2(x_1 + y_1)} \rho_t f_y \qquad (9\text{–}41)$$

The flexural strength, M_n, is given by the well-known equation for singly reinforced members:

$$M_n = A_s f_y jd \qquad (9\text{–}42)$$

where d is the effective depth in flexure and jd is the lever arm. Letting ρ_s be the total volume percentage of flexural steel with respect to the gross cross section, xy, the flexural strength can be written in terms of ρ_s:

$$M_n = xyjd\,\rho_s f_y \qquad (9\text{–}43)$$

Equating T_n and M_n from Eqs. (9–41) and (9–43), we can find the ratio of the percentage of torsional steel to the percentage of flexural steel:

$$\frac{\rho_t}{\rho_s} = \frac{2jd(x_1 + y_1)}{x_1 y_1} = 2\frac{(jd)}{y_1}\left(1 + \frac{y_1}{x_1}\right) \qquad (9\text{–}44)$$

In Eq. (9–44), jd/y can be taken approximately as unity. Therefore:

$$\frac{\rho_t}{\rho_s} = 2\left(1 + \frac{y_1}{x_1}\right) \qquad (9\text{–}45)$$

Equation (9–45) shows that $\rho_t/\rho_s = 4$ and 8 for $y_1/x_1 = 1$ and 3, respectively. This means that for normal cross sections with a height-to-width ratio from 1 to 3, the efficiency of flexural reinforcement is 4 to 8 times better than that of the torsional reinforcement.

The above comparison of the efficiency of torsional steel and flexural steel takes into account only the amount of materials required. It should be realized that half of the torsional steel consists of closed stirrups which are much more expensive to manufacture and to install than the longitudinal flexural steel. Therefore, in terms of cost, the torsional steel will be in excess of 4 to 8 times more expensive than the flexural steel.

In short, the redistribution of moments after torsional cracking can be utilized to achieve great economy. The torsional reinforcement in the spandrel beam could be greatly reduced with the addition of only a small amount of flexural steel to the floor slab or joists. This economic advantage was recognized by the 1977 ACI Code, which adopted for the first time the torsional limit design method.

Torsional limit design is not only more economical than the elastic analysis, but it is also considerably simpler to use. There is no need to perform computations as required in Section 9.2. All that is required is to design torsional reinforcement for a spandrel beam to resist a specified cracking torque at the critical section that will ensure torsional ductility and serviceability. This specified cracking torque is given as $\phi(4\sqrt{f'_c} \ \Sigma x^2 y/3)$ at a distance d from the support. The experimental justification for this cracking torque is given in the following sections.

9.3.2 Moment Redistribution

The redistribution of moments from a spandrel beam to floor systems after torsional cracking will be illustrated by a three-dimensional structural frame, shown in Fig. 9.7. The portion of the frame shown includes four columns, two spandrel beams, and three floor beams. The floor beam at the center is supported by the two spandrel beams. When a uniform load w is applied on the floor beam, it will produce a rotation at the ends that in turn induces a torsional moment in the spandrel beam. The interaction of the floor beam and the spandrel beam can be studied using the T-shape specimen indicated by heavy lines. This specimen includes a spandrel beam between two inflection points and a floor beam from the junction to an inflection point. The three

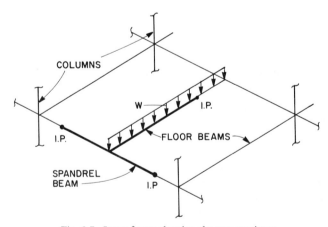

Fig. 9.7. Space frame showing the test specimen.

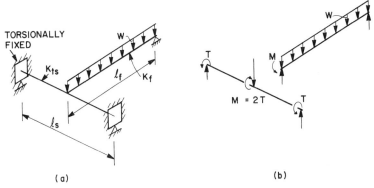

Fig. 9.8. Test specimen.

inflection points, indicated by solid dots, can be simulated by hinges in the test.

Figure 9.8 (a) shows the test specimen resting on three hinges and the floor beam loaded by a uniform load w. The ends of the spandrel beams are maintained torsionally fixed. This condition is more severe, and therefore more conservative, than that existing in the frame. It is adopted to simplify the analysis and the testing procedures.

The load w will create a negative bending moment at the continuous end of the floor beam due to the restraint of the spandrel beam. If we separate the floor beam from the spandrel beam as shown in Fig. 9.8 (b), this negative bending moment is designated as M acting on the end of the floor beam. The reaction of this moment M becomes a twisting moment acting at the midspan of the spandrel beam, thus creating a uniform torque T in the spandrel beam. The torque T is equal to $M/2$. The bending moment, shear, and torque diagrams are drawn in Fig. 9.9 from equilibrium conditions. It can be seen that all three diagrams can be plotted if the moment M is known.

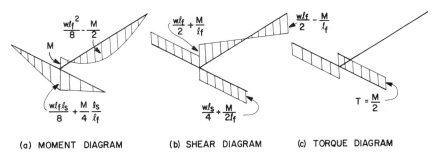

(a) MOMENT DIAGRAM   (b) SHEAR DIAGRAM   (c) TORQUE DIAGRAM

Fig. 9.9. Moment, shear, and torque diagrams of test specimen.

This moment, which is a bending moment on the floor beam and a twisting moment on the spandrel beam, will be called the joint moment.

We shall first derive the joint moment, M, based on elastic analysis. The flexural stiffness of the floor beam, K_f, and the torsional stiffness of the spandrel beam, K_{ts}, are defined as follows:

$$K_f = \frac{4EI}{\ell_f}$$

$$K_{ts} = \frac{GC}{\ell_s}$$

where

EI = flexural rigidity of the floor beam
GC = torsional rigidity of the spandrel beam
ℓ_f = length of the floor beam in the test specimen
ℓ_s = length of the spandrel beam in the test specimen

The joint moment, M, can be determined by the compatibility of rotation at the joint. The flexural rotation of the floor beam, ϕ_f, and the torsional rotation of the spandrel beam, ϕ_s, at the joint can be expressed as follows:

$$\phi_f = -\frac{M\ell_f}{3EI} + \frac{w\ell_f^3}{24EI} = -\frac{4M}{3K_f} + \frac{w\ell_f^2}{6K_f}$$

$$\phi_s = \frac{M\ell_s}{4GC} = \frac{M}{4K_{ts}}$$

Equating ϕ_f and ϕ_s:

$$M = \frac{\frac{1}{8}w\ell_f^2}{1 + \left(\frac{3}{16}\right)\dfrac{K_f}{K_{ts}}} \tag{9-46}$$

Notice that the joint moment M is a function of the stiffness ratio K_f/K_{ts}. If $K_f/K_{ts} = 0$, $M = (1/8)w\ell_f^2$. This is the special case in which the spandrel beam is infinitely rigid in torsion, and the joint moment becomes the fixed end moment. If $K_f/K_t = \infty$, $M = 0$. This gives the other extreme case, in which the spandrel beam has no torsional stiffness. The floor beam is then simply supported, and the joint moment must be zero.

It should be mentioned that an assumption was made in Eq. (9–46) that the vertical deflection of the spandrel beam is small and can be neglected. An exact analysis of this T-shape specimen, taking into account the vertical deflection, was demonstrated in Example 8.10 in Chapter 8. It was found that this refinement will increase the accuracy by only a few percent in normal cases. At the same time, the complexity involved will obscure the clear concept of the interaction of the floor beam and the spandrel beam given in Eq. (9–46). This refinement, therefore, is not adopted here.

The joint moment, M, calculated by Eq. (9–46) using the uncracked elastic stiffness properties, should be applicable to the specimen before cracking. After cracking, however, K_{ts} drops drastically to a small fraction of the value before cracking (say 5%), while K_f only drops to roughly 50% of the precracking value. Therefore the K_f/K_{ts} ratio increases in the order of ten times after cracking. According to Eq. (9–46), this change will cause a large decrease of the joint moment, M, compared to the uncracked elastic value. A decrease of M means that the torque in the spandrel beam is reduced, accompanied by a corresponding increase of flexural moment in the floor beam near the midspan. This phenomenon, which can be viewed as a redistribution of moments from the spandrel beam to the floor beam, is known as "moment redistribution after cracking."

Moment redistribution after cracking has been observed by several groups of investigators.[4-9] It can be utilized to achieve economy and is the basis of the torsional limit design method. This phenomenon will be illustrated by three T-shape test specimens, B1, B2, and B5, taken from series B in Ref. 4. A typical specimen is photographed in Fig. 9.10, subjected to the test condition described in Fig. 9.8 (a). The spandrel beam was 6 in. wide, 12 in. high, and 9 ft long, while the floor beam was 6 in. wide, 9 in. high, and 9 ft long. The top surface of the floor beam was level with the top surface of the spandrel beam.

The test program and ultimate loads for these three beams are summarized in Table 9.1. The specimens were all subjected to a load of 25.6 k uniformly distributed along the floor beam. The design torques, however, are different from each of the specimens. Specimen B1 was designed by elastic analysis based on an uncracked section. This method requires a nominal torsional stress, $\tau_n = 8.9\sqrt{f'_c}$, resulting in a high web reinforcement index, rf_y, of 475 psi. In contrast, specimens B2 and B5 were designed by the limit design method using nominal torsional stresses of $4.4\sqrt{f'_c}$ and $2.7\sqrt{f'_c}$, respectively. These stresses, in turn, give corresponding stirrup indices of 165 psi and 60 psi, respectively. The resulting reinforcement requirements are given in Table 9.2.

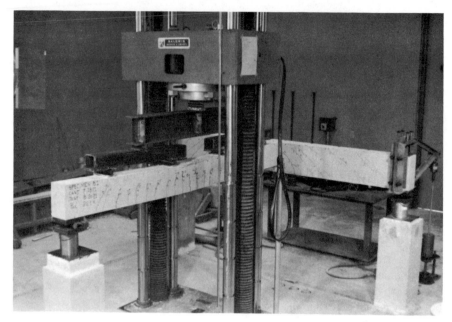

Fig. 9.10. Testing of a T-shape specimen.

Figure 9.11 shows the variation of torque with increase of load w for all three specimens. It can be seen that all three test specimens behave in a reasonably elastic manner up to cracking. Upon cracking, the torques remain essentially constant while the load is increased. This confirms the existence of the moment redistribution after cracking. This redistribution can also be observed in Fig. 9.12, showing the torque–twist relationship of the three test specimens. Before cracking the torque–twist curve is approximately linear. After cracking, however, the torque–twist curves reach a horizontal plateau, where the angle of twist increases under a constant torque. In other words, a plastic hinge is formed in the spandrel beam after cracking. When sufficient amounts of stirrups were provided as in specimens B1 and B2, the plastic hinge in the spandrel beam permitted the floor beam to bend until the longitudinal bars yielded near midspan.

In specimen B1, where an excessive amount of stirrups was provided, the plastic hinge in the spandrel beam not only allowed the floor beam to yield, but also allowed the torque again to increase rapidly under load (see Fig. 9.11). This phenomenon can be considered as a redistribution of moments from the floor beam to the spandrel beam and is known as the "second redistribution of moment after yielding." Although this second redistribution

Table 9.1 Test program and results of specimens B1, B2 and B5 (Ref. 4)

SPECIMEN	SPANDREL BEAM				MEASURED τ_n (psi)	DESIGN ULTIMATE LOAD (kips)	MEASURED ULTIMATE LOAD (kips)	MODE OF FAILURE
	METHOD OF DESIGN		$rf_y{}^*$ (psi)	τ_n (psi)				
B1	Elastic analysis (uncracked section)		475	$8.9\sqrt{f_c'}$	$6.9\sqrt{f_c'}$	25.6	26.1	Flexure
B2	Limit design		165	$4.4\sqrt{f_c'}$	$4.3\sqrt{f_c'}$	25.6	28.2	Flexure
B5	Limit design		60	$2.7\sqrt{f_c'}$	$2.3\sqrt{f_c'}$	25.6	24.5	Torsion

$^*rf_y$ = web reinforcement index, where r is the ratio of web reinforcement (A_w/bs) and f_y is the yield strength of web reinforcement.

Table 9.2   Details of specimens B1, B2 and B5

SPECI-MEN	REINFORCEMENT			f_y ksi	f'_c psi
	FLOOR BEAM, MIDSPAN	FLOOR BEAM, JOINT	SPANDREL BEAM		
B1	2 #2 #2 @ 4" 2 #5	2 #4 #2 @ 4" 2 #5	2 #3 #3 @ 5-3/8 2 #6	#2 50.0 #3 69.9	4380
B2	2 #2 #2 @ 4" 2 #5 + 1 #3	2 #3 #2 @ 4" 2 #5	2 #2 #2 @ 5" 2 #5	#2 50.0 #3 69.9	4120
B5	2 #2 #2 @ 4" 2 #5 + 1 #4	2 #2 #2 @ 4" 2 #5 + 1 #4	2 D1 D1 @ 4" 5 #3*	D1 72.0 #3 73.8	3610

Cross section of spandrel beam 6 in. x 12 in. Stirrup dimensions 5 in. x 11 in. o. to o. (½" cover)
Cross section of floor beam 6 in. x 9 in. Stirrup dimensions 5 in. x 8 in. o. to o. (½" cover)
Yield strength of steel: #6 − 60.0 ksi, #5 − 62.5 ksi, #4 − 80.0 ksi. Others indicated.
(1 ksi = 6.89 MN/m²; 1 psi = 6.89 kN/m²; 1 in. = 25.4 mm)
* 3 #3 bars are cut-off according to ACI requirements

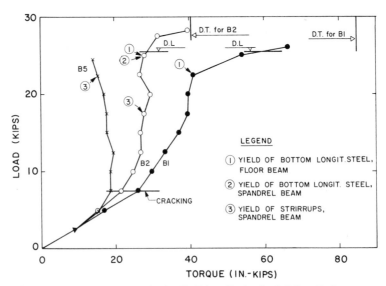

Fig. 9.11. Load-torque curves (series B). DL = Design load; DT = Design torque.

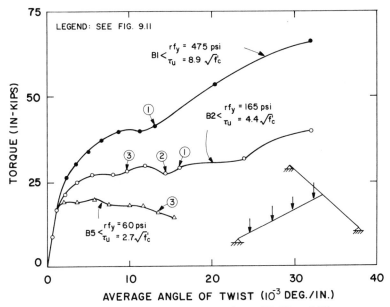

Fig. 9.12. Torque–twist curves, specimens B1, B2, and B5.

may allow us to utilize the ultimate torsional strength of the web reinforcement in the spandrel beam, Fig. 9.11 shows that the increase of load is small. This small gain in load also occurs at very large angles of twist, as indicated in Fig. 9.12. Such large deformation is undersirable from a serviceability point of view. In short, from an economical and practical point of view, this second redistribution of moment after yielding should not be utilized. The excessive amount of web reinforcement required by the elastic analysis is therefore wasteful and unnecessary.

In specimen B5, an insufficient amount of web reinforcement was provided. Figure 9.12 shows that torques decrease after cracking. Figure 9.13 also shows the brittle torsional failures that occurred in the spandrel beam before the yielding of longitudinal bars in the floor beams. Therefore, the ultimate design load cannot be reached (see Table 9.1). Such a design would be unsafe from the viewpoint of both strength and ductility.

Comparison of the three specimens shows clearly that specimen B2 has the most desirable design, assuming a torsional design stress τ_u of about $4\sqrt{f_c'}$. This specimen behaves in a ductile manner, can reach its design strength, and can satisfy the serviceability condition on crack width control. It is also very economical. Based on such tests, the 1977 and 1983 ACI

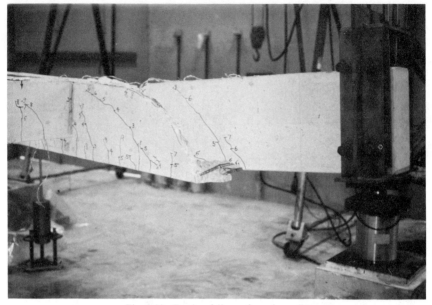

Fig. 9.13. Torsion failure of specimen B5.

Codes, Section 11.6.3, allows the design of spandrel beams using a maximum factored torque of $\phi(4\sqrt{f'_c}\ \Sigma x^2 y/3)$.

9.3.3 Interaction of Torsion and Shear

Figure 9.14 shows a reinforced concrete space frame, where three floor beams are framing into each span of a continuous spandrel beam. The T-shaped test specimen described in Section 9.3.2 and Fig. 9.8 (a) is indicated by heavy dotted lines. The spandrel beam in this T-shaped specimen represents the portion of the spandrel beam near the midspan where the flexural shear stresses are small. The flexural shear stresses in all three test specimens, B1, B2, and B5, are less than $2\sqrt{f'_c}$, the contribution of concrete. The critical sections, however, are near the column faces where large shear stresses as well as large torsional stresses occur simultaneously. Hence, the portion of the space frame that we shall now study is the U-shaped specimen consisting of the negative moment region of the spandrel beam and the two floor beams, as shown by the heavy solid lines.

The behavior of this U-shaped specimen, shown in Fig. 9.15 (a), can also be simulated by a T-shaped specimen similar to that given in Fig. 9.8 (a), but with an additional concentrated load at the midspan of the spandrel

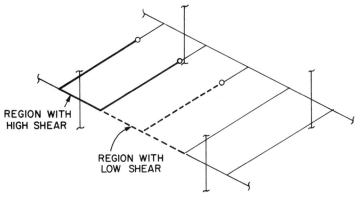

Fig. 9.14. High shear region of spandrel beam in frames.

beam. Such a T-shaped specimen is shown in Fig. 9.15 (b). In the U-shaped specimen the external load on the spandrel beam P_s and the uniform load on the floor beam w are acting at the ends of the spandrel beam. The reaction R, representing the difference between the axial loads of columns above and below, is acting at the midspan. In the T-shaped specimen, however, all the external loads P_s and w are acting at the midspan, while the reaction R is divided and is placed at the ends of the spandrel beam. This interchange

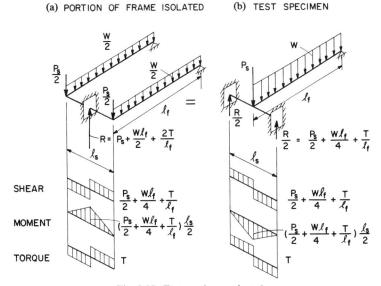

Fig. 9.15. Test specimen selected.

of load and reaction reverses all the directions of the bending moments, shear forces, and torsional moments in the spandrel beam. These reversals of forces are evident if we compare the shear, moment and torque diagrams in Fig. 9.15 (a) to those in Fig. 9.15 (b). It can be seen that the magnitude of these forces (or moments) and the relationship among them have not been changed. As far as the spandrel beam is concerned, the T-shaped specimen is identical to the U-shaped specimen.

Four T-shaped specimens with concentrated load P_s, taken from series C in Ref. 5, will be used to illustrate the interaction of high shear and high torque in spandrel beams. The test program and the main test results of these four specimens C1, C2, C3, and C4 are summarized in Table 9.3. The details of the specimens and the reinforcement are given in Table 9.4. As shown in Table 9.3, the design floor beam load was 25.6 kips for all four T-specimens C1 through C4. The spandrel beam loads were chosen to be 0, 14.0 kips, 38.9 kips, and 24.7 kips for specimens C1, C2, C3, and C4, respectively. As a result, the longitudinal reinforcement ratio of the spandrel beams varies from 0.5% to 1.88% and covers the typical reinforcement ratios of a continuous spandrel beam near the column faces.

Specimens C1 and C2 were designed by the limit design method. A design torsional stress of $4\sqrt{f_c'}$ was assumed for specimen C2 as adopted currently in the ACI Code. The design of shear and torsional reinforcement also followed the code method using the elliptical interaction curve of Eqs. (4–36) and (4–37). The torsional web reinforcement calculated was added to the shear web reinforcement, and the torsional longitudinal reinforcement was added to the flexural longitudinal steel. An identical method was also used for specimen C1, except that a larger design torsional stress of $6\sqrt{f_c'}$ was assumed. This larger stress was adopted in order to examine whether there was any improvement in the torsional ductility and the crack width control. The stirrup index of 261 psi in Specimen C1 was designed to be the same as that in specimen C2 by reducing the load P_s on the spandrel beam to zero.

Specimens C3 and C4 were designed by an alternate method based on a specified minimum amount of torsional web reinforcement for a spandrel beam. This minimum reinforcement will govern if it is greater than the web reinforcement required by flexural shear. On the other hand, if the required flexural shear is greater than this minimum torsional reinforcement, then the flexural shear reinforcement alone will govern. In other words, the alternate method permits us to calculate the torsional reinforcement and the shear reinforcement separately, and the greater of the two will govern. Specimen C3 was designed such that the flexural shear reinforcement governed, and the flexural shear alone required a stirrup index of 270 psi (close to those

Table 9.3   Test program and results of specimens C1, C2, C3 and C4 (Ref. 5)

| | TEST PROGRAM | | | | | TEST RESULTS | | | |
SPECIMEN	DESIGN TORSIONAL STRESS, τ_n (psi)	STIRRUP INDEX, rf_y (psi)	LONGIT. STEEL %	FLOOR BEAM LOAD (kips)	SPANDREL BEAM LOAD (kips)	FLOOR BEAM LOAD (kips)	SPANDREL BEAM LOAD (kips)	ULTIMATE REACTION FORCE, R (kips)	MODE OF FAILURE
C1	$6.0\sqrt{f'_c}$	261	0.50	25.6	0	32.5	0	17.1	F_f
C2	$4.0\sqrt{f'_c}$	261	0.94	25.6	14.0	32.0	17.8	34.5	$F_s + F_f$
C3	—	270	1.88	25.6	38.9	25.0	37.5	50.7	$(S + T)_s$
C4	—	142	1.27	25.6	24.7	24.4	19.8	31.6	$(S + T)_s$

F_f = flexural failure of floor beam (ductile).
F_s = flexural failure of spandrel beam (ductile).
$(S + T)_s$ = shear and torsion failure of spandrel beam (brittle).

Table 9.4   Details of specimens C1, C2, C3 and C4

SPECI-MEN	REINFORCEMENT			f_y (ksi)	f'_c (psi)
	FLOOR BEAM MIDSPAN	FLOOR BEAM JOINT	SPANDREL, OR SIMPLE BEAM		
C1	2#3 #2@4"* 2#5+1#4	2#3 #2@4"* 2#5	2#2 #2@4" 3#3	#2* – 80.3 #2 – 62.8 #3 – 60** #4 – 73.8 #5 – 64.9	4333
C2	2#3 #2@4"* 2#5+1#4	2#3 #2@4"* 2#5	2#2 #2@4" 2#5	#2* – 80.3 #2 – 62.8 #3 – 60** #4 – 73.8 #5 – 64.9	4875
C3	2#3 #2@4"* 2#5+1#4	2#3 #2@4"* 2#5	2#2 #2@4" 4#5	#2* – 81.1 #2 – 64.7 #3 – 60** #4 – 73.8 #5 – 60.9	4032
C4	2#3 #2@4"* 2#5+1#4	2#3 #2@4"* 2#5	2#2 2D2@4" 2#5+2#3	D2 Annealed Deformed Wire – 42.5 #2* – 84.2 #2 – 64.6 #3 – 60** #4 – 77.8 #5 – 69.4	4067

* #2 plain bar

** $f_y = 52.33$ ksi at $\epsilon_y = 2.4 \times 10^{-3}$ in./in.; $f_y = 63.8$ ksi at $\epsilon_y = 4.5 \times 10^{-3}$ in./in.

in specimens C1 and C2). In constrast, specimen C4 was designed such that the minimum torsional web reinforcement governed. This minimum torsional web reinforcement was assumed to have a stirrup index of 142 psi in this particular case.

The torque–twist curves for these four specimens are given in Fig. 9.16. Specimen C2 behaved as a specimen with optimum web reinforcement in the spandrel beam. After diagonal cracking in the spandrel beam, a torsional plastic hinge was formed. The large increase of the angle of twist allowed the moment to be redistributed from the spandrel beam to the floor beam. Failure was initiated by the yielding of stirrups and longitudinal steel in the spandrel beam (represented by the circled numbers 2 and 3 in Fig. 9.16). This was followed by the yielding of the bottom longitudinal steel near the midspan of the floor beam (represented by the circled number 1), and a ductile collapse of the specimen. Two observations are evident: (1) the development of the torsional hinges was accompanied by a slight decrease of torque after the yielding of the steel in the spandrel beam, and (2) the bottom longitudinal reinforcement in the floor beam just reached its yield strain at collapse. In view of these observations, it can be concluded that the web reinforcement of the spandrel beam designed by the ACI limit design method

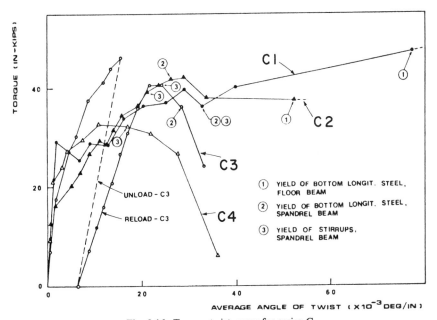

Fig. 9.16. Torque–twist curve for series C.

using a torionsal stress of $4\sqrt{f'_c}$ was just enough to provide the required ductility, and that the most economical design had been achieved.

Specimen C1 behaved as a specimen with excessive web reinforcement in the spandrel beam. The process of moment redistribution and the mode of failure were similar to those of specimen C2. However, the development of the torsional plastic hinge in the spandrel beam was accompanied by a continuous increase of torque, and some of the web reinforcement in the spandrel beam did not yield at failure. These observations indicate that the full capacity of the web reinforcement had not yet been reached. This spandrel beam behaved satisfactorily from the viewpoint of the limit design concept, but the assumed torsional stress of $6\sqrt{f'_c}$ is obviously uneconomical.

In contrast, specimens C3 and C4, which were designed by the alternate method, behaved as specimens with inadequate web reinforcement in the spandrel beam. Instead of a flexural failure in the floor beam, collapse of these specimens was caused by a premature shear–torsion failure of the spandrel beam. As shown in Fig. 9.16, the torques reached their maximum at the yielding of stirrups in the spandrel beam and then decreased rapidly. Since the angle of twist of the torsional plastic hinges was insufficient to allow a redistribution of moment from the spandrel beams to the floor beams, no yielding was observed in the bottom longitudinal reinforcement of the floor beams. These behaviors of specimens C3 and C4 mean that the web reinforcement designed by the alternate method was inadequate to ensure the development of torsional plastic hinges and to utilize fully the flexural capacity of the floor beam.

In terms of ultimate strength, Table 9.3 also shows that the test values for specimens C1 and C2 were about 25% higher than the design values. On the other hand, specimens C3 and C4 could not reach their design capacities. This again emphasizes that the ACI limit design criterion was satisfactory, whereas the alternate design method must be improved.

In view of these tests, we can conclude that the ACI limit design method using a torsional stress of $4\sqrt{f'_c}$ is applicable to high shear regions of a continuous spandrel beam near the columns. In these regions of high torsion and high shear, the torsional web reinforcement calculated from this specified torsional stress should be added to the web reinforcement required by flexural shear. The torsional longitudinal reinforcement should also be added to the longitudinal steel required by bending moments.

9.3.4 ACI Torsional Limit Design Method

A complete ACI limit design method is stated very concisely in Section 11.6.3 of the 1977 or 1983 ACI Codes as follows:

11.6.3—In a statically indeterminate structure where reduction of torsional moment in a member can occur due to redistribution of internal forces, maximum factored torsional moment T_u may be reduced to $\phi(4\sqrt{f'_c} \ \Sigma x^2 y/3)$.

11.6.3.1—In such a case the correspondingly adjusted moments and shears in adjoining members shall be used in design.

11.6.3.2—In lieu of more exact analysis, torsional loading from a slab shall be taken as uniformly distributed along the member.

Section 11.6.3.1 reminds us that when the ACI limit design method is used, the bending moments and shear forces in the floor beam must be calculated using a negative end moment at the joint, M, which is in equilibrium with the specified torque of $\phi(4\sqrt{f'_c} \ \Sigma x^2 y/3)$ at the torsional plastic hinges. These plastic hinges in the spandrel beam are considered to be located at the critical section a distance d from the faces of the columns.

In the case of a spandrel beam supporting a slab or very closely spaced joists, Section 11.6.3.2 of the ACI Code allows us to consider the torsional moment from the slab as uniformly distributed along the member. In other words, the torque in the spandrel beam can be assumed to vary from a maximum of $\phi(4\sqrt{f'_c} \ \Sigma x^2 y/3)$ at the critical section to zero at the midspan. The negative end moment of the slab would be $\phi(4\sqrt{f'_c} \ \Sigma x^2 y/3)$ divided by $l_s/2$, giving a unit of moment per unit length.

The ACI limit design method will be illustrated by an example problem in Section 9.4.

9.3.5 Alternate Design Method

An alternate method of spandrel beam design was also suggested by ACI Committee 438, Torsion (now merged into ACI/ASCE Committee 445, Shear and Torsion). This method prescribes a minimum torsional reinforcement including both web and longitudinal reinforcement. In view of the tests described in Section 9.3.3, this minimum torsional reinforcement should be added to those required by shear and flexure.

The minimum torsional web reinforcement is specified as:

$$A_{t,\,min} = \frac{250 s A_c}{f_y p_c} \tag{9-47}$$

where

$A_{t, min}$ = area (one leg) of minimum closed stirrups (in.^{2})
A_c = area enclosed by the outer perimeter of the concrete section (in.^{2})
p_c = outer perimeter of the concrete section (in.^{2})
s = spacing of stirrups (in.)
f_y = yield strength of reinforcement (psi)

Equation (9–47) was derived by equating the ultimate strength of a reinforced concrete hollow beam, T_u, to the cracking strength of the same beam, T_{cr}.

The ultimate torsional strength of a reinforced concrete hollow beam is expressed by Rausch's Eq. (3–13) and the cracking strength by Bredt's Eq. (1–72). Taking $T_n = T_{cr}$ gives:

$$\frac{2AA_t f_y}{s} = 2Ah\tau_{cr} \qquad (9\text{–}48)$$

where τ_{cr} is the cracking stress of a hollow beam, corresponding to the cracking torque, T_{cr}. Theoretically A on the left-hand side is the area enclosed by the center line of the stirrups, while A on the right-hand side is the area enclosed by the center line of the wall thickness. The two can be taken as the same, since they are very close, and since the wall thickness h has to be defined empirically. Hence:

$$\frac{A_t f_y}{s} = h\tau_{cr} \qquad (9\text{–}49)$$

The wall thickness h can be expressed by:

$$h = K\frac{A_c}{p_c} \qquad (9\text{–}50)$$

where K is an empirical coefficient. Substituting h into Eq. (9–49) gives:

$$A_t = \frac{(K\tau_{cr})s}{f_y}\frac{A_c}{p_c} \qquad (9\text{–}51)$$

Assuming $(K\tau_{cr}) = 3.5\sqrt{f'_c}$, and $f'_c = 5,000$ psi, then $(K\tau_{cr}) = 247$ psi. When $(K\tau_{cr})$ is rounded out to 250 psi, we arrive at Eq. (9–47) for minimum torsional web reinforcement. This constant of 250 psi was substantiated by tests.[5, 9]

An equal volume of torsional longitudinal steel should also be included as part of the specified minimum reinforcement. From Eq. (3–17):

$$\hat{A}_l = \frac{u}{s} A_{t,\,min} \tag{9–52}$$

where u is the perimeter of the area bounded by the center line of a complete hoop bar.

To simplify the design, the specified minimum torsional reinforcement, Eqs. (9–47) and (9–52), should be required throughout the beam in all cases. This is in contrast to the ACI limit design method (Section 11.6.3.2 of the ACI Code) that allows the torsional moments to reduce linearly from a maximum torque at the critical sections to zero at the midspan, if the spandrel beam supports a slab or closely spaced joists.

The alternate design method is simpler to use than the ACI limit design method in two ways. First, the minimum torsional steel at the critical sections can be calculated directly by Eqs. (9–47) and (9–52), without involving the torsional moments. Second, the stirrups required by torsion are added directly to those required by shear without involving an interaction equation. This alternate method will also be illustrated by the design example in Section 9.4.

9.3.6 Design of Joint

The design of joints between the floor beam and the spandrel beam will be illustrated by the failure of specimen A4, taken from Ref. 4. Specimen A4 was designed to fail at the joint, as no shear or torsional reinforcement was provided in the spandrel beam at the joint. Figure 9.17 shows in perspective the joint and the shaded failure surface. It can be seen that a vertical crack first occurred between the floor beam and the spandrel beam due to a negative moment at the end of the floor beam. This crack extended about halfway down the depth and caused failure to occur suddenly along the shaded failure surface indicated. A secondary failure followed, tearing the bottom layer of concrete along the longitudinal bars. The failure surface consists of a horizontal plane and two inclined planes. The horizontal projection of each inclined plane can be conservatively taken as one-half of the effective depth of the spandrel beam, $d_s/2$.

In order to prevent this type of joint failure, stirrups in the spandrel beam must be provided within $d_s/2$ of the spandrel beam on either side of the

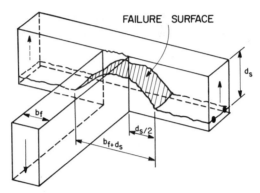

Fig. 9.17. Failure surface at joint of A4.

floor beam (see Fig. 9.17). The number of stirrups should be sufficient to carry the reaction from the floor beam. Thus:

$$V_n = \frac{b_f + d_s}{s} A_v f_y \qquad (9\text{--}53)$$

where

V_n = nominal strength to resist the joint force transferred from the floor beam to the spandrel beam
b_f = width of the floor beam
d_s = effective depth of the spandrel beam
s = spacing of stirrups within the failure zone of length $b_f + d_s$
A_v = area of one stirrup (including two legs if they exist)
f_y = yield strength of stirrups

Equation (9–53) was used to design the joint of other specimens in Ref. 4 and was found to be adequate. It is believed that Eq. (9–53) should be applicable when the depths of the floor beam and the spandrel beam are about the same.

A suggestion has been made that closely spaced stirrups be placed within the width b_f of the joint to "hang up" 100% of the load transmitted from the floor beam. If this suggestion is followed, it is often necessary to place several stirrups in the joint. Apparently such a severe requirement is not supported by tests in Ref. 4.

A practical rule should also be observed in detailing the joint. When longitudinal steel bars intersect each other at the joint, those coming into the joint from the floor beam should be placed above those from the spandrel beam. This rule is simply common sense in engineering.

It has been suggested by ACI Committee 426, Shear[10] that the requirement of Eq. (9–53) can be waived under the following three conditions: (1) if the shear stresses at the end of the floor beam are less than $3\sqrt{f'_c}$, as in the case for one-way joists; (2) if the lower face of the spandrel beam is more than 0.5 times the depth of the floor beam below the bottom of the floor beam; and (3) if the spandrel beam is supported on its lower face at the joint. In cases (2) and (3) the inclined compressive thrust that develops in the floor beam can be resisted without tearing away the bottom of the spandrel beam.

9.4 DESIGN EXAMPLE 9.1

9.4.1 Design Frame and Loading

To illustrate the application of the ACI torsion design criteria, an interior span of spandrel beam is chosen, as shown in Fig. 9.18 (a). Column spacing is 27 ft by 30 ft (8.23 m by 9.15 m). The 18 in. by 18 in. (45.7 cm by 45.7 cm) columns are 11 ft (3.35 m) long and are assumed to be fixed at the base. The columns support the spandrel beams, which, in turn, support the transverse joists and the slab. The cross sections of the joists and spandrel beams are shown in Fig. 9.18 (b) and (c), respectively. The 4-in. (10.2-cm) slab supports a uniform live load of 127 psf (6.08 kN/m²). Its 16-in. (40.6-

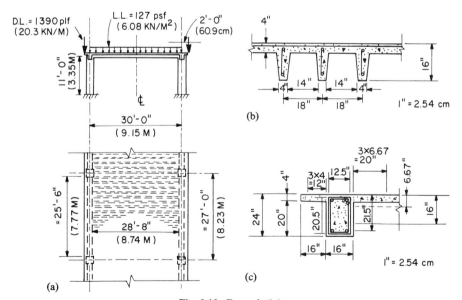

Fig. 9.18. Example 9.1.

cm) cantilever also supports a linear dead load of 1,390 plf (20.3 kN/m) at the edge. A design reinforcement with yield strength of 60 ksi (414 MN/m²) and concrete compressive strength of 4,000 psi (27.6 MN/m²) is adopted. For this design, use the 1977 or 1983 ACI Code.

9.4.2 Flexural and Shear Design

Loading:
Dead load:
Assuming the joist to have an equivalent thickness of 6.67 in.
(16.9 cm):

$$\text{Joist} \qquad \frac{6.67}{12}(14.33)(150) = 1{,}194 \text{ plf}$$

$$\text{Beam} \qquad \frac{16(24)}{144}(150) = \quad 400$$

$$\text{Cantilever} \qquad \frac{4(16)}{144}(150) = \quad 66$$

$$\text{D.L. on cantilever} \qquad = \underline{1{,}390}$$
$$3{,}050 \text{ plf}$$

Factored dead and live load (w_u):

$$\begin{aligned} \text{Dead load} \qquad & 1.4(3{,}050) = 4{,}270 \text{ plf} \\ \text{Live load} \qquad & 1.7(127)(15) = \underline{3{,}240} \\ & w_u = 7{,}510 \text{ plf} \end{aligned}$$

Flexural Design:
Maximum negative moment:

$$-M_u = \frac{1}{11} w_u \ell_n^2 = \frac{1}{11}(7.51)(25.5)^2(12) = 5{,}330 \text{ in.k}$$

Negative steel area:
Assuming $a = 6$ in.

$$A_s = \frac{M_u}{\phi f_y\left(d - \dfrac{a}{2}\right)} = \frac{5{,}330}{0.9(60)\left(21.5 - \dfrac{6}{2}\right)} = 5.34 \text{ in.}^2$$

$$a = \frac{A_s f_y}{0.85 f'_c b} = \frac{5.34(60)}{0.85(4)(16)} = 5.9 \text{ in.} \approx 6 \text{ in.} \qquad \text{O.K.}$$

Maximum positive moment:

$$+M_u = \frac{1}{16} w_u l_n^2 = \frac{1}{16}(7.51)(25.5)^2(12) = 3,660^{\text{in.k}}$$

Positive steel area:
   Assuming $a = 1.5$ in.

$$A_s = \frac{M_u}{\phi f_y \left(d - \dfrac{a}{2}\right)} = \frac{3,660}{0.9(60)\left(21.5 - \dfrac{1.5}{2}\right)} = 3.27 \text{ in.}$$

$$a = \frac{A_s f_y}{0.85 f'_c b} = \frac{3.27(60)}{0.85(4)(48)} = 1.2 \approx 1.5 \text{ in.} \qquad \text{O.K.}$$

The factored design moment envelope is shown in Fig. 9.19 (a). It is obtained from graph A.11, p. 625, of. Ref. 11, except that the negative moment envelope has been approximated by a straight line.

Shear Design:
The factored shear at the column face is:

$$V_u = \frac{w_u l_n}{2} = \frac{7.51(25.5)}{2} = 95.7 \text{ kips}$$

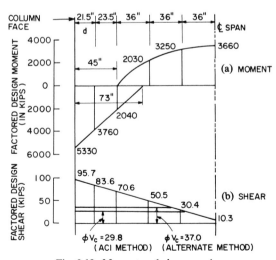

Fig. 9.19. Moment and shear envelopes.

The factored shear at midspan is conservatively obtained by placing a live load on one half of an assumed simple span:

$$V_u = 3,240 \left(\frac{25.5}{2}\right)\frac{1}{4} = 10.3 \text{ kips}$$

The factored shear envelope shown in Fig. 9.19 (b) is taken as a straight line between the value at the column and that at midspan. The critical section for design is taken at a distance d from the column face. At this location the factored shear is 83.6 kips.

9.4.3 Elastic Torque

The factored elastic torque at the face of the column is computed by the elastic method given in Section 9.2, using the following formula:

$$T_u = C_1 \frac{l_n}{2} M_F \qquad \text{where } C_1 = f(P,Q)$$

We shall first calculate the fixed end moment M_F from the loads and then find the coefficient C_1 from the properties of the frame. The loads that induce torsion in the spandrel beam are:

Joist dead load $\qquad 1.4 \left(\frac{6.67}{12}\right)(150) = 117 \text{ lb/ft}^2$

Live load on floor $\qquad\qquad 1.7(127) = \underline{216 \text{ lb/ft}^2}$

$\qquad\qquad\qquad\qquad\qquad\qquad w_u = 333 \text{ lb/ft}^2$

Hence:

$$M_F = \frac{1}{12} w_u \, l_s^2 - N l_c$$

$$= \frac{1}{12} 0.333(30)^2(12) - [1.39(24) + 0.05(16)]$$

$$= 265 \text{ in.-k/ft}$$

To find C_1 we calculate the frame and material constants. The moment of inertia of joist I_s was calculated with the aid of the *ACI Ultimate Strength Design Handbook,* Vol. 1, SP 17(73), p. 255.

$$I_s = K_{i\,4}\left(\frac{b_w h^3}{12}\right)\left(\frac{1}{b}\right) = 1.82\,\frac{4(16)^3}{12}\left(\frac{1}{18}\right) = 138 \text{ in.}^4/\text{in.}$$

$$C = \Sigma\beta x^3 y = 0.196(16)^3(24) + 0.264(4)^3(12) + 0.264(6.67)^3(20)$$
$$= 21{,}038 \text{ in.}^4$$

$$I_c = \frac{1}{12}(18)^4 = 8{,}748 \text{ in.}^4$$

$$\frac{E}{G} = 2(1+\nu) = 2(1+0.2) = 2.4$$

Then:

$$P = \ell\sqrt{\frac{2EI_s}{\ell_s GC}} = 27(12)\sqrt{\frac{2(2.4)(138)}{30(12)(21{,}038)}} = 3.03$$

$$Q = \frac{\ell I_s h}{\ell_s I_c} = \frac{27(138)(11)(12)}{30(8{,}748)} = 1.87$$

From Fig. A1 in Appendix A, $C_1 = 0.385$. Now:

$$T_u = C_1\frac{\ell_n}{2}\,M_F = 0.385\left(\frac{25.5}{2}\right)(265) = 1{,}300 \text{ in.k}$$

$$\tau_u = \frac{3T_u}{\Sigma x^2 y} = \frac{3(1{,}300{,}000)}{7{,}226} = 540 \text{ psi} = 8.5\sqrt{f'_c} > 4\sqrt{f'_c}$$

Since τ_u is considerably greater than $4\sqrt{f'_c}$, much economy can be achieved by using the ACI torsional limit design method given in Section 9.3.

9.4.4   ACI Torsional Limit Design Method

Flexural Shear and Torque (at a distance $d = 21.5$ in. from the column face):

$$V_u = 95.7 - \frac{95.7 - 10.3}{(12.75)}\left(\frac{21.5}{12}\right) = 83.6 \text{ kips}$$

$$\Sigma x^2 y = 16^2(24) + 4^2(12) + 6.67^2(20) = 7{,}226 \text{ in.}^2$$

$$T_u = \phi 4\sqrt{f'_c}\,\Sigma x^2 y/3 = 518{,}000 \text{ in.-lb}$$

$$C_t = \frac{b_w d}{\Sigma x^2 y} = \frac{(16)(21.5)}{7{,}226} = 0.0476\,\frac{1}{\text{in.}}$$

$$T_c = \frac{0.8\sqrt{f'_c}\,\Sigma x^2 y}{\sqrt{1 + \left(\dfrac{0.4 V_u}{C_t T_u}\right)^2}} = \frac{0.8\sqrt{4,000}\,(7,226)}{\sqrt{1 + \left[\dfrac{0.4(83.6)}{0.0476(518)}\right]^2}} = 217,000 \text{ in.-lb}$$

$$T_{n,\,\text{max}} = 5 T_c = 5(217,000) = 1,085,000 \text{ in.-lb} > T_u/\phi \qquad \text{O.K.}$$

$$V_c = \frac{2\sqrt{f'_c}\,b_w d}{\sqrt{1 + \left(\dfrac{C_t T_u}{0.4 V_u}\right)^2}} = \frac{2\sqrt{4,000}\,(16)(21.5)}{\sqrt{1 + \left[\dfrac{0.0476(518)}{0.4(83.6)}\right]^2}} = 35,000 \text{ lb}$$

Shear Reinforcement (at a distance d from the column face):

$$V_u - \phi V_c = 83.6 - 0.85(35.0) = 53.8 \text{ kips}$$

$$\frac{A_v}{s} = \frac{V_u - \phi V_c}{\phi d f_y} = \frac{53.8}{0.85(21.5)(60)} = 0.0490 \text{ in.}^2/\text{in.}$$

$$\left(\frac{A_v}{s}\right)_{\text{min}} = \frac{50b}{f_y} = \frac{50(16)}{60,000} = 0.0133 \text{ in.}^2/\text{in.}$$

$$< 0.0490 \text{ in.}^2/\text{in.} \qquad \text{O.K.}$$

$$V_s = (V_u - \phi V_c)/\phi = 53.8/0.85 = 63.3 \text{ kips}$$

$$4\sqrt{f'_c}\,b_w d = 4\sqrt{4,000}\,(16)(21.5) = 87,000 \text{ lb} > V_s$$

$$s_{\text{max}} = \frac{d}{2} = 10.75 \text{ in.}$$

Use $s_{\text{max}} = 10.75$ in. when $T_u \leqslant \phi 0.5\sqrt{f'_c}\,\Sigma x^2 y = 194,000$ in.-lb

Torsional Web Reinforcement (at distance d from the column face):

Assuming 1½ in. cover and No. 4 stirrups:
$$x_1 = 12.5 \text{ in. and } y_1 = 20.5 \text{ in.}$$
$$\alpha_t = 0.66 + 0.33 y_1/x_1 = 1.20 < 1.5 \qquad \text{O.K.}$$
$$T_u - \phi T_c = 518 - 0.85(217) = 334 \text{ in.-kips}$$
$$\frac{A_t}{s} = \frac{T_u - \phi T_c}{\phi \alpha_t x_1 y_1 f_y} = \frac{334}{0.85(1.20)(12.5)(20.5)(60)} = 0.0212 \text{ in.}^2/\text{in.}$$
$$s_{\text{max}} = \frac{x_1 + y_1}{4} = \frac{12.5 + 20.5}{4} = 8.25 \text{ in.} \qquad \text{governs}$$
$$s_{\text{max}} = 12 \text{ in.}$$

Use stirrups at 8 in. maximum spacing. When $T_u \leqslant 194,000$ in.-lb, a maximum spacing of 10.75 in. may be used.

Total Web Reinforcement (at distance d from the column face):

$$\frac{A_t}{s} + \frac{A_v}{2s} = 0.0212 + \frac{0.0490}{2} = 0.0457 \text{ in.}^2/\text{in.}$$

For No. 4 stirrups:

$$s = 0.20/0.0457 = 4.38 \text{ in.}$$

The factored design torque diagram is assumed to be a straight line as shown in Fig. 9.20 (a). Web reinforcement at 3 ft, 6 ft, and 9 ft from midspan is calculated proportionately according to ($V_u - \phi V_c$) and ($T_u - \phi T_c$) in Table 9.5 Closed stirrups are required and provided in the spandrel beam as shown in Fig. 9.21 (a). The final arrangement of web reinforcement is given in Fig. 9.22

Torsional Longitudinal Reinforcement (at distance d from the column face):

$$\hat{A}_l = 2\left(\frac{A_t}{s}\right)(x_1 + y_1) = 2(0.0212)(12.5 + 20.5) = 1.40 \text{ in.}^2 \qquad \text{governs}$$

$$\hat{A}_l = \left[\frac{400x}{f_y}\left(\frac{T_u}{T_u + \dfrac{V_u}{3C_t}}\right) - 2\frac{A_t}{s}\right](x_1 + y_1)$$

$$= \left[\frac{400(16)}{60,000}\left(\frac{518}{518 + \dfrac{83.6}{3(0.0476)}}\right) - 2(0.0212)\right](12.5 + 20.5)$$

$$= 0.253 \text{ in.}^2$$

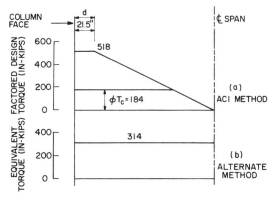

Fig. 9.20. Torque envelopes.

Table 9.5 Summary of reinforcement (ACI method)

ITEM	UNITS	AT COLUMN FACE	AT DISTANCE d FROM COL. FACE	DISTANCE FROM CENTER OF SPAN			
				9 FT	6 FT	3 FT	0
Stirrups							
$V_u - \phi V_c$	kips	—	53.8	40.8	20.7	0.62	0
$T_u - \phi T_c$	in.-kips	—	334	241	99.2	0	0
$A_v/2s$	in.²/in.	—	0.0245	0.0185	0.0094	0.0002	0
A_t/s	in.²/in.	—	0.0212	0.0153	0.0063	0	0
$A_v/2s + A_t/s$	in.²/in.	—	0.0457	0.0338	0.0157	0.0002	0
s	in.	—	4.38[1]	5.92[1]	7.01[2]	—	—
Top Longitudinal Bars							
$-M_u$	in.-kips	−5,330	−3,760	−2,040	0	—	—
$T_u - \phi T_c$	in.-kips	—	334	241	99.2	0	0
A_s	in.²	5.34	3.57	1.84	0	—	—
$A_{l,top}$	in.²	0.39	0.39	0.28	0.12	—	—
$A_s + A_{l,top}$	in.²	5.73	3.96	2.12	0.12	—	—
Use		4#9 + 2#8	4#9 + 2#8	4#9	2#9	2#9	2#9
Bottom Longitudinal Bars							
$+M_u$	in.-kips	—	—	0	2,030	3,250	3,660
$T_u - \phi T_c$	in.-kips	—	334	241	99.2	0	0
A_s	in.²	—	—	0	1.82	2.90	3.27
$A_{l,bot}$	in.²	0.39	0.39	0.28	0.12	—	—
$A_s + A_{l,bot}$	in.²	0.39	0.39	0.28	1.94	2.90	3.27
Use		2#7	2#7	2#7	2#8 + 3#7	2#8 + 3#7	2#8 + 3#7

Note: The values underlined have been calculated previously. The values at other sections were obtained from the underlined values by proportion.

1: #4 stirrups; 2: #3 stirrups.

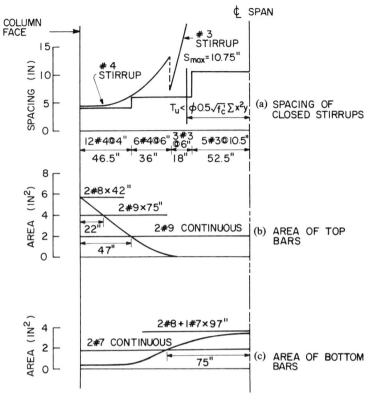

Fig. 9.21. Reinforcement required and provided (ACI method).

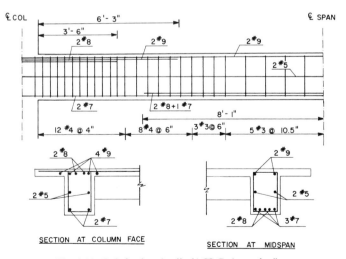

Fig. 9.22. Reinforcing details (ACI Code method).

Maximum spacing of longitudinal bars $s'_{max} = 12$ in. Provide longitudinal bars at mid-depth; then $s' = 9.5$ in. < 12 in. O.K. Assume equal areas of longitudinal steel at the three levels.

$$\frac{\hat{A}_l}{3} = \frac{1.40}{3} = 0.467 \text{ in.}^2$$

Use two No. 5 intermediate bars, $(\hat{A}_l/3) = 0.62$ in.$^2 > 0.467$ in.^{2}:

$$A_{l,top} = A_{l,bot} = \frac{1}{2}(1.40 - 0.62) = 0.39 \text{ in.}^2$$

Total Longitudinal Reinforcement:
The area of top longitudinal bars required to resist bending at the column face has been found to be 5.34 in.^{2} The steel areas required for other sections are also calculated and are recorded in Table 9.5. The area of top longitudinal bars required to resist torsion is 0.39 in.^{2} at a distance d from the column face. This required area is also used at the column face. The steel areas at the other sections are proportional, according to $(T_u - \phi T_c)$. The total areas of top reinforcement for both bending and torsional moments are also summarized in Table 9.5 and plotted in Fig. 9.21 (b). The cut-off points of steel bars have been determined according to the ACI Code.

Similarly, the total areas required for the bottom bars at various sections are summarized in Table 9.5. Figure 9.21 (c) shows the steel required and that provided. All the longitudinal reinforcement details are given in Fig. 9.22.

9.4.5 Alternate Design Method

Flexural Shear Web Reinforcement (at a distance d from the column face):

$$V_u = 83.6 \text{ kips}$$
$$V_c = 2\sqrt{f'_c}\,b_w d = 2\sqrt{4,000}(16)(21.5) = 43,500 \text{ lb}$$
$$8\sqrt{f'_c}\,b_w d = 8\sqrt{4,000}(16)(21.5) = 174,000 \text{ lb} > V_u \qquad \text{O.K.}$$
$$V_u - \phi V_c = 83.6 - 0.85(43.5) = 46.6 \text{ kips}$$
$$\frac{A_v}{s} = \frac{V_u - \phi V_c}{\phi df_y} = \frac{46.6}{0.85(21.5)(60)} = 0.0425 \text{ in.}^2/\text{in.}$$
$$\left(\frac{A_v}{s}\right)_{min} = \frac{50b}{f_y} = \frac{50(16)}{60,000} = 0.0133 \text{ in.}^2/\text{in.}$$
$$< 0.0425 \text{ in.}^2/\text{in.} \qquad \text{O.K.}$$
$$4\sqrt{f'_c}\,b_w d = 4\sqrt{4,000}(16)(21.5) = 87,000 \text{ lb} > V_s\ (= V_u/\phi - V_c)$$
$$\text{Use } s_{max} = d/2 = 10.75 \text{ in.}$$

Shear web reinforcement at other locations was proportioned in Table 9.6 according to the $V_u - \phi V_c$ diagram. The shear diagram V_u is assumed to be a straight line that has a maximum at the critical section and zero at the midspan.

Table 9.6 Summary of reinforcement (alternate method)

ITEM	AT COLUMN FACE	AT DISTANCE, d, FROM COLUMN FACE	DISTANCE FROM CENTER OF SPAN			
			9 FT	6 FT	3 FT	0
(1)	(2)	(3)	(4)	(5)	(6)	(7)
(a) Stirrups						
$V_u - \phi V_c$, in kips	—	46.6	33.6	13.5	0	0
A_v/s, in square inches per inch	—	0.0425	0.0307	0.0123	0	0
$2A_t/s$, in square inches per inch	—	0.0400	0.0400	0.0400	0.0400	0.0400
$2A_t/s + A_v/s$ in square inches per inch	—	0.0825	0.0707	0.0523	0.0400	0.0400
s, in inches	—	4.85	5.66	7.65	10.0	10.0
(b) Top Longitudinal Bars						
$-M_u$, in in.-kips	−5,330	−3,760	−2,040	0	—	—
A_s, in square inches	5.34	3.57	1.84	0	—	—
$A_{l,top}$, in square inches	0.35	0.35	0.35	0.35	0.35	0.35
$A_s + A_{l,top}$, in square inches	5.69	3.92	2.19	0.35	0.35	0.35
Use	4 No. 9 + 2 No. 8	4 No. 9 + 2 No. 8	4 No. 9	2 No. 9	2 No. 9	2 No. 9
(c) Bottom Longitudinal Bars						
$+M_u$, in in.-kips	—	—	0	2,030	3,250	3,660
A_s, in square inches	—	—	0	1.82	2.90	3.27
$A_{l,bot}$, in square inches	—	0.35	0.35	0.35	0.35	0.35
$A_s + A_{l,bot}$, in square inches	—	0.35	0.35	2.17	3.25	3.62
Use	2 No. 7	2 No. 7	3 No. 8 + 2 No. 7	3 No. 8 + 2 No. 7	3 No. 8 + 2 No. 7	3 No. 8 + 2 No. 7

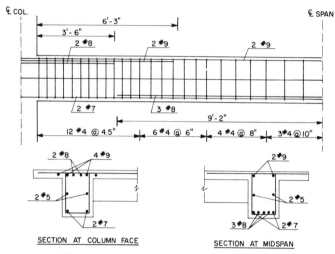

Fig. 9.23. Reinforcing details (alternate method).

Torsional Web Reinforcement:
The minimum amount of torsional web reinforcement is calculated by Eq. (9–47):

$$\frac{A_t}{s} = \frac{250 A_c}{f_y p_c} = \frac{250(16)(24)}{60,000(16 + 24)2} = 0.0200 \text{ in.}^2/\text{in.}$$

This torsional web reinforcement is used throughout the beam.

Total Web Reinforcement (at a distance d from the column face):

$$\frac{2A_t}{s} + \frac{A_v}{s} = 2(0.0200) + 0.0425 = 0.0824 \text{ in.}^2/\text{in.}$$

For No. 4 stirrups, $s = 0.4/0.0825 = 4.85$ in. Web reinforcement at other locations is calculated in Table 9.6. The detailed arrangement of web reinforcement in the spandrel beam is shown in Fig. 9.23.

Torsional Longitudinal Reinforcement:
From Eq. (9–52):

$$\hat{A}_1 = \frac{A_t}{s} 2(x_1 + y_1) = 0.0200(2)(12.5 + 20.5) = 1.32 \text{ in.}^2/\text{in.}$$

Assuming equal areas of steel at three levels gives:

$$\frac{\hat{A}_l}{3} = 0.44 \text{ in.}^2/\text{in.}$$

Use two No. 5 intermediate bars, $(\hat{A}_l/s) = 0.62$ in.$^2 > 0.44$ in.^{2}:

$$A_{l,\text{top}} = A_{l,\text{bot}} = \frac{1}{2}(1.32 - 0.62) = 0.35 \text{ in.}^2$$

Total Longitudinal Reinforcement:
Both top and bottom flexural reinforcements are calculated at various locations, and they are added to the torsional longitudinal steel as shown in Table 9.6. The final arrangement of longitudinal steel is given in Fig. 9.23.

9.4.6 Comparison

Figures 9.22 and 9.23 give the reinforcing details designed by the ACI torsional limit design method and the alternate method, respectively. The steel requirements per beam for those two designs are given in Table 9.7. This table also includes the steel requirement for the same spandrel beam designed according to uncracked elastic analysis (Ref. 3, Chapter 4). When compared to the elastic method, the ACI torsional limit design method saves about 30% in longitudinal steel and 56% in stirrups, while the alternate method saves about 27% in longitudinal steel and 51% in stirrups.

Table 9.7. Comparison of reinforcement designed by three methods

	ACI TORSIONAL LIMIT DESIGN	ALTERNATE METHOD	ELASTIC ANALYSIS (REF. 3, CHAPTER 4)
Longitudinal steel (lb)	613	641	882
Stirrups (lb)	173	193	397

Comparing the alternate method to the ACI torsional limit design method, it can be seen that the former requires 5% more in longitudinal steel and 11% more in stirrups. However, this extra steel is compensated by the simplicity of the design method, as discussed in Section 9.3.5.

REFERENCES

1. Saether, K. and N. M. Prachand, "Torsion in Spandrel Beams," *Journal of the American Concrete Institute,* Proc., Vol. 66, No. 1, January 1969, pp. 24–30.
2. Salvadori, M. G., "Spandrel–Slab Interaction," *Journal of the Structural Division, ASCE,* Vol. 96, No. ST1, January 1970, pp. 89–106.
3. Shoolbred, R. A. and E. P. Holland, "Investigation of Slab Restraint of Torsional Moments in Fixed-Ended Spandrel Girders," *Torsion of Structural Concrete,* SP-18, American Concrete Institute, Detroit, 1968, pp. 69–88.
4. Hsu, T. T. C. and K. Burton, "Design of Reinforced Concrete Spandrel Beams," *Journal of the Structural Division, ASCE,* Proc., Vol. 100, No. ST1, January 1974, pp. 209–229.
5. Hsu, T. T. C. and C. S. Hwang, "Torsional Limit Design of Spandrel Beams," *Journal of the American Concrete Institute,* Proc., Vol. 74, No. 2, February 1977, pp. 71–79.
6. Behera, U., K. S. Rajagopalan, and P. M. Ferguson, "Reinforcement for Torque in Spandrel L-Beams," *Journal of the Structural Division, ASCE,* Proc., Vol. 96, No. ST2, February 1970, pp. 371–380.
7. Collins, M. P. and P. Lampert, "Redistribution of Moments at Cracking—The Key to Simpler Torsion Design?" Publication No. 71–21, Dept. of Civil Engineering, University of Toronto, February 1971, 49 pp.
8. Pillai, S. U., "Moment–Torque Distribution in Reinforced Concrete Frames," Research Bulletin, Dept. of Civil Engineering, Regional Engineering College, Calicut, Kerala, India, July 1974, 44 pp.
9. Mansur, M. A. and B. V. Rangan, "Torsion in Spandrel Beams," *Journal of the Structural Division, ASCE,* Proc., Vol. 104, No. ST7, July 1979, pp. 1061–1075.
10. ACI-ASCE Committee 426, Shear, "Suggested Revisions to Shear Provisions for Building Codes," American Concrete Institute, Detroit, 1979, 82 pp.
11. Winter, G., and A. H. Nilson, *Design of Concrete Structures,* 9th Ed, McGraw Hill Book Co., N.Y., 1979, 647 pp.

Appendix A   Coefficients for Maximum Torque in Elastic Analysis of Spandrel Beams ($C_1, C_2, \ldots C_7$)

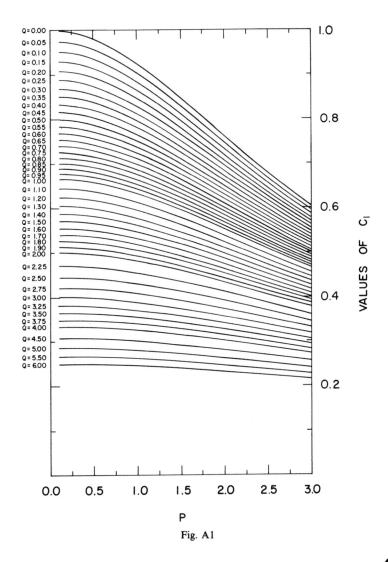

Fig. A1

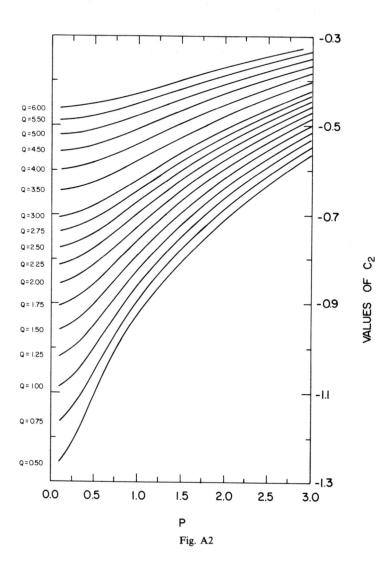

Fig. A2

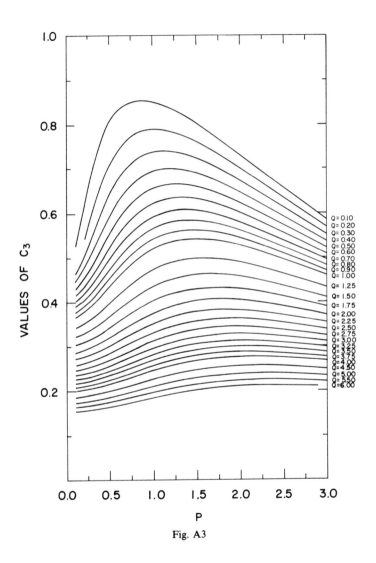

Fig. A3

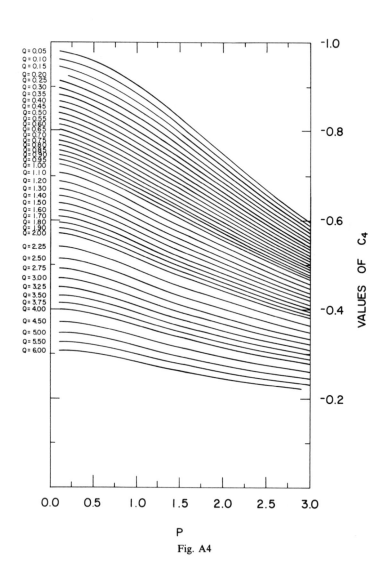

Fig. A4

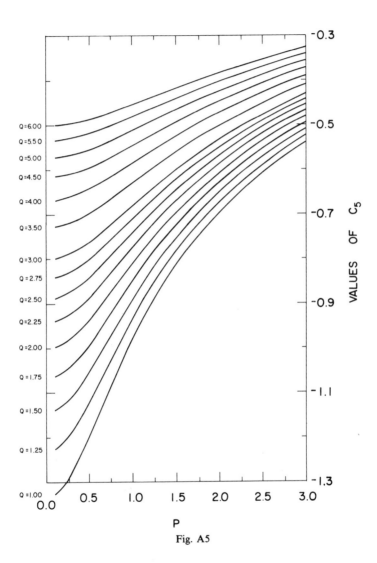

Fig. A5

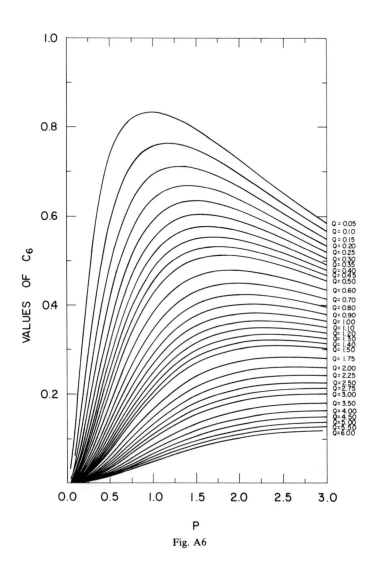

Fig. A6

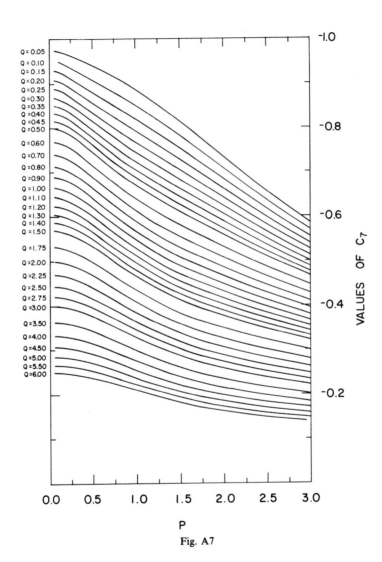

Fig. A7

Appendix B Tentative Recommendations for the Design of Prestressed and Nonprestressed Members to Resist Torsion

These Tentative Recommendations are generalizations of the torsion provisions in the 1977 and 1983 ACI Codes. In order to made the comparison easier, the section numbers used in these codes are retained. Furthermore, the differences are highlighted by the asterisk symbols (*) at the right margin of the pages.

11.0—Notation

A_g = gross area of section, sq. in.

A_l = total area of longitudinal reinforcement to resist torsion, sq. in.

A_{ps} = area of prestressed reinforcement in tension zone, sq. in.

A_s = area of nonprestressed tension reinforcement, sq. in.

A_t = area of one leg of a closed stirrup resisting torsion within a distance s, sq. in.

A_v = area of shear reinforcement within a distance s.

b = width of compression face of member, in.

b_t = width of that part of cross section containing the closed stirrups resisting torsion

b_w = web width, in.

C_t = factor relating shear and torsional stress properties

  = $b_w d / \Sigma x^2 y$

d = distance from extreme compression fiber to centroid of longitudinal tension reinforcement, but need not be less than $0.80\,h$ for prestressed members, in. (For circular sections, d need not be less than the distance from extreme fiber to centroid of tension reinforcement in opposite half of member.)

f'_c = specified compressive strength of concrete, psi

$\sqrt{f'_c}$ = square root of specified compressive strength of concrete, psi

f_y = specified yield strength of nonprestressed reinforcement, psi

h = wall thickness of box section, in.   *

s = spacing of shear or torsion reinforcement in direction parallel to longitudinal reinforcement

T_c = nominal torsional moment strength provided by concrete in com-   *
bined loading

T_{c0} = nominal torsional moment strength provided by concrete under   *
pure torsion

T_n = nominal torsional moment strength

T_s = nominal torsional moment strength provided by torsion reinforcement

T_u = factored torsional moment at section

V_c = nominal shear strength provided by concrete in combined loading   *

V_{c0} = nominal shear strength provided by concrete without torsion   *

V_{ci} = nominal shear strength provided by concrete when diagonal cracking results from combined shear and moment

V_{cw} = nominal shear strength provided by concrete when diagonal cracking results from excessive principal tensile stress in web

V_n = nominal shear strength

V_s = nominal shear strength provided by shear reinforcement

V_u = factored shear force at section

x = shorter overall dimension of rectangular part of cross section

y = longer overall dimension of rectangular part of cross section

$\Sigma x^2 y$ = torsional section property

x_1 = shorter center-to-center dimension of closed rectangular stirrup

y_1 = longer center-to-center dimension of closed rectangular stirrup

α_t = coefficient as a function of y_1/x_1

ϕ = strength reduction factor

γ = a prestress factor = $\sqrt{1 + 10\sigma/f'_c}$ *

γ_2 = a prestress factor = $(1 - 0.833\sigma/f'_c)\sqrt{1 + 10\sigma/f'_c}$ *

σ = average prestress in member, psi

  *

11.6—Combined shear and torsion strength for prestressed and nonprestressed   * members with rectangular or flanged sections

11.6.1—Torsion effects shall be included with shear and flexure where factored torsional moment T_u exceeds $\phi\gamma(0.5\sqrt{f'_c}\ \Sigma\ x^2y)$. Otherwise, tor-   * sion effects may be neglected.

11.6.1.1—For members with rectangular or flanged sections, the sum Σx^2y shall be taken for the component rectangles of the section, but the overhanging flange width used in design shall not exceed three times the flange thickness.

11.6.1.2—A rectangular box section may be taken as a solid section provided wall thickness h is at least $x/4$. A box section with wall thickness less than $x/4$, but greater than $x/10$, may be taken as a solid section except that Σx^2y shall be multiplied by $4h/x$. When h is less than $x/10$, stiffness of wall shall be considered. Fillets shall be provided at interior corners of all box sections.

11.6.2—If the factored torsional moment T_u in a member is required to maintain equilibrium, the member shall be designed to carry that torsional moment in accordance with Sections 11.6.4 through 11.6.9.

11.6.3—In a statically indeterminate structure where reduction of torsional moment in a member can occur due to redistribution of internal forces, maximum factored torsional moment T_u may be reduced to $\phi\gamma(4\sqrt{f'_c}\ \Sigma x^2 y/3)$. *

11.6.3.1—In such a case the correspondingly adjusted moments and shears in adjoining members shall be used in design.

11.6.3.2—In lieu of more exact analysis, torsional loading from a slab will be taken as uniformly distributed along the member.

11.6.4—Sections located less than a distance d from face of support may be designed for the same torsional moment T_u as that computed at a distance d.

11.6.5—Torsional moment strength
Design of cross sections subject to torsion shall be based on

$$T_u \leqslant \phi T_n \qquad\qquad (11\text{--}20)$$

where T_u is factored torsional moment at section considered and T_n is nominal torsional moment strength computed by

$$T_n = T_c + T_s \qquad\qquad (11\text{--}21)$$

where T_c is nominal torsional moment strength provided by concrete in accordance with Section 11.6.6, and T_s is nominal torsional moment strength provided by torsion reinforcement in accordance with Section 11.6.9.

11.6.6—Torsional moment strength provided by concrete

11.6.6.1—Torsional moment strength T_c and shear strength V_c shall be *
computed by *

$$T_c = \frac{T_{c0}}{\sqrt{1 + \left(\dfrac{T_{c0}}{V_{c0}}\dfrac{V_u}{T_u}\right)^2}} \qquad\qquad (11\text{--}22) \quad *$$

$$V_c = \frac{V_{c0}}{\sqrt{1 + \left(\dfrac{V_{c0}}{T_{c0}}\dfrac{T_u}{V_u}\right)^2}} \qquad (11\text{–}5) \quad *$$

where

$T_{c0} = 0.8\sqrt{f'_c} \; \Sigma x^2 y \; (2.5\gamma - 1.5)$ *

$V_{c0} = V_{ci}$, Eq. (11–11) or V_{cw}, Eq. (11–13), whichever is less.   *

For nonprestressed concrete $T_{c0} = 0.8\sqrt{f'_c} \; \Sigma x^2 y$, $V_{c0} = 2\sqrt{f'_c} \; b_w d$ and   *
$T_{c0}/V_{c0} = 0.4/C_t$ *

11.6.6.2—For members subject to significant axial tension, torsional reinforcement shall be designed to carry the total torsional moment, unless a
more detailed calculation is made in which T_c given by Eq. (11–22) and
V_c given by Eq. (11–5) shall be multiplied by $(1 + N_u/500A_g)$, where
N_u is negative for tension.

11.6.7—Torsion reinforcement requirements

11.6.7.1—Torsion reinforcement, where required, shall be provided in addition to reinforcement required to resist shear, flexure, and axial forces.

11.6.7.2—Reinforcement required for torsion may be combined with that
required for other forces, provided the area furnished is the sum of individually required areas and the most restrictive requirements for spacing and
placement are met.

11.6.7.3—Torsion reinforcement shall consist of closed stirrups, closed ties,
or spirals combined with longitudinal bars.

11.6.7.4—Design yield strength of torsional reinforcement shall not exceed
60,000 psi.

11.6.7.5—Stirrups and other bars and wires used as torsion reinforcement
shall extend to a distance d from extreme compression fiber and shall be
anchored according to Section 12.14 to develop the design yield strength
of reinforcement.

11.6.7.6—Torsion reinforcement shall be provided at least a distance
$(d + b_t)$ beyond the point theoretically required.

11.6.8—Spacing limits for torsion reinforcement

11.6.8.1—Spacing of closed stirrups shall not exceed the smaller of $(x_1 + y_1)/4$ or 12 in.

11.6.8.2—Spacing of longitudinal bars, not less than #3, distributed around the perimeter of the closed stirrup, shall not exceed 12 in. At least one longitudinal bar shall be placed in each corner of the closed stirrups.

11.6.9—Design of torsional reinforcement

11.6.9.1—Where factored torsional moment T_u exceeds torsional moment strength ϕT_c, torsional reinforcement shall be provided to satisfy Eqs. (11–20) and (11–21), where torsional moment strength T_s shall be computed by

$$T_s = \frac{A_t \alpha_t x_1 y_1 f_y}{s} \tag{11–23}$$

where A_t is the area of one leg of a closed stirrup resisting torsion within a distance s, and $\alpha_t = (0.66 + 0.33 y_1/x_1)$ but not more than 1.50. Longitudinal bars distributed around the perimeter of the closed stirrups A_t shall be provided in accordance with Section 11.6.9.3.

11.6.9.2—Where factored torsional moment T_u exceeds $\phi \gamma (0.5 \sqrt{f'_c} \, \Sigma x^2 y)$, minimum area of closed stirrups shall be computed by

$$A_v + 2A_t = 50 \frac{b_w s}{f_y} \left(1 + 12 \frac{\sigma}{f'_c}\right) \leqslant 200 \frac{b_w s}{f_y} \tag{11–16}$$

In Eq. (11–16) $(50 b_w s/f_y)(1 + 12\sigma/f'_c)$ shall be multiplied by $(4h/x)$ for box sections and by $\sqrt{d/b_w}$ for I-sections.

11.6.9.3—Required area of longitudinal bars $\hat{A}_l$ distributed around the perimeter of the closed stirrups A_t shall be computed by

$$\hat{A}_l = 2A_t \frac{x_1 + y_1}{s} \tag{11–24}$$

or by

$$\hat{A}_l = \left[\frac{400xs}{f_y} \left(\frac{T_u}{T_u + \dfrac{V_u}{3C_t}} \right) - 2A_t \right] \frac{x_1 + y_1}{s} \tag{11–25}$$

whichever is greater. Value of $\hat{A}_l$ computed by Eq. (11–25) need not exceed that obtained by substituting

$$\frac{50b_w s}{f_y}(1 + 12\sigma/f'_c) \text{ for } 2A_t \qquad *$$

11.6.9.4—The factored torsional moment T_u shall not exceed $\qquad *$

$$\frac{4\sqrt{f'_c}\gamma_2\,\Sigma x^2 y}{\sqrt{1 + \left(\dfrac{0.4\gamma_2 V_u}{C_t\,T_u}\right)^2}} \qquad *$$

where $\gamma_2 = (1 - 0.833\,\sigma/f'_c)\sqrt{1 + 10\sigma/f'_c}$. For nonprestressed concrete $\quad *$
$\gamma_2 = 1.0$. $\qquad\qquad\qquad\qquad\qquad\qquad\qquad\qquad\qquad\qquad\qquad\quad *$

Index